BOUNDARY LAYER STUDIES AND APPLICATIONS

A Special Issue of Boundary-Layer Meteorology in honor of Dr. Hans A. Panofsky (1917–1988)

Reprinted from Boundary-Layer Meteorology
Vol. 47, Nos. 1–4 (1989)

KLUWER ACADEMIC PUBLISHERS
DORDRECHT / BOSTON / LONDON

ISBN 978-94-010-6928-1   ISBN 978-94-009-0975-5 (eBook)
DOI 10.1007/ 978-94-009-0975-5

Published by Kluwer Academic Publishers,
P.O. Box 17, 3300 AA Dordrecht, The Netherlands.

Kluwer Academic Publishers incorporates
the publishing programmes of
D. Reidel, Martinus Nijhoff, Dr W. Junk and MTP Press.

Sold and distributed in the U.S.A. and Canada
by Kluwer Academic Publishers Group,
101 Philip Drive, Norwell, MA 02061, U.S.A.

In all other countries, sold and distributed
by Kluwer Academic Publishers,
P.O. Box 322, 3300 AH Dordrecht, The Netherlands.

BOUNDARY LAYER STUDIES AND APPLICATIONS

*A Special Issue of Boundary-Layer Meteorology
in honor of Dr. Hans A. Panofsky (1917–1988)*

HANS PANOFSKY, 1917–1988

JOHN C. WYNGAARD

National Center for Atmospheric Research[1], *Boulder, Colorado 80307, U.S.A.*

The editor of *Boundary-Layer Meteorology*, R. E. Munn, asked me to write a dedication for a special memorial volume for Hans A. Panofsky. Having known Hans and his beloved wife and companion Nancy for nearly 25 years, having fallen victim to their charm and joy of life, and having witnessed first hand Hans' constructive influence on our discipline, I could not but agree.

We lost an outstanding leader with Hans' death on February 28, 1988 in San Diego, CA. HAP, as he signed his name and as many affectionately called him, developed strong, lasting friendships throughout our community.

I asked several of Hans' students and colleagues to share their memories of him, knowing that there was a story there. I was not prepared for the depth of their response, however. I am pleased to convey their reminiscences to readers of this special volume of *Boundary-Layer Meteorology* dedicated to his memory.

From Germany to Princeton to New York University

Alfred K. Blackadar, now retired from the Department of Meteorology at the Pennsylvania State University, and his wife Betty are long-standing friends of the Panofskys. Al has provided this review of some of the early history of the Panofsky family in the United States:

> The original Panofsky family left Germany in 1934 and settled in Princeton, New Jersey. Hans, along with his younger brother Wolfgang, were accepted as undergraduates in Princeton. Hans majored in astronomy, and graduated in 1938 with a nearly perfect *A* record. He continued his astronomy study at the University of California at Berkeley where he completed his Ph.D. in 1941. He spent that summer at UCLA teaching Army cadets how to draw weather maps – which he was just trying to do himself. After three months of this, he made the decision not to try to be a meteorologist.
>
> I first met HAP in late 1941 when he visited the Princeton Observatory where I too was studying astronomy. Our paths crossed the following summer at New York University, where I was an army cadet. He had been hired as a meteorology instructor, having reconsidered his earlier decision about becoming a meteorologist.

[1] The National Center for Atmospheric Research is sponsored by the National Science Foundation.

Boundary-Layer Meteorology **47**: 1–14, 1989.
© 1989 *Kluwer Academic Publishers*.

Hans had a monumental impact on the field we now call boundary-layer meteorology. Another student from the New York University years, Morton L. Barad, now president of Barad Consultants, Inc. and previously Chief of the Meteorology Laboratory of the Air Force Cambridge Research Laboratories, reveals a little-known story about Hans' entry into the field:

In 1948, as I was beginning my search for a doctoral thesis subject, Bernard Haurwitz, then chairman of the Department of Meteorology at New York University, informed me that the services of a meteorologist would be required in a new air quality study at N.Y.U. to be sponsored by the Consolidated Edison Company of New York. I joined the research team and soon had a thesis topic in the field of air quality research to propose to the department.

However, since none of the department's faculty members had done research in this field, it was necessary for me to find a willing thesis adviser. Hans Panofsky became that person. Thirty-three years later, while attending a meeting of the American Meteorological Society, HAP introduced me to two of his current graduate students as the fellow who was responsible for his switching his research interest from the atmosphere of Jupiter to the boundary layer of Earth. Since the meteorological world now knows of his many accomplishments in the field he adopted so many years ago, I feel good knowing that something I did for myself has had such positive consequences.

Robert Fleagle, a professor of Atmospheric Sciences at the University of Washington and another of Hans' Ph.D. students at N.Y.U., reflects on some of the qualities that even in those early years marked Hans as a special individual:

Hans combined an exceptionally keen and versatile mind with remarkable personal qualities. Together these made him an outstanding teacher who inspired by example. Hallmarks of his personal and professional life were honesty, modesty, and generosity. He was totally without guile and was incapable of believing in another's selfish motive.

Hans was invariably curious about things he didn't know. He could undertake a subject entirely new to him and in a very short time develop a graduate course or a textbook which combined interpretations of background material with original, fresh insights. He did this in such fields as theory of climate, oceanography, planetary atmospheres, atmospheric statistics, and atmospheric turbulence, long before these subjects became major interests in geophysics. He was the ideal colleague, interested in the work of those around him, capable of contributing new ideas or questions, and totally unselfish in sharing credit.

Hans loved teaching and did it well. His particular gift was in making teaching a collaboration of teacher and student. In his hands difficult material became clear, and he encouraged students to believe in their own abilities. Research was the natural emanation of his style of teaching, and many of his students and colleagues will always treasure the experience of working with him on a research problem.

The Penn State years

By 1950 Hans was greatly esteemed at N.Y.U., and the position he held there seemed to be secure with every promise of providing him with a comfortable, stimulating environment. He was conveniently close to Princeton, where he was associated actively with John von Neumann and the exciting numerical prediction project that was underway at that time. He was also close to Brookhaven where there was an active scientific group and facility that provided him with associations and challenges that fueled his imagination and activities. Nonetheless, Hans accepted an appointment in the Department of Meteorology at Penn State University in 1951. Blackadar recalls:

It is really quite interesting that he made the decision to come to Penn State. When in 1950 he wrote to Hans Neuberger inquiring about the possibility of coming, he had already turned down handsome offers of secure positions at Chicago and Seattle. At that time the Penn State program was small and obscure, at least in comparison with other institutions of that time. Penn State had only meager resources and was struggling for a critical mass that would enable it to carry on a viable curriculum and research program. The salary he was offered by Penn State was less than what he had been earning at N.Y.U., and half of that would depend on his own ability to secure research support from outside sources. The offer letter also described the work that would be expected of him. It was more than two pages long and would have discouraged any but the stoutest of souls. But he made the decision to come and all of us would agree, I think, that his decision was a crucial turning point for meteorology at Penn State.

Hans maintained his ties with Brookhaven after moving to Penn State. Meteorology had become a part of Brookhaven's program originally because of the need to understand what might happen to the plume that was to be released from the 350-ft stack associated with the research reactor. They built an elaborate facility, consisting of 420-ft and 150-ft towers, 900 feet apart, with a cable-cart unit between them at 100 feet above ground. These towers were fully instrumented with excellent temperature and wind equipment, including bivanes. All of these were spaced logarithmically in the vertical and recorded in a central building. By 1951, the Brookhaven meteorology group had performed sufficient diffusion experiments to fulfill its original mission, but the laboratory decided that it would be appropriate to have the group continue its work in diffusion and in micrometeorology and turbulence as well, even though the central focus of the laboratory program was nuclear. The leader of the Brookhaven meteorology group at the time, Maynard Smith, writes:

> At this time, HAP was beginning his intense interest in low-level tur-
> bulence, and he was developing a strong meteorological program at Penn
> State. A marriage between the unusual theoretical strength at Penn State and
> the unique instrumental program at BNL was almost inevitable, especially
> since HAP and I had known each other since undergraduate days and Irving
> Singer had studied with him as a graduate student. Included in this close
> relationship was Robert McCormick who was then head of a U.S. Weather
> Bureau research station located with Brookhaven's Meteorology Group.
>
> My most vivid memories of HAP during this period (mid fifties through
> late sixties) were of his boundless enthusiasm and insight into problems that
> were then new to us all, and of his remarkable ability to transmit his ideas
> and enthusiasm to others. He had a very strong influence on our program at
> BNL, especially in relation to our work on turbulence power spectra and the
> specific effects of turbulence on gas and particle diffusion. He made exten-
> sive use of BNL data in conjunction with the Penn State program, and it
> served as a part of more theses than I can possibly remember.
>
> I do recall a characteristic that is not exactly common among eminent
> scientists. HAP remained open-minded about our field throughout his career
> as far as I know. He was always willing to explore a new thought from
> whatever source. I recall a member of the audience at one meeting chiding
> him for having just proposed something radically different from the ideas
> presented in an earlier paper. Panofsky's retort was: "That was what I
> thought last year – it was wrong."
>
> Sometimes his open-mindedness worked to his disadvantage. We had an
> electro-mechanical marvel at BNL, consisting of a host of self-synchronous
> motors and differential gears designed to provide a direct recording of the
> differences among wind speeds at three levels on the main tower (remember
> that no computers existed at that time). In any case, on one cloudy day with
> a sturdy wind and neutral turbulence conditions, HAP noticed that this
> machine was reporting almost no differences in wind speed over the 420-ft
> height of the tower. Undaunted, HAP busily developed fragmentary ideas as
> to how this could happen, even for periods as long as an hour. He was a bit
> taken aback when I pointed out that the instrument had three sensitivity
> scales, and there was no possibility of seeing the usual differences on the
> coarse scale that the technician had set it on!
>
> During the BNL-Penn State association, I personally learned how effective
> a teacher HAP could be. He was so capable and patient that he could
> transmit his ideas successfully to almost any sort of group.

Another of Hans' early students, Frank Gifford, first met him during the 5th
wartime course in meteorology at N.Y.U. 46 years ago. After the war Gifford

returned to N.Y.U. and took, as he recalls, every course that Hans offered, after which he followed him to Penn State to do graduate work under him. Frank, who later became Director of the Atmospheric Turbulence and Diffusion Laboratory at Oak Ridge, Tennessee, writes:

> We lived just a few doors down from the Panofskys in State College. Since my two boys are about the same age as his two girls, our families have been friends ever since. Two of my strongest impressions of Hans, though only two among many, are the following.

> Hans attracted anecdotes the way a dog attracts fleas. Get any group of his ex-students together, and the HAP stories will surely follow in abundance. Some have become classics. A good example – one of my favorites – is the story of the cheese demonstration. Hans had a kind of delicatessen approach to teaching dynamic meteorology. He used a jelly sandwich to illustrate subsidence, divergence, and mass continuity, for example. Phillips' graphic description of large, flat oceanic eddies as "blini" pleased Hans greatly, and he frequently used it in discussing the atmospheric case. Hans was teaching an introductory survey course in meteorology at Penn State, and he decided to illustrate the geometry of warm and cold fronts by means of the shapes of cheeses: a wedge of cheddar became a warm front and a gouda or some similar cheese a cold front. He picked up suitable examples at the grocery store and put them in the refrigerator until class time. Just before the lecture he went to the refrigerator to retrieve the cheeses, only to discover that a person or persons unknown had helped themselves to a snack, thereby ruining the frontal structure demonstration. It is a measure of Hans's good nature that he was the first one to tell this story and thought it a fine joke on himself.

> Many will testify to Han's superb teaching skills; his excellence as a research worker; in short, his greatness as a scientist. I would like to add that he was also an extremely kind, pleasant, and invariably helpful person whose friendship never stopped at the office or classroom door.

Another Penn State Ph.D. student during this time was George McVehil, now a principal with McVehil-Monnet Associates Inc. George arrived at Penn State in the fall of 1958 to begin doctoral studies after receiving a masters degree in meteorology from MIT. He recalls:

> My first meeting with HAP is still most memorable. On the first day of classes, I waited in a classroom with eight to ten other graduate students for HAP to appear for the first lecture of one of his courses. When he arrived, he quickly scanned the room of mostly familiar faces and immediately walked back to where I was seated. He stuck out his hand and said "You must be McVehil. Hi!" I was astounded. Such informality had not existed in

the MIT meteorology department. Though the classes were small there, I don't think I ever saw a professor greet a student individually in a classroom setting. At the time, I believe I was flattered to think that Professor Panofsky even knew my name. But in retrospect the striking fact was his warmth and sincere interest in his students and associates. It was not long until, like most of the graduate students, I was invited to HAP's home for dinner with him and Nancy.

There are many stories about HAP's great lectures, his unique capability for presenting ideas in an original way, and his seemingly casual lecture preparation. As all probably remember, his lecture notes usually consisted of a crumpled-up sheet of paper pulled from a jacket pocket. I remember at least one occasion when his notes were, quite literally, scribbled on the back of a used envelope. But on some later occasions, when I substituted for him in giving undergraduate lectures, I had the opportunity to review the material to be presented with him. There was never any question as to what he had covered or the topics to be treated in the next lecture.

When I was doing research for my doctoral thesis, Al Blackadar was my official advisor. But since HAP was interested in the subject, he quickly became an unofficial advisor. He would stop by my office on a nearly daily basis to ask how the work was progressing. After a minute or two of scanning my data or listening to my explanation of problems, he would toss out two or three suggestions about analyses to try or ways to solve the problem. Then he was off on his way to class. Since I was very methodical and many steps behind HAP in insight, I almost never had fully implemented his suggestions when he returned (usually that afternoon or the next morning). He seldom expressed any irritation at my slow progress; he simply added one or two new suggestions – he had thought about my problem in the interim – and was gone. I think that my desire to have something positive to show HAP on his next visit was as important to my eventual success in completing the research as his scientific contributions.

I think that these thoughts best typify HAP to me. We all recognize his scientific contributions and teaching skill, but I most appreciate his warm human qualities and his dedication to giving something more than technical knowledge to each of his students.

About this time, Jim Deardorff and Chandran Kaimal were Ph.D. candidates at the University of Washington. Jim credits Hans for encouraging him, while still in graduate school, to enter the field of small-scale meteorology and turbulence:

He reviewed my first paper, dealing with wind fluctuations, which I submitted to the old *Journal of Meteorology*. His review indicated I would need to cut out half the paper, but that the other half could be made suitable. Ordinarily this much alteration is cause for rejection. However, by allowing the greatly shortened paper to go forward, he helped me towards a career in boundary-layer turbulence. The fact that Hans often felt no need to be anonymous in his reviews testifies to his outgoing personal nature.

Chandran writes:

To Hans Panofsky, more than anyone else, I owe my interest in atmospheric turbulence and spectra. His earliest papers and reports drew me to that field, and I have been hooked ever since. To me, a graduate student, he appeared a distant and formidable figure. Only later, as a research scientist at the Air Force Cambridge Research Laboratories, did I come to know him as a colleague, when our paths crossed frequently at conferences within the U.S. and abroad. But it is the last twelve years that I remember most fondly – when we were family friends. Hans and Nancy often stayed with my family when they visited Boulder and we spent long evenings talking. But the debt I owe Hans was never expressed – at least not explicitly. He would have brushed it aside with characteristic modesty. I acknowledge it now to my friends and colleagues.

1964 saw a milestone, the publication of the Lumley–Panofsky monograph. John Lumley tells the story behind this remarkable effort:

In the early sixties, Hans invited me to join him in writing what was to become *The Structure of Atmospheric Turbulence*. I was a very young assistant professor and was quite flattered at the attention of such a distinguished man. At the same time, I did not know what I was getting into because I had never worked closely with anyone outside may own little specialty. At about the same time, Hans started sitting in on my turbulence course and dragged along with him whatever other people from the meteorology department could not escape.

As a graduate student, I had been introduced to the idea of a faculty member's going to another department and auditing a colleague's course; my thesis advisor did it all the time, asking lots of pointed questions that often brought the lecture to a halt. To a certain extent, this prepared me for the experience of having Hans in the class although I found it a bit different from the other side of the lecture platform. I was appalled to find that many of the background subjects that seemed to me the stock-in-trade of any educated person were news to Hans. At the same time, he seemed vitally concerned about all sorts of things that I had never heard of and thought were probably irrelevant. I recall two questions, probably because I have heard them a hundred times since from other people in meteorology and oceanography: first, how can we use a statistical approach when we have only one realization, the earth?; and, how can we use stationary processes with integral scales as paradigms, when the weather is not stationary and integral scales do not exist experimentally? (Readers who are still having trouble with these points should send a self-addressed envelope.) Although I still tend to feel that the things I think are important are the ones that are really important, in the fullness of time I have come to realize how parochial we all are and how rare are the people, like Hans, who actively seek out fields about which they know little, in hopes of learning more, and brave the difficulties of trying to establish communication across the gaping chasms.

The book was a lot of fun to write. In addition to totally different backgrounds, our approaches are very different. Hans had in mind a book primarily covering Monin–Obukhov similarity theory, then relatively new and unfamiliar in the West. I had in mind something to alleviate what seemed to me the shocking ignorance in the meteorological community. Each of us wrote a first draft of his own part, and then we exchanged parts. Then the telephone calls started; Hans used to call me at 7:00 A.M. on Sunday morning (and every other time of day also – it's just those particular calls that stick in the mind). Why had I said this, and why that? What did this mean? And so forth. He was concerned, and rightly so, that meteorologists would not understand what I, still wet behind the ears, had written. I tried hard to moderate the tone and simplify the presentation. On the other hand, I found his part filled with things that seemed to me logically unfounded. When each of us had had enough of this, the manuscript was sent to two colleagues and mentors, one of mine and one of HAP's. Predictably, HAP's colleague found nothing to comment on in HAP's part, other than its unusual clarity, but found my part unnecessarily obscure, mathematical and difficult. My colleague found nothing to comment on in my part, other than *its* clarity, but found HAP's part untidy, confused, reaching unjustified conclusions, etc. In light of these comments, we revised again. By this process we crept toward each other, finally arriving just close enough to touch. It is, of course, still clear that the two halves were written by very different sorts of people, something I have managed to avoid in subsequent efforts with other coauthors. But then, these subsequent coauthors may not have been such different people. Hans was much more of a naturalist and not at all a mathematician/physicist, and I am practically the inverse.

My mother always said that things that were hard work or were painful were probably good for you in the long run. My experience with Hans certainly epitomized this. As a result of my interaction with him, although I have staunchly remained a fluid mechanic, I have a lifelong professional interest in small-scale meteorology and oceanography, as well as in the Russian school of Kolmogorov. Perhaps harder to pin down, but more useful to me in the long run, is a *general* enthusiasm for interaction with people in many other fields. I have Hans to thank for the appreciation that there are lots of interesting problems in other people's fields, and that these people are happy to talk to you about them if you are half-way civilized. Sadly, very few of them are as nice as Hans.

In the early 1960s Gary Briggs arrived as a Ph.D. student in the Meteorology Department at Penn State. Gary, now a research scientist at the Meteorology Division of the Environmental Protection Agency in Research Triangle Park, North Carolina, recounts:

Luckily for me, my days as a graduate student at Penn State began with a desk in HAP's great "oval office" (actually, keyhole-shaped), just above the entrance and below the rotunda of the old Earth and Mineral Sciences building. HAP shared it with his talented "woman Friday", Lisa Mares, secretary, draftswoman, technical editor, and operator of a 90-lb calculator machine, and with two or three graduate students at a time. It was a stimulating atmosphere, with HAP inquiring about both our work *and* our well-being rather often, tossing us little suggestions about our data analyses, and then asking us for *our* opinion. It was he who first asked my opinion of our war in Vietnam. I had no opinions at all at that time, but he started me thinking about it.

I even found the distraction of eavesdropping on his phone calls rather educational. "Hello, is this the mechanic working on the Studebaker? This is *Mr.* Panofsky . . . ." As his daugher Ruth said in an eloquent eulogy of her father at the campus chapel last spring, HAP never exalted himself or let on to anyone that he was famous or anyone out of the ordinary. Of particular interest to *BLM* readers, I remember his immediate reaction as he read the announcement soliciting papers for the start-up of this publication. "What, another journal? We have too many already!" Of course, he later published much of his own voluminous boundary-layer research in *BLM*, starting with the second issue.

That paper, with Roger Pielke, and others are full of "Panofsky diagrams", scatter plots with a best-fit line "fitted by eye". Not many scientists these days dare to venture outside the safe respectability of least-squares fits. HAP certainly could talk, teach, or apply statistics when it seemed useful, but he was completely down-to-earth in assuming that if there was no relationship in the data evident to the eye, then fancy statistics certainly were not going to help it. He was right, and I wish that his practical attitudes and emphasis on intuitive understanding of phenomena would permeate teaching everywhere. Fortunately, many of us were lucky to have learned from this humble, great teacher.

Steve Hanna, now Vice President of Sigma Research Corporation in Westford, Massachusetts, was also a Ph.D. candidate under Hans at this time. He recalls:

My best memories of Hans are from the period when I shared an office with him and others in the room above the Mineral Industries Building Library at Penn State in 1965. He was constantly doing various statistical analyses that required squaring large numbers of data, and in the days before the widespread use of computers, he would do many of these exercises in his head. I believe he knew the squares of all numbers up to at least 100. He used this same method to remember telephone numbers – for example, if the last four digits of someone's number were 4346, he would remember this as 66^2–10.

It never occurred to me at the time, but his willingness to share an office with five research associates and graduate students is an indication of his unselfish dedication to his students. As an Evan Pugh Research Professor, he could have commanded a one-person office with a potted plant, but he preferred to work in the trenches with the students.

Another graduate student from the sixties was David Smith, now Vice President and Manager of the Environmental Division of Chas. T. Main, Inc. of Boston. Smith writes:

I was struggling with a particularly difficult theoretical problem while working towards my masters degree. Several with whom I shared my frustrations suggested that I see Hans Panofsky who might be able to help. I admired him as an instructor in some of my courses, but I didn't really know him personally.

Hans greeted me enthusiastically and, after a discussion on my predicament, said he had some ideas for a master's thesis for a student who was supported by an air pollution control grant program, as I was. Hans was guiding an air pollution research program in a polluted, industrial valley town nearby and wanted to test some hypotheses about the behavior of pollution in such a valley. He thought it might be interesting to actually test how pollutant emissions behaved and relate that behavior both to meteorological parameters in the valley, which he had been measuring, and to the new diffusion typing schemes recently advanced by his friend Frank Pasquill in England. It sounded like just the challenge I needed and certainly a problem that could be readily tackled within my constraints.

While I didn't realize it at the time, I later came to appreciate that Hans recognized my strong inclinations towards application rather than theory. He helped me identify and approach a problem which allowed me to take advantage of my strengths. Later, when my thesis fell victim to writer's block as I was struggling to meet a deadline, he inspired me to complete it on schedule. Finally, his urging prompted me to publish a paper on the results.

As a result of Hans' keen insight into people as well as research problems plus his inspirational role as mentor and teacher, my foundering graduate program was rescued, and my career path redirected, both at a critical juncture. From the perspective of some 20 years later, I owe a great deal to Hans for the challenges and rewards that followed.

Meanwhile, Henk Tennekes had arrived at the Aerospace Engineering Department of Penn State in 1964 and soon became well acquainted with Hans. Henk reflects on those days:

How do I remember Hans Panofsky? I recall a pleasant Sunday morning in central Pennsylvania, enjoying tea and crackers in bed and lazily browsing through scattered sections of the massive newspaper that was dropped on the front porch half an hour earlier. What is more relaxing than being able to take your time and to allow the process of waking up its due course?

Suddenly the telephone rang. Who has the temerity to shatter my domestic bliss? Since one can never be sure that there is no emergency on the other side of the line (I wish the phone company would invent a special blinker or buzzer for the purpose!), I stumbled out of bed and picked up the receiver.

"Hello, this is Hans Panofsky. Are you awake yet?" (No, I was rather successful in avoiding exactly that). "I am reviewing a manuscript that deals with the Lagrangian coherence of lateral velocity fluctuations in nearly neutral turbulence over rough terrain. The author claims that the exponential decay of the correlation function scales with . . . ."

Doing my best to be polite, I listened to the case and attempted to phrase a few halfway intelligent questions. Before too long – Panofsky could be very charming and very persistent simultaneously, with not a trace of cunning – I found myself promising to have a look at the paper first thing Monday morning and to participate in writing the referee report before the week was over.

Hans Panofsky was wholly without pretense. The "minor" Panofsky – father Erwin revolutionized art history, brother Wolfgang was in charge of the Stanford Linear Accelerator – always shared everything he knew and didn't know. Not always at the most opportune time, but unfailingly free of the encumbrance of shame and pride. That is how I remember him best.

Beginning with Niels Busch's visit to Penn State in 1965, a strong relationship grew between Hans and Denmark's Riso National Laboratory. Leif Kristensen, one of the Riso scientists who worked with Hans at Penn State and at Riso, recalls the experience in this way:

HAP was extremely cooperative with young scientists, in particular newcomers who had just learned the basics. He would always treat them "on the level", never "top-down", to use a phrase of contemporary boundary-layer physics. The young scientist who worked with HAP would soon be captivated by his intuitive approach which most often would lead to sound results. If at some point HAP's young colleague suggested what he considered an elegant formulation involving a display of mathematical *haute école*, HAP would soon get absent-minded and at the first opportunity return to *the physics* of the problem under consideration. Many of us who became familiar with HAP's views learned a lot of interesting physics as a result. Perhaps it was his interaction with newly hatched scientists that has contributed most of all to the international conformity in the approach to problems concerning the atmospheric boundary layer. Worldwide, scientists in our field seem to speak the same language and to use the same basic approach.

Hans' monograph with John Lumley, *Structure of Atmospheric Turbulence*, was required reading for a generation of students. Eventually it fell victim to progress and needed to be updated. In 1984 the Panofsky–Dutton text *Atmospheric Turbulence* appeared. John Dutton remembers Hans:

> My memories of Hans Panofsky center on his boundless curiosity and his inexhaustable kindness.
>
> HAP was relentless in pursuing answers to the questions that puzzled him. He insisted on understanding the physical world in his terms and would contemplate a new issue until he was satisfied. He would query colleagues, often by long-distance phone and sometimes far too early in the morning, seeking answers to the current questions. If he were asked a question to which he did not know the answer, he would puzzle over that new challenge and when he had a solution, reveal it enthusiastically to his questioner. This curiosity and his innate quickness of mind were the key to the breadth of his scientific contributions.
>
> HAP was kind to everyone he knew and profoundly concerned about their welfare. He was deeply distressed by the misfortunes of his friends and students. Especially with young people, he would commit himself to helping to alleviate their distress. As if his own good fortune troubled him, he would attend to those less lucky in life, and he often made a difference in providing a new option or a new opportunity for someone who needed it.

Hans supervised the thesis work of more than 30 doctoral and 70 master degree candidates at Penn State and New York University. One of these, Dennis A. Trout, now an administrator with the Environmental Protection Agency, writes:

> HAP enthusiastically shared his knowledge, insight, and philosophy with his students and colleagues. Though I am but one of many, I feel most privileged to have been one of HAP's students, advisees, co-authors, and friends. HAP was well loved and respected, and will certainly be missed and well remembered.

While I never took a course from Hans, I did meet him early in my graduate student days at Penn State. I was in mechanical engineering but soon became a Ph.D. student of John Lumley, then a young staff member in aerospace engineering. In the fall of 1962 I began studying turbulence under Lumley. I soon discovered Hans, but not as a professor; Hans was the liveliest student in Lumley's turbulence course! In a subsequent course we were privileged to be the "test students" for a draft version of *The Structure of Atmospheric Turbulence*.

In the summer of 1964 I periodically visited fellow graduate student Steve Hanna in the office he shared with Hans, Gary Briggs, and a few others. Steve, David Smith, and I were enrolled in visiting professor Frank Pasquill's diffusion course. Gary would have been as well, but he had a summer job with Frank Gifford's group at Oak Ridge. Based on those office visits, I can testify that the atmosphere in that office, if stimulating, was also close to bedlam. Nonetheless, great things happened.

When I finished my Ph.D. degree in 1967 and turned to face the real world, I was armed with my enthusiasm, my new knowledge, Henk Tennekes' sage advice was to seek a job in geophysical rather than engineering turbulence, but there were no job leads. I asked Hans for advice. Hans immediately put together a list of prospects, complete with addresses and phone numbers. I followed up each and eventually accepted one with Duane Haugen's Boundary-Layer Branch (which was then preparing for the 1968 Kansas experiments) at the Air Force Cambridge Research Laboratories. This was part of Mort Barad's Meteorology Laboratory there; Mort's name had been at the top of Hans' list. Hans pointed me in the right direction.

Our family saw the Panofskys frequently over the next 20 years. Hans and I both found ourselves on sabbatical at the University of Washington in the spring of 1973. We took family trips together, including a memorable ferry ride from Tsawwassen (I wouldn't have remembered that name except for Hans' fascination with it). We had dinners together, at his urging usually at Ivar's Salmon House (the salmon is cooked over alder wood, he would remind us) where he chose the meatloaf. The salmon was for Nancy and us.

I last saw Hans at the American Meteorological Society Short Course on Air Pollution Modeling in San Diego in March of 1986. He enrolled as a student and as usual took a front-row seat. In my mind it was 1962 again, with Hans keeping the lecturer alert with his penetrating questions and comments.

Hans A. Panofsky shaped the thinking of two generations of students, many of whom went on to very successful careers in our field. His wealth of contributions to our science are securely recorded in the archival literature. While HAP touched our lives in many ways, his colleagues' reminiscences in this dedication are poignant indicators that his human qualities lie foremost in our memories. Alfred Blackadar sums it up eloquently:

> But the subject that he loved best was people – his friends, his colleagues, his students. They can be found everywhere, in every country, in all walks of life. He loved us very much and was so generous in the way he shared his great talents and understanding with us. In his modest and unassuming way, he let us take credit for many new ideas that were fostered out of his

boundless imagination. No one will ever know how many people were touched and influenced by his life and the warmth of his friendship. The richness of his life remains in our memories and in the accomplishments of the many people he influenced. We were indeed fortunate to have had him as our teacher and our friend.

PUBLICATIONS BY PROFESSOR H. A. PANOFSKY IN
BOUNDARY-LAYER METEOROLOGY

Turbulence Characteristics Along Several Towers, R A. Pielke and H. A. Panofsky, Vol. 1 (1970) 115–130.

Variances and Spectra of Vertical Velocity Just Above the Surface Layer, H. A. Panofsky and C. Mazzola, Vol. 2 (1971) 30–37.

Book Review: *Statistical Methods and Instrumentation in Geophysics*, A. G. Kjelaas (ed.), Teknologisk Forlag, Oslo, Norway (1971) 337 pp., H. A. Panofsky, Vol. 3 (1973) 397.

Turbulence Structure in the Planetary Boundary Layer, N. E. Busch, H. Tennekes and H. A. Panofsky, Vol. 4 (1973) 251–264.

Satellite Radiances and Clear Air Turbulence Probabilities, J. A. Woods and H. A. Panofsky, Vol. 4 (1973) 361–375.

Horizontal Coherence of Wind Fluctuations, C. F. Ropelewski, H. Tennekes and H. A. Panofsky, Vol. 5 (1973) 353–363.

Book Review: *The Planetary Boundary Layer of the Atmosphere*, F. Wippermann, Deutscher Wetterdienst (1973) 346 pp., H. A. Panofsky, Vol. 7, 236.

Two-Point Velocity Statistics Over Lake Ontario, H. A. Panofsky, D. W. Thomson, D. A. Sullivan and D. E. Moravek, Vol. 7 (1974) 309–321.

A Model for Vertical Diffusion Coefficients in a Growing Urban Boundary Layer, H. A. Panofsky, Vol. 9 (1975) 235–244.

Horizontal Coherence and Pasquill's Beta, Hans A. Panofsky and Tateke Mizuno, Vol. 9 (1975) 247–256.

The Validity of Taylor's Hypothesis in the Atmospheric Surface Layer, T. Mizuno and H. A. Panofsky, Vol. 9 (1975) 375–380.

Comments on 'On the Representation of Frequency Spectra in Meteorology', (Correspondence), H A. Panofsky, Vol. 10 (1976) 227.

The Characteristics of Turbulent Velocity Components in the Surface Layer Under Convective Conditions, H. A. Panofsky, H. Tennekes, D. H. Lenschow and J. C. Wyngaard, Vol. 11 (1977) 355–361.

Reply to Comments by B. B. Hicks, H. A. Panofsky, J. C. Wyngaard and D. H. Lenschow, Vol. 15 (1978) 259–260.

On Characteristics of Wind Direction Fluctuations in the Surface Layer, H. A. Panofsky, C. A. Egolf and R. Lipschutz, Vol. 15 (1978) 439–446.

Reply to Comments by J. Lacaze, (Correspondence), H. A. Panofsky and J. C. Wyngaard, Vol. 15 (1978) 527.

Wind Characteristics at the Boulder Atmospheric Observatory, (A Preliminary Report), Steven Schotz and Hans A. Panofsky, Vol. 19 (1980) 155–164.

Lateral Coherence of Longitudinal Wind Components in Strong Winds, Leif Kristensen, Hans A. Panofsky and Stuart D. Smith, Vol. 21 (1981) 199–205.

Wind Profiles at the Boulder Tower, Ann Korrell, H. A. Panofsky and R. J. Rossi, Vol. 22 (1982) 295–312.

Vertical Cross-Spectra of Horizontal Velocity Components at the Boulder Observatory, Rick Soucy, Robert Woodward and H. A. Panofsky, Vol. 24 (1982) 57–66.

Wind Profiles Over Complex Terrain, Zhao Ming, H. A. Panofsky and Robert Ball, Vol. 25 (1983) 221–228.

Wind Fluctuations in Stable Air at the Boulder Tower, Zhou Leyi and H. A. Panofsky, Vol. 25 (1983) 353–362.

A Suggested Refinement for O'Brien's Convective Boundary Layer Eddy Exchange Coefficient Formulation, (Research Note), R. A. Pielke, H. A. Panofsky and M. Segal, Vol. 26 (1983) 191–195.

Boundary-Layer Meteorology 47: 15–16, 1989.

Vertical Coherence and Phase Delay Between Wind Components in Strong Winds Below 20 m, A. J. Bowen, R. G. J. Flay and H. A. Panofsky, Vol. **26** (1983) 313–324.

Vertical Variation of Roughness Length at the Boulder Atmospheric Observatory, (Research Note), H. A. Panofsky, Vol. **28** (1984) 305–308.

Book Review: *Large-Scale Oceanographic Experiment in the WCRP*, Report of the Study Conference in Tokyo, May 1982, 2 volumes, H. A. Panofsky, Vol. **28** (1984) 413–414.

THE STRUCTURE OF THE STABLY STRATIFIED
INTERNAL BOUNDARY LAYER IN OFFSHORE
FLOW OVER THE SEA

J. R. GARRATT and B. F. RYAN

CSIRO Division of Atmospheric Research, Private Bag No. 1, Mordialloc, Vic., 3195 Australia

(Received 13 May, 1988)

Abstract. Observations obtained mainly from a research aircraft are presented of the mean and turbulent structure of the stably stratified internal boundary layer (IBL) over the sea formed by warm air advection from land to sea. The potential temperature and humidity fields reveal the vertical extent of the IBL, for fetches out to several hundred of kilometres, geostrophic winds of 20–25 m s^{-1}, and potential temperature differences between undisturbed continental air and the sea surface of 7 to 17 K. The dependence of IBL depth on these external parameters is discussed in the context of the numerical results of Garratt (1987), and some discrepancies are noted.

Wind observations show the development of a low-level wind maximum (wind component normal to the coast) and rotation of the wind to smaller cross-isobar flow angles. Potential temperature (θ) profiles within the IBL reveal quite a different structure to that found in the nocturnal boundary layer (NBL) over land. Over the sea, θ profiles have large positive curvature with vertical gradients increasing monotonically with height; this reflects the dominance of turbulent cooling within the layer. The behaviour is consistent with known behaviour in the NBL over land where curvature becomes negative (vertical gradients of θ decreasing with height) as radiative cooling becomes dominant.

Turbulent properties are discussed in terms of non-dimensional quantities, normalised by the surface friction velocity, as functions of normalised height using the IBL depth. Vertical profiles of these and the normalised wavelength of the spectral maximum agree well with known results for the stable boundary layer over land (Caughey *et al.*, 1979).

1. Introduction

There has been extensive research on internal boundary layers and the effects of abrupt changes in surface conditions on turbulent boundary layers, much of this work with direct application to atmospheric boundary-layer problems (see Mulhearn, 1981; Garratt, 1987 for a summary and a list of relevant references). The range of such problems includes coastal pollution from industrial sites, the formation of surface-based advective radar ducts and the propagation of low-level prefrontal internal gravity waves or bores. Because of its more significant practical implications, greater attention has been given to the case of onshore flow, particularly that of cool air over warm land. Raynor *et al.* (1979) described light aircraft observations of the mildly convective IBL overland, and compared these with an empirical model for predicting the height of the IBL. They also discussed earlier work, almost all confined to the onshore flow problem. Venkatram (1977, 1986) has considered the modelling aspects of IBL development with onshore flow and has recently examined methods to estimate the height of the coastal IBL under such conditions.

Boundary-Layer Meteorology **47**: 17–40, 1989.
© 1989 *Kluwer Academic Publishers*.

There has been far less work on the offshore flow problem, both in terms of cold continental air over the sea (unstable IBL) and warm continental air flowing out over a cooler sea (stable case). The present paper is concerned exclusively with the stable case. Growth of the stable thermal IBL, for which growth rates are small with fetches of several hundreds of kilometres required to develop an IBL several hundreds of metres deep, has recently been studied by appeal to some historic data and dimensional analysis (Mulhearn, 1981; Hsu, 1983). More recently, Garratt (1987) studied the IBL structure, and factors determining its growth, using a two-dimensional mesoscale numerical model and a simple physical model. In the steady-state situation, the results showed that the IBL could be characterised by a critical layer-flux Richardson number, with the overall behaviour of the mean and turbulent profiles being similar to that of the horizontally homogeneous stable boundary layer overland. The depth of the IBL, h, could be described by

$$h = \alpha U x^{1/2} (g \Delta \theta / \theta)^{-1/2} \tag{1}$$

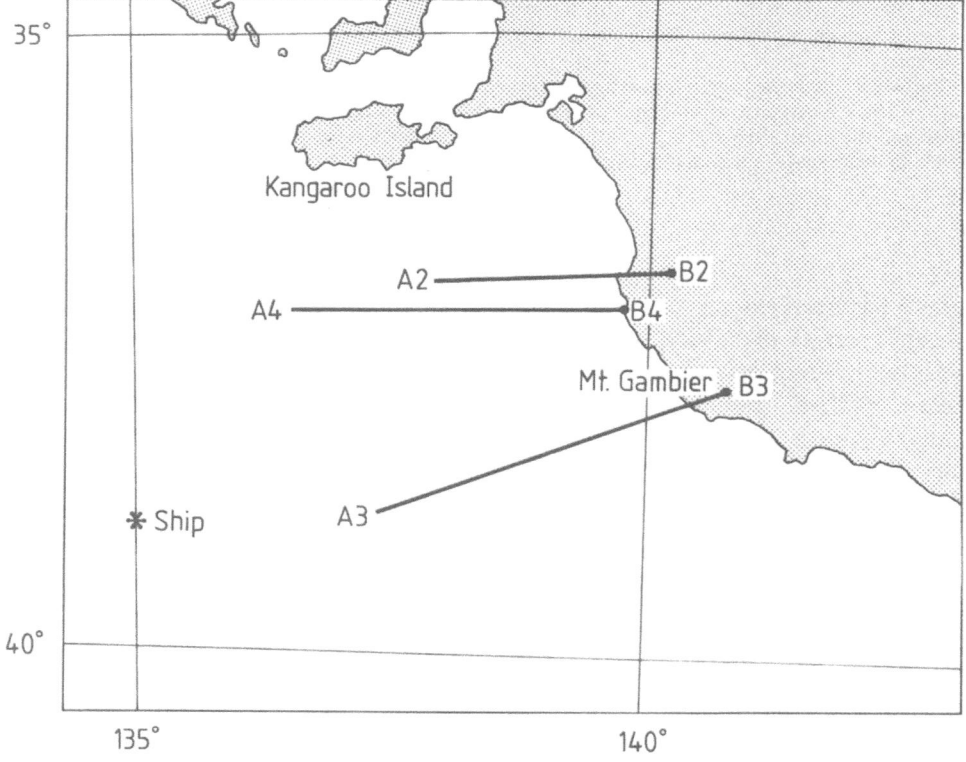

Fig. 1.   Region of southeast Australia, showing: (a) Location of ship HMAS Kimbla (Case 1); (b) Aircraft flight path for Case 2 with coast-parallel winds; (c) Aircraft flight path for Case 3 with coast-normal winds; (d) As in (c) for Case 4.

with $\alpha = 0.014$, where x is the distance from the coast, U is the x component of the geostrophic wind, with the x-axis normal to the coastline, $\Delta\theta$ is the potential temperature difference between continental mixed-layer air and sea surface, θ is the mean potential temperature and g is the acceleration due to gravity. In the diurnally-varying situation, mean profiles within the IBL showed only small differences from the steady-state, although diurnal variations, particularly in the wind maximum, were evident within a few hundreds of kilometres of the coast.

The stimulus for the study of Garratt (1987) and the present work arose indirectly from a major study of Australian summertime cold fronts described for example in Ryan *et al.* (1985). The most vigorous of such cold fronts as they approach the coastline of southeast Australia (Figure 1) are often associated with a strong, offshore prefrontal flow, where hot continental air with mixed-layer potential temperatures of up to 310 to 315 K flows over the ocean with a surface temperature of about 288 K. Consequently a stably stratified IBL forms over the ocean whose structure also reflects the large change in aerodynamic roughness between the land and sea surface. In support of the numerical and physical model study, we present here an analysis of mainly research aircraft observations, the first of their kind we believe to be published on this problem. Most observations were made in prefrontal (cold front) flow conditions offshore, but within several hundreds of kilometres of the coast, in the Mt. Gambier region of southeast Australia (Figure 1).

2. Observations

a. GENERAL

Suitable data were available from both ship (case 1) and aircraft (cases 2 to 4) platforms; the ship, HMAS Kimbla, was stationed at 39° S, 135° E during Phase 3 of the Cold Fronts Research Programme (Ryan *et al.*, 1985) whilst the CSIRO F-27 research aircraft (Figure 2) was used during this and a subsequent experiment to make overwater sorties out of Mt. Gambier (see Figure 1). Ship observations were confined to radiosonde releases made every 2 hours before and during frontal passages. In contrast, aircraft observations consisted of detailed wind and thermodynamic fields, both mean and turbulent, measured on several occasions over time periods of 4 hours approximately (see Sections 2 and 3).

Four situations will be discussed; surface analysis charts are given in Figure 3 to illustrate the generally offshore flow in southeast Australia where measurements were made. Fetches (for air flow from land to sea) lay in the range 0 to 600 km; low-level geostrophic winds were between 15 and 30 m s^{-1}. General information on platform and data types, and on low-level winds can be found in Tables I and II.

Fig. 2.   CSIRO F-27 research aircraft showing instrumented nose cone.

b. Research Aircraft – Instrumentation and Data Analysis

Figure 2 shows the CSIRO F-27 research aircraft with the nose cone carrying the turbulence instruments; during a research flight, the aircraft speed is close to $80 \, \text{m s}^{-1}$. On board, a computer-controlled data acquisition system records data from slow response instruments (at 5 hz) and fast response turbulence instruments (at 50 hz). A hard-copy digital output of data each 20 s allows for onboard preliminary analysis whilst all data are recorded onto magnetic tape.

Standard observations included –

  (i) Dry- and wet-bulb temperatures (platinum–resistance thermometers);
 (ii) Static and dynamic pressure (pitot tubes);
(iii) Surface temperature, T_s (Barnes' radiometer); this was converted to a sea-surface potential temperature θ_s using $\theta_s = T_s(1000/p)^{R/c_p}$ with p in mbar. Values of p for each case were taken from Figure 3.
 (iv) Aircraft altitude (radio altimeter – below 600 m);
  (v) Mean winds from
        (a) Doppler radar wind-finding unit (2-min averages);
        (b) Aircraft inertial navigation system.

Turbulence instrumentation consisted of the following –

  (i) Damped pressure vanes of the type described by Johnson *et al.* (1978) giving angles of sideslip and attack;

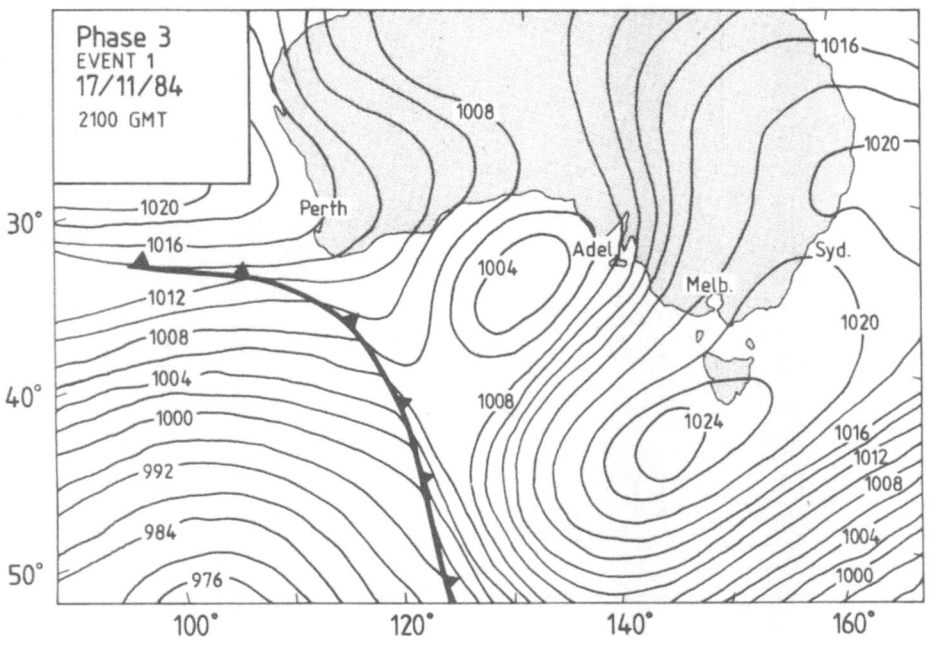

Fig. 3a.

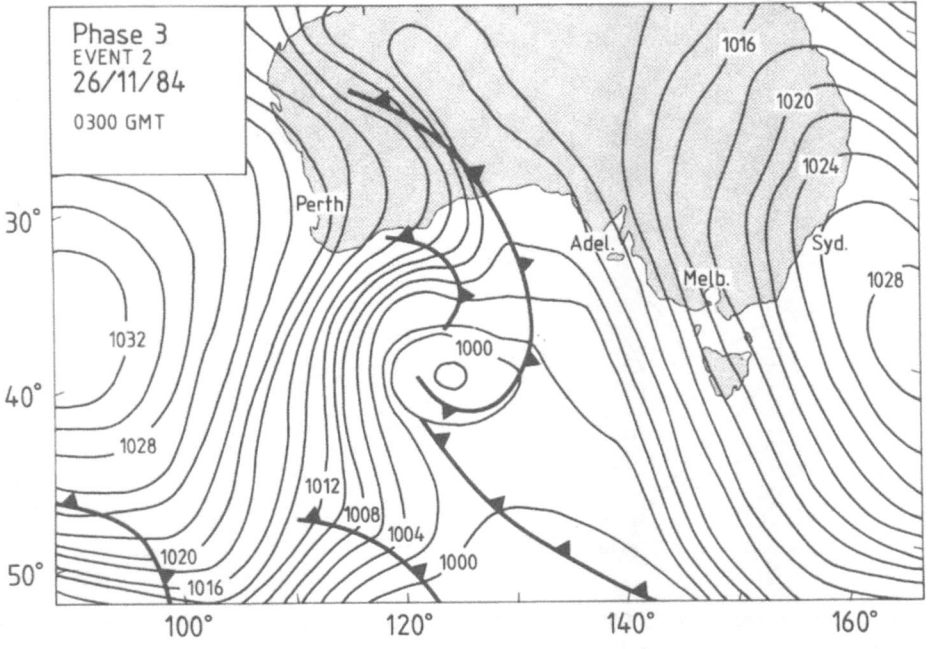

Fig. 3b.

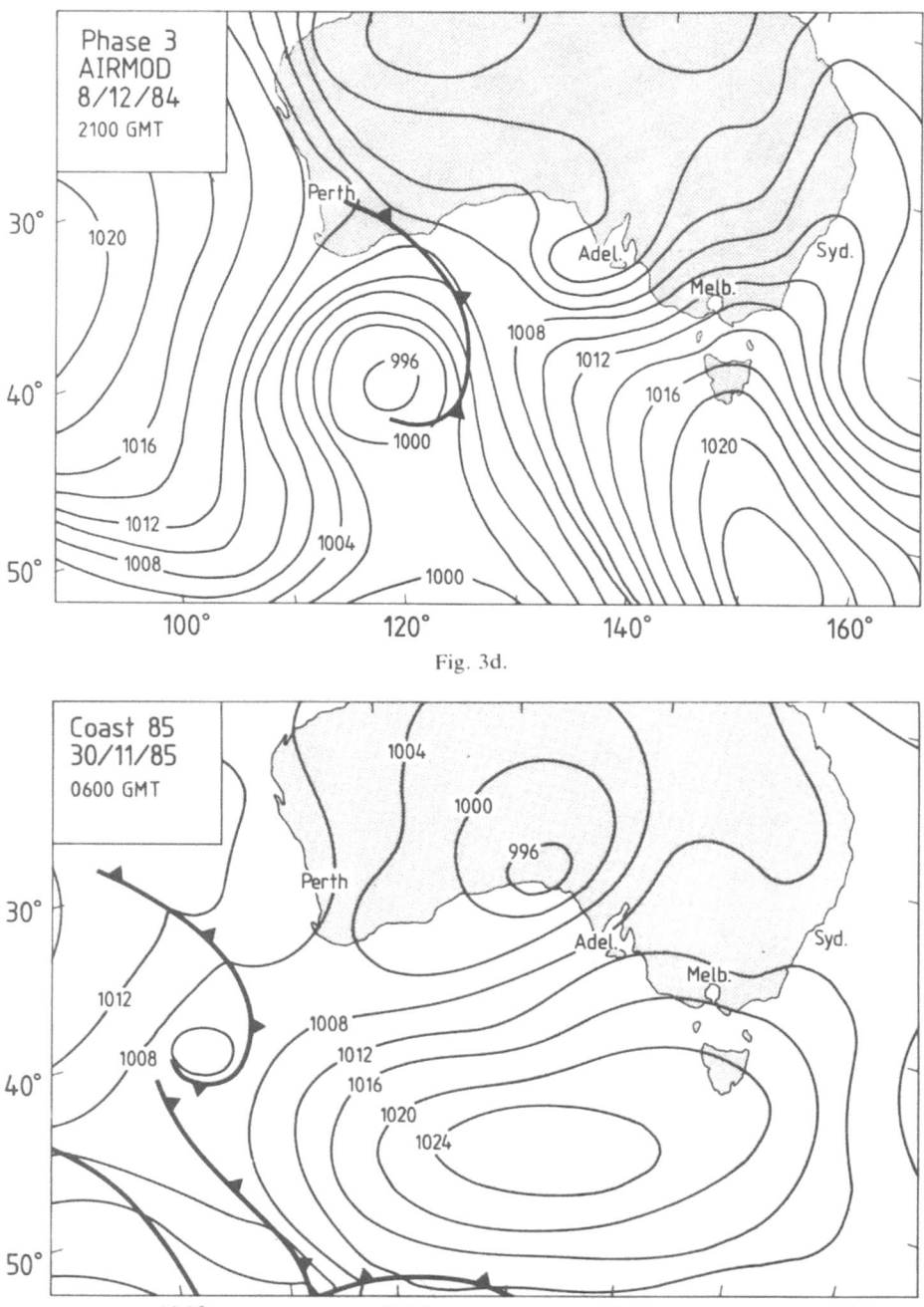

Fig. 3d.

Fig. 3a–d. Surface analysis charts for the four cases under consideration: (a) Case 1 – 2100 GMT, 17 November, 1984; (b) Case 2 – 0300 GMT, 26 November, 1984; (c) Case 3 – 2100 GMT, 8 December, 1984; (d) Case 4 – 0600 GMT, 30 November, 1985.

TABLE I

General information for the 4 cases under study

Case	Date	Platform	Flow	Data
1	18 Nov. 84	Ship	Coast normal	Mean
2	26 Nov. 84	Aircraft	Coast parallel	Mean
3	9 Dec. 84	Aircraft	Coast normal	Mean and turbulent
4	30 Nov. 85	Aircraft	Coast normal	Mean and turbulent

TABLE II

Summary of low-level winds for the 4 cases under study

Case	Location	Time (GMT)	Height (m)	Speed (m s^{-1})	Direction
1	Kimbla	1810 (17 Nov.)	100	14.5	049
		0028 (18 Nov)	100	15.5	066
	Mt. Gambier	1600 (17 Nov.)	200	5.4	068
		2200	200	9.9	031
		0000 (18 Nov.)	200	7.5	037
2	Aircraft (point A2)	0250	150	21.0	345
3	Aircraft (point B3)	0453	150	6.5	066
	(point A3)	0541	150	17.0	072
4	Aircraft (point B4)	0447	150	7.5	115
	(point A4)	0525	150	15.0	096

(ii) Static and dynamic pressure transducers located at the tip of the nose-boom.

Winds were calculated by combining vane, pressure transducer and INS data.

(iii) Dry- and wet-bulb fine-wire resistance thermometers.

All data were analysed using a suite of programs developed by Dr C. E. Coulman of the CSIRO Division of Radiophysics. 5 hz data were summarised as 1-min averages to give mean fields, whilst 50 hz data were used to give variance and covariances, spectra and cospectra for selected portions of the flight legs (usually about eight 3-min runs per flight).

c. IBL OBSERVATIONS, WITHOUT TURBULENCE DATA

Case 1. An anticylone (see Figure 3a) well to the southeast of the ship (Figure 1) directed a NEly flow from continent to the ocean, with the 900 mb over-water fetch at the ship equal to about 500 km. Sea-surface temperatures at the ship were approximately 287 K (θ_s = 286.3 K).

Figure 4a shows a time-height cross-section of potential temperature in the prefrontal and frontal regions to illustrate the shallow prefrontal IBL extending well ahead of the leading edge of the frontal transition zone. The deep, well-mixed region about the IBL is the result of strong heating over the land (during the day) from where the air has originated. Figure 4b gives two sets of vertical profiles which overall reveal an IBL of depth about 300 m.

Case 2. Anticyclonic flow (Figure 3b) directed a NNW flow over southeastern Australia giving coast-parallel flow in the Mt. Gambier region. Air arriving at point A2 farthest from land on the flight path (Figure 1) has travelled some 100–150 km over the water from Kangaroo Island; thus Figure 4c shows a set of vertical profiles of temperature and humidity based on aircraft soundings made at that point which reveals a shallow, stable IBL of depth $\approx$ 200 m. Winds at 200 m (22 m s^{-1} from 343 deg) and 1000 m (19 m s^{-1} from 311 deg) suggest the existence of a low-level maximum at, or just above, 200 m height.

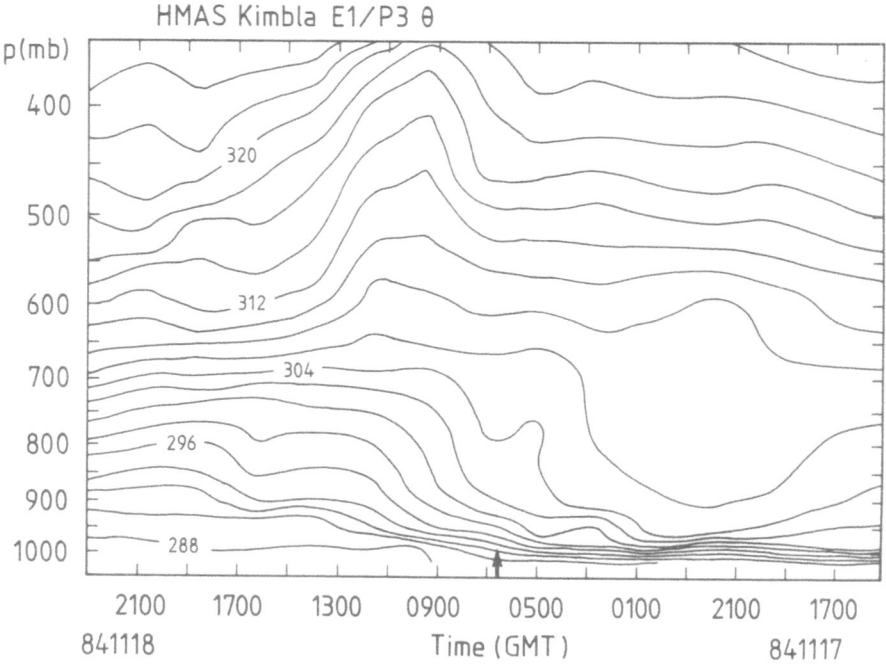

Fig. 4a.

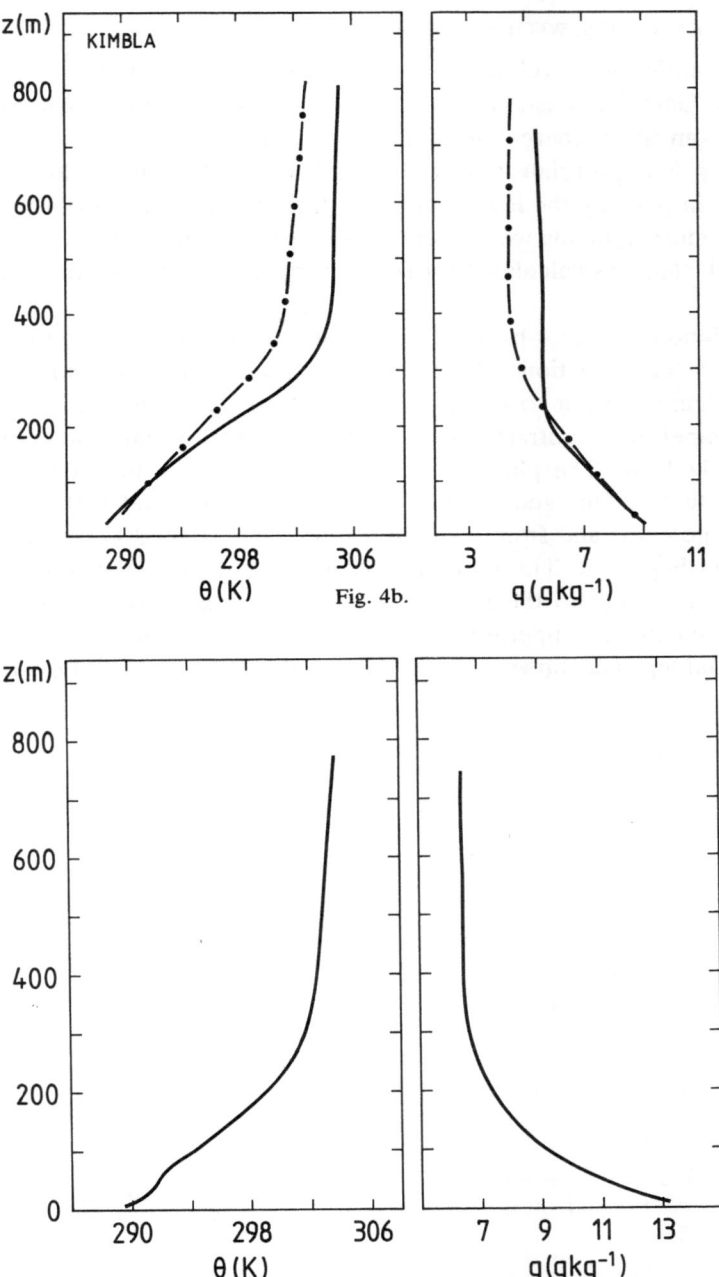

Fig. 4a–c. (a) Time-height cross-section of potential temperature from HMAS Kimbla on 17 and 18 November, 1984 (Case 1). The vertical arrow denotes the position of the surface cold front. (b) Vertical profiles of potential temperature and specific humidity at 1810 GMT, 17 November, 1984 (pecked curve) and 0028 GMT, 18 November, 1984 (full curve) (Case 1). (c) Vertical profiles of potential temperature and specific humidity based on aircraft sounding at point A2 (Figure 1) for Case 2 – 0250 GMT, 26 November, 1984.

d. IBL OBSERVATIONS, WITH TURBULENCE DATA

Case 3. An anticyclone well to the south of Tasmania directed an E to NE flow over the Mt. Gambier coastal region (Figure 3c). Surface geostrophic winds were calculated from special barometer data available in this area; a value of 27 m s^{-1} from 055 deg is appropriate midway through the flight. The flight pattern was determined, in part, by the low-level winds (Figure 1) and is shown in Figure 5. Turbulence measurements were made at the three nominal levels – 20, 150 and 300 m – with statistics calculated for seven 3-min runs indicated in Figure 5 (see later).

Time variations on the 4-hr flight were small so all data were used to construct the composite cross-sections of potential temperature (Figure 6a) and specific humidity (Figure 6b) out to distances of about 300 km from the coastline. The potential temperature contrast between the continental air and the sea surface is approximately 14 K. The plane of the cross-sections lies along 070 deg, which compares with a surface geostrophic wind along 055 deg and 150 m winds from 066 deg at the coast and from 072 deg at point A3, some 300 km offshore (see Figure 1 and Figure 5). The sections reveal a well-defined IBL growing slowly with fetch (x), similar to the numerical results described in Garratt (1987; his Figure 2a). Potential temperature profiles constructed from aircraft soundings and horizontal legs are shown in Figure 6c for three values of x. Low-level wind

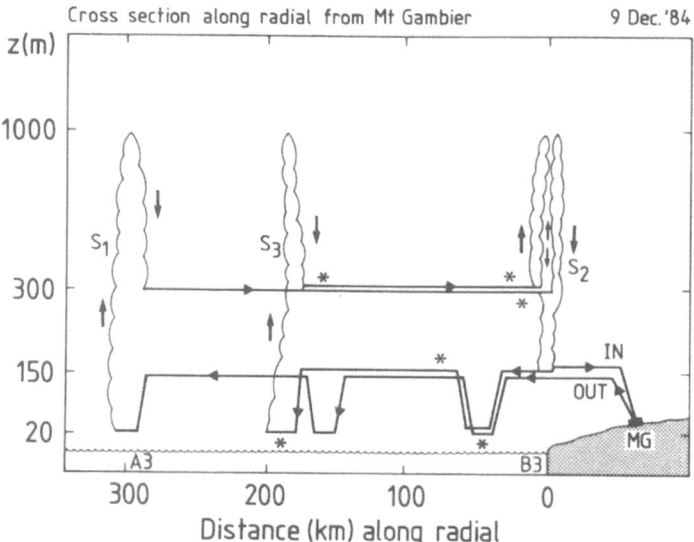

Fig. 5.   Cross-section of flight path on 9 December, 1984 along a radial out of Mt. Gambier; this shows soundings (S) and horizontal legs. The asterisks denote locations of 3-min runs where turbulence statistics have been evaluated.

profiles of u and v (these are defined relative to the radial along 070 deg, with positive u along the radial towards the sea, and positive v normal to the radial and rotated anticlockwise on positive u) are shown in Figure 7 at three values of x. Overland v values are large and negative, indicative of significant cross-isobar flow at low levels. The trend of v towards zero as x increases is consistent with anticlockwise rotation due to both a decrease in frictional effects over the sea and to an inertial oscillation of period about 24 hr for this latitude. The change in the u profile as x increases indicates significant advective acceleration, as would be expected in flow from rough to a smooth surface. Over land, the u profile shows some evidence of a low-level maximum; over the sea, this feature is much more prominent and lies near the top of the IBL (see later).

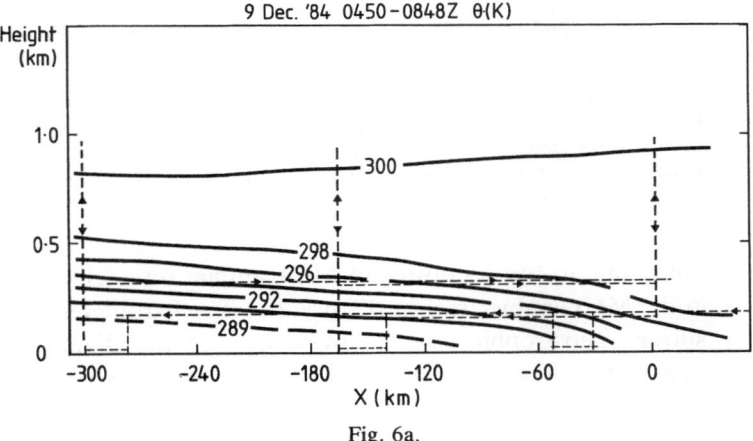

Fig. 6a.

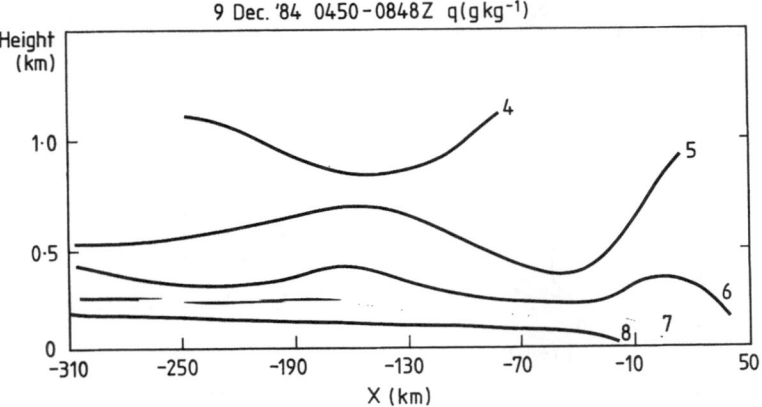

Fig. 6b.

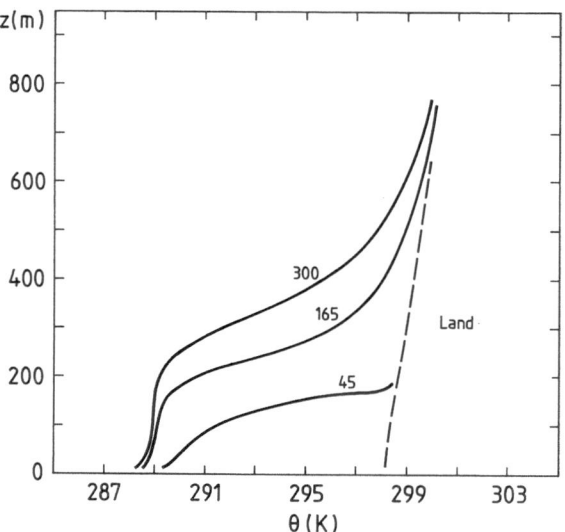

Fig. 6a–c. (a) Cross-section (x, z) of potential temperature for Case 3–9 December, 1984. Pecked lines show aircraft path. (b) As in (a) for specific humidity (units of $g\,kg^{-1}$). (c) Vertical profiles of potential temperature at three values of x (km), $x = 165$ and $300\,km$ corresponding to soundings S3 and S1 (Figure 5) and $x = 45\,km$ based on a low-level leg and the composite of other horizontal legs (Figure 6a). The pecked curve is based on the sounding S2 at $x = 0\,km$.

Case 4. An anticyclone to the southwest of Mt. Gambier directed an E to ENE flow offshore in the Mt. Gambier region (Figure 3d). The barometer array in this case gave a surface geostrophic wind midway through the flight of $19\,m\,s^{-1}$ from 080 deg. As in the previous case, the aircraft pattern was partly based on

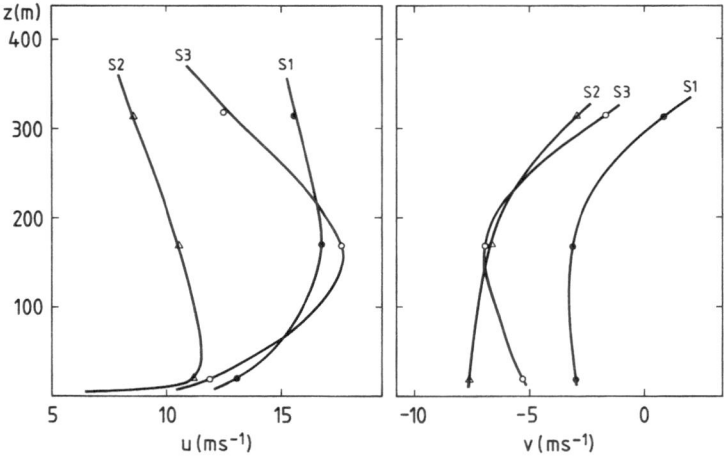

Fig. 7. Vertical profiles of wind components u and v for the three soundings shown in Figure 5 (Case 3). Sounding S1 at point A3 ($x = 300\,km$); sounding S2 at point B3 ($x = 0$) and sounding S3 at $x = 165\,km$ (refer Figure 5).

low-level winds; nominal flight levels were 20, 150, 225 and 300 m with turbulence statistics being calculated for nine 3-min runs. On this occasion, time
variations were significant – we therefore took overlapping soundings or legs,
made an estimate of the time trend and adjusted data at later times based on the
inferred time rate of change of properties. Cross-sections of potential temperature (Figure 8a) and specific humidity (Figure 8b), and vertical profiles of
potential temperature based on aircraft soundings at 3 values of x (Figure 8c),

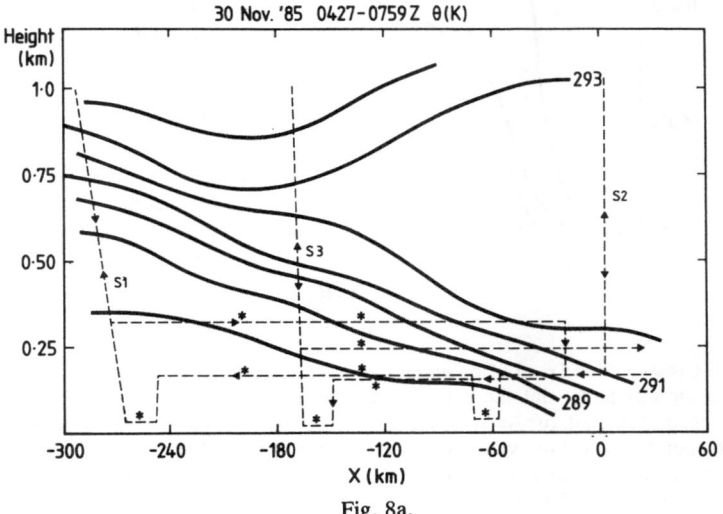

Fig. 8a.

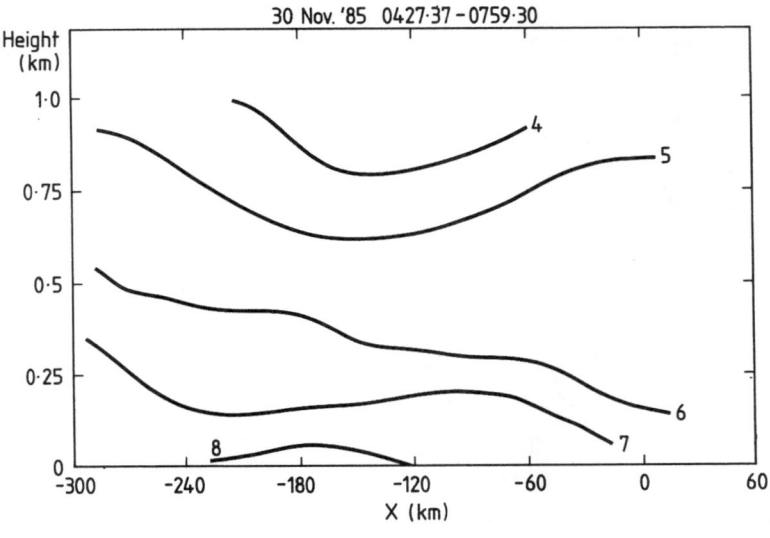

Fig. 8b.

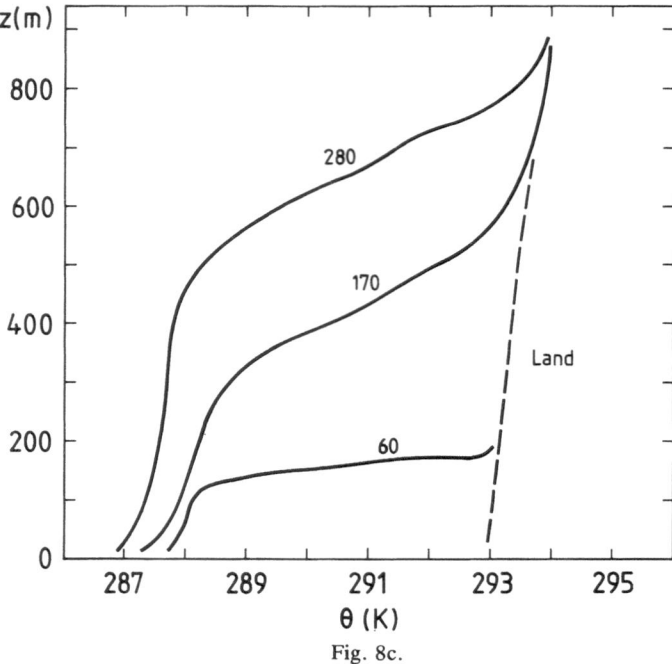

Fig. 8c.

Fig. 8a–c. (a) Cross-section (x, z) of potential temperature for Case 4–30 November, 1985. Pecked lines show the aircraft path, and the asterisks denote locations of 3-min Runs where turbulence statistics have been evaluated. (b) As in (a) for specific humidity (units of $g\,kg^{-1}$). (c) Vertical profiles of potential temperature for three values of x (km); $x = 170$ and 280 km based on soundings S3 and S1 respectively (Figure 8a) and $x = 60$ km based on a low-level leg and a composite of other horizontal legs. The pecked curve is based on the sounding S2 at $x = 0$ km.

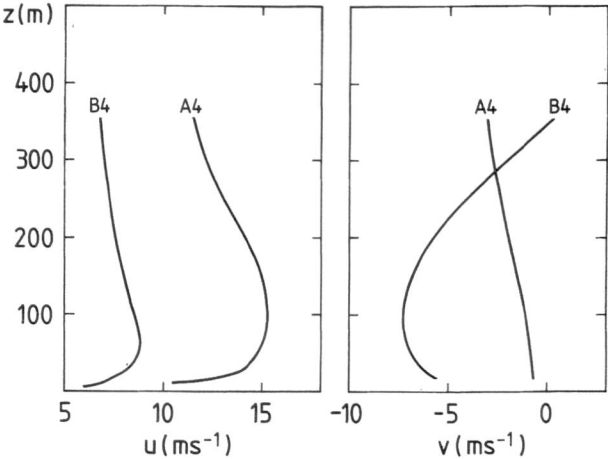

Fig. 9.   Vertical profiles of wind components u and v for Case 4, over the land (near point B4 at $x = 0$) and at point A4 (Figure 1) some 280 km offshore (Case 4).

again reveal an IBL, somewhat deeper than in Case 3, presumably related to the weaker temperature difference between continental air (potential temperature of 293 K) and the sea (potential temperature of 286 K), i.e., 7 K compared with 14 K for case 3. On this occasion, the plane of the cross-section lies along 090 deg, with the surface geostrophic wind from 080 deg and low-level (50 m) winds from 025 deg at the coast and 110 deg at point A4 (Figure 1), some 300 km from the coastline. Wind profiles shown in Figure 9 have features very similar to Case 3, with offshore acceleration in u and development of the maximum, and decreases in the v component due to rotation towards smaller cross-isobar flow.

Wind structure in both cases 3 and 4, with the implied rotation in the flow over the sea and the low-level wind maximum close to the top of the IBL, is consistent with the numerical results of Garratt (1987; his Figure 4d).

3. Turbulence Properties Within the IBL

a. FLUCTUATIONS

Values of variances and covariances were calculated for several 3-min runs for each of Flights 3 and 4 to cover a range in x and z. Samples of calibrated output from various instruments are shown in Figure 10 – these illustrate the relatively large fluctuations in T at 150 m height overland prior to crossing the coastline (Figure 10a), and in Figure 10b the relatively small temperature fluctuations at 25 m height over the sea characteristic of a near-neutral or strongly stably-stratified layer (Case 3). In Figure 10c we show fluctuations over the sea for Case 4 (not so stable) together with velocity fluctuations. These data illustrate the strong correlation between dry- and wet-bulb temperature variations close to the sea surface, even in stable conditions, and the much larger fluctuations in horizontal than in vertical velocity. Statistics of these data records, together with spectral decomposition, will be discussed in Section 3c.

b. LOW-LEVEL STATISTICS

Table III gives a range of quantities measured at the lowest flight levels for Cases 3 and 4, typically for z between 20 and 40 m, as far as 300 km offshore.

Sea-surface temperatures (SST), identified as T_s and converted to a potential temperature θ_s, were measured with a Barnes PRT-5 radiometer. They show little systematic variation with x, with a range of only 0.9 K in Case 3 and of 1.3 K in Case 4; potential temperature differences do show a systematic decrease in both cases with increase in x, and although turbulence quantities and mean wind exhibit an increase with increasing x for Case 4, this is not seen in Case 3. Sensible heat fluxes into the sea surface are negligible in the more stable case, and in Case 4 are comparable with values typically seen in the nocturnal

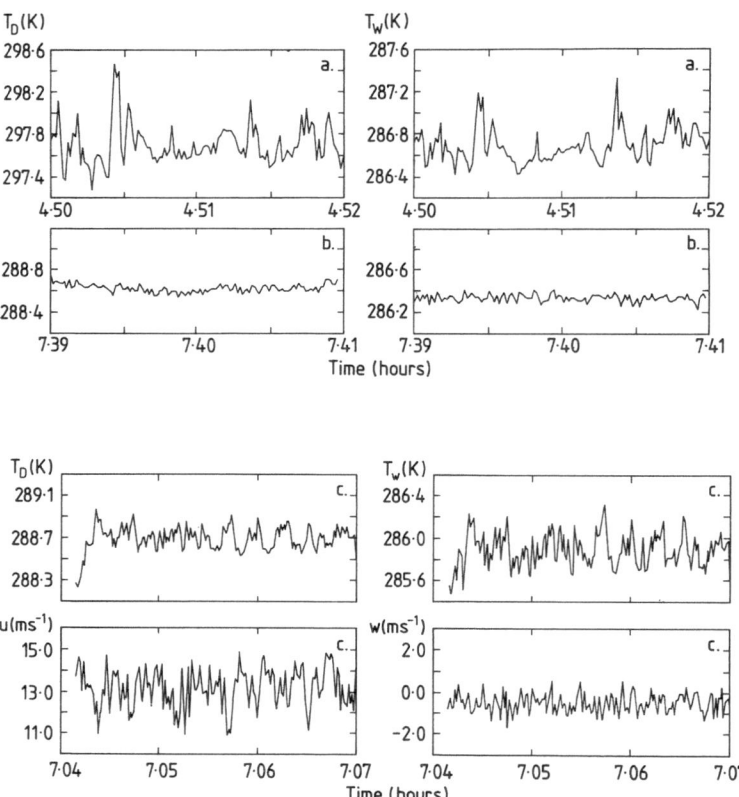

Fig. 10. (a) Case 3 – sample dry- and wet-bulb temperature fluctuations over land at a height of 150 m. (b) As in (a) at a height of 25 m over the sea ($x = 165$ km). (c) Case 4 – sample dry- and wet-bulb temperature fluctuations, together with u and w wind components, at a height of 25 m above the sea ($x = 160$ km).

boundary layer overland (e.g., see Garratt, 1982). Evaporative fluxes are much larger than H as is to be expected over the sea; the strong correlation seen in Figure 10c between dry- and wet-bulb temperatures is verified in Table III with reference to the correlation coefficient $r_{\theta q}$; with one exception, its values are in excess of 0.66, typical of flow over the ocean (Friehe *et al.*, 1975).

Drag coefficients, reduced to a reference level of 10 m by assuming a logarithmic wind profile, are typically about 0.8×10^{-3}; even allowing for the effects of the stable stratification, these values seem small compared to neutral values (e.g., Garratt, 1977) – at a wind speed $V = 12.5$ m s^{-1}, these are about 1.5×10^{-3}. In contrast, the ratio σ_w/u_* has values (about 1.35, with one exception) quite comparable with those found elsewhere over the land and the sea (e.g., Pasquill and Smith, 1983; Chapter 2).

TABLE III

Low-level mean and turbulent quantities for flights 3 and 4. The identification of relevant 3-min runs is also shown (R1, R2 etc.). Here $\delta\theta$ is the potential temperature difference between the aircraft and the sea surface

Time (GMT) Case 3	x (km)	z (m)	θ_s (K)	$\delta\theta$ (K)	V	u_* m s^{-1}	σ_w	C_D $\times 10^3$	σ_w/u_*	H W m^{-2}	E	$r_{\theta q}$
0502	40	20	287.3	3.04								
0719 (R2)	40	21	287.3	2.71	15.5	0.41	0.56	0.80	1.37	−2	30	0.29
0512	150	20	286.7	2.31								
0740 (R4)	150	19	286.4	2.50	13.6	0.34	0.48	0.71	1.41	−1	12	0.66
0545	280	20	287.2	1.70								
Case 4												
0454 (R1)	65	40	285.4	1.50	12.1	0.24	0.44	0.51	1.83	−6	53	0.68
0706 (R6)	165	24	286.6	1.57	13.2	0.36	0.48	0.86	1.33	−18	55	0.78
0529 (R4)	265	35	286.7	0.81	14.7	0.40	0.53	0.92	1.33	−22	89	0.86

c. VERTICAL VARIATION OF TURBULENCE QUANTITIES

Studies of the stably stratified boundary layer over land show a systematic and well ordered decrease of turbulence quantities with height, when height and turbulence parameters are suitably normalised. For velocity, this is done with the surface friction velocity u_{*0}, and for height with boundary-layer depth, h (see Caughey et al., 1979). In the time-evolving boundary layer over land, normalised turbulence quantities are a function of z/h and of time, but where the time evolution is small (as is usually the case well after sunset for example), z/h becomes the dominant dependent variable. In the case of the quasi-steady IBL, but with evolution in distance downstream present, a dependence upon both z/h and x/h must be assumed. Our data cover an x/h range from 300 to about 1800 (see Figure 13 later and related text) but are too few to examine any x/h dependence; in a slowly evolving IBL, however, turbulence profiles readily adjust to the mean conditions and any x/h effect is likely to be small. This is implied in the numerically based K profiles in Figure 4e of Garratt (1987), which show only small changes from $x = 50$ to 500 km (x/h range of 250 to 2000).

Values of u_* in Table III have been averaged for each of Case 3 and 4 to give an estimate of the 'surface' value u_{*0}. This is done to provide a more statistically

TABLE IV

Curvature of potential temperature profiles in stable boundary layers; γ is the profile curvature factor defined by André and Mahrt (1982) as $\gamma = 1 - 2(\theta(h/2) - \theta_s)/\Delta\theta$

Source	Surface	Turbulent cooling	Radiative cooling	γ	η
Andre and Mahrt (1982)	Land (Wangara)	small	large	−0.74	0.5
	Land (Voves)	medium	medium	0.17	>1
Garratt and Brost (1981)	Land	medium	nil	0.14	>1
(Numerical)	Land	medium	medium	−0.14	<1
These data	Sea	medium	small	0.5	2

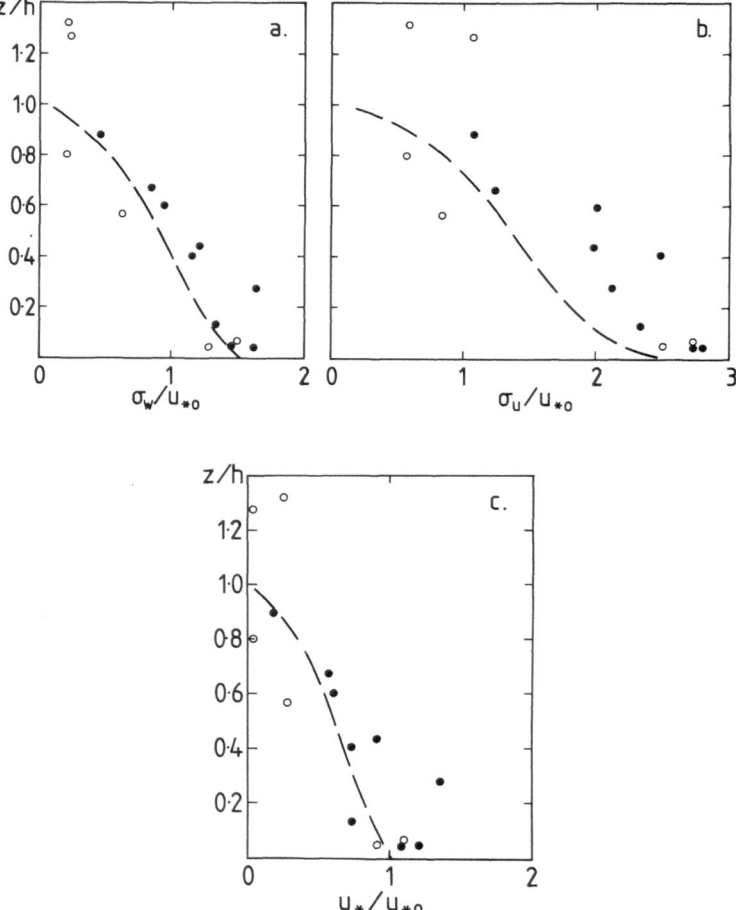

Fig. 11.   Normalised turbulence velocity quantities as a function of z/h – data from Case 3 (○) and Case 4 (●); curves are from Caughey et al. (1979).

reliable value of the surface friction velocity used as the basic scaling velocity here (a 3-min leg at $80\ \mathrm{m\ s^{-1}}$ is equivalent to a 20 min averaging time at a single point for a $12.5\ \mathrm{m\ s^{-1}}$ wind – Wyngaard, 1973). Values of h in Table IV were estimated mainly from the potential temperature cross-sections (Figures 6a and 8a) based on horizontal legs and vertical profiles; h was taken as the level at which there is a sharp decrease in the vertical gradient of θ. In presenting turbulence statistics, we have concentrated, though not exclusively so, on velocity data since temperature variances and hence spectral intensities tended to be small. Figure 11 shows normalised velocity quantities (x/h between 125 and 330) σ_u/u_{*0}, σ_w/u_{*0} and u_*/u_{*0} plotted against z/h. Also shown are mean curves from Caughey *et al.* (1979) for the evolving NBL over land; given that we are using only 3-min runs, there is a satisfactory degree of agreement. Residual (decaying) turbulence existing above the IBL from advection of the continental mixed layer probably contributes to non-zero variance at $z/h > 1$.

We have not shown any spectra or cospectra but prefer to examine the behaviour of the spectral peaks, suitably normalised, for all quantities as was done by Caughey *et al.* They found (as is intuitively to be expected) that λ_m, the wavelength of the spectral maximum, depends on z in the boundary layer and on the boundary layer depth. Their curves are shown in Figure 12, together with our own data, for q and θ (12a), u and v (12b) and w (12c); again there is broad correspondence, though we have not taken into account any h/L dependence (L being the Monin–Obukhov length) as done by Caughey *et al.* As is the case over land, both the scalars and horizontal velocity components have $\lambda_m > h$ near the top of the IBL.

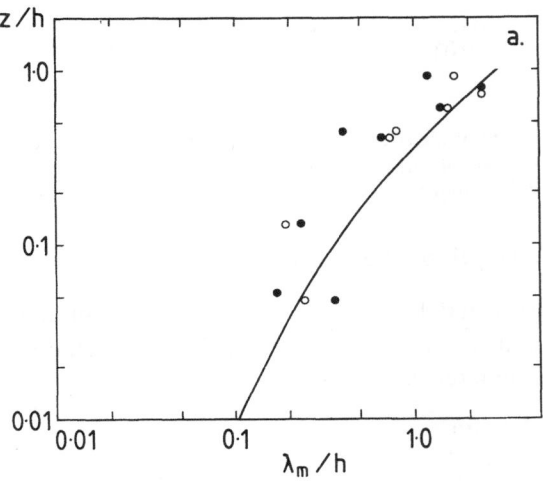

Fig. 12a.

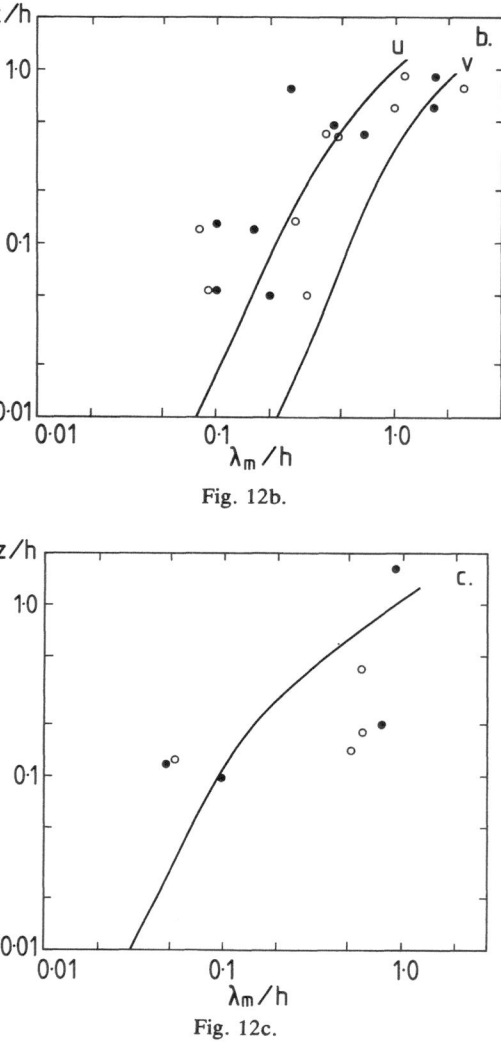

Fig. 12b.

Fig. 12c.

Fig. 12a–c.   The wavelength of the spectral maximum, normalised by the IBL depth h, as a function of z/h – data from Cases 3 and 4: (a) Potential temperature (O) and specific humidity (●); (b) u (●) and v (O) wind components; (c) w component, for Case 3 (O) and Case 4 (●).

4. Depth of the IBL and its Parameterisation

Equation 1 was suggested by a physical model, with numerical model results giving $\alpha = 0.014$ (Garratt, 1987 – his Figure 10); the relation was at least valid for the ranges of parameter values as follows,

$$x:\ 100\text{–}900\ \text{km}$$
$$U:\ 10\text{–}30\ \text{m s}^{-1}$$
$$\Delta\theta:\ 5\text{–}16\ \text{K}\ .$$

The numerical coefficient represented by α (see Appendix) actually depends on the θ profile shape (through the quantity A_0), layer flux Richardson number (R_f), geostrophic drag coefficient (C_G) and β, the angle the wind vector above the IBL makes with the coastline. For the values of fetch, x, used for the 4 cases, this angle is probably within the range $+10$ deg. For cases 3 and 4, geostrophic drag coefficients have values close to 0.25×10^{-3}, and for case 4, for x between 165 and 300 km where H values are significant, the layer flux Richardson number as defined in Garratt (1987) has a value of 0.13. Both parameters are thus comparable with the numerical results.

The shape factor A_0 is not too sensitive to the profile shape; this can be seen by setting

$$(\theta - \theta_s)/\Delta\theta = (z/h)^n \tag{2}$$

and evaluating Equation (A2) with the assumption that wind and θ profile shapes are the same (this is not a critical assumption). The result gives A_0 a value of 3 for $n = 1$ and 2.3 for $n = 3$. Potential temperature profiles for Cases 1, 3 and 4 are shown in Figures 4b, 6c and 9c for several values of x. Normalised profiles are shown in Figure 13 for a range in x/h from 300 to 1800; data are generally consistent with a value of $n = 2$, and hence with $A_0 = 2.7$, with no significant x/h dependence. This value of n implies positive curvature in the θ profile (concave to the surface) and contrasts for example with the large negative curvature found within the Wangara NBL; Yamada (1979) analysed data from 5 nights and, though using a different profile form, his results imply $n = 0.5$. This behaviour was confirmed by André and Mahrt (1982) who discussed the dependence of profile curvature upon the relative contributions of radiative and turbulent

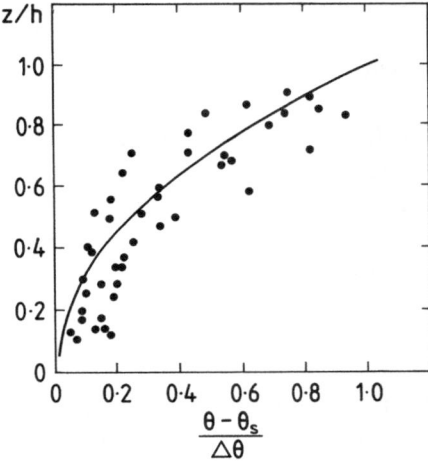

Fig. 13. Normalised temperature profiles based on data contained in Figures 4, 6 and 8 (Case 2 excluded). The mean curve represents Equation 2 with $n = 2$.

TABLE V

Quantities used for the evaluation of Equation 1

Case	x (km)	$\Delta\theta$ (K)	U (m s^{-1})	h (m) obs	α	h (m) calc
1	560	17	20	300	0.015	280
1	560	14.5	20	350	0.016	305
2	200	16	20	300	0.023	175
3	45	12	25	150–200	0.020	120
3	165	11	25	350–500	0.024	240
3	300	11	25	450–550	0.021	320
4	50	7	19	150–200	0.018	125
4	170	7	19	500–650	0.032	230
4	280	8	19	700–850	0.037	275

cooling. They introduced a profile curvature factor, γ, and compared values for the Wangara site and a French site where turbulent cooling was more important. These results and others are shown in Table IV; in particular, the numerical results of Garratt and Brost (1981) showed positive curvature when radiative cooling was set to zero, but slightly negative curvature when radiative and turbulent cooling were comparable.

Table V gives a summary of the quantities relevant to the IBL depth and an estimate of the IBL depth using Equation 1 with $\alpha = 0.014$ as given in Garratt (1987). The most significant result is that, in most cases, Equation 1 with $\alpha = 0.014$ underestimates h considerably, i.e., the implied value of α using the observed heights varies between 0.015 and 0.035 for the range of x considered (which lies within the range of the numerical results); the mean value of α is 0.024. Hsu (1983) described some data covering the x range predominantly of 5–200 km; his values of α can be inferred from his Table I and cover the range 0.01–0.03 approximately with a mean value of 0.015. Mostly his data are for $x < 50$ km.

Equation 1 is of course based on several bold assumptions, including steady state conditions, the absence of geostrophic wind shear and no turbulence above the IBL, all of which may not be valid in the cases summarised in Table IV. In addition, the parameters A_0, R_f and C_D are assumed constant with x in the model although this may not be the case in the real world.

5. Final Comment

The stably stratified IBL in offshore flow over the sea has a number of features common to the NBL over land. These include the low-level wind maximum near the boundary-layer top and the vertical variation, using z/h as the vertical coordinate, of normalised turbulent velocity quantities and the wavelength of the spectral maximum. One major feature concerns the shape of the mean potential temperature profiles; the IBL observations reveal a profile with large positive curvature (concave to the surface – refer Figure 13) consistent with profiles found in the NBL over land when turbulent cooling dominates over radiative cooling. Where radiative cooling dominates, profiles have large negative curvature.

The IBL depth h is found to depend upon several external parameters as described by, e.g., Garratt (1987) and in Equation 1. The mean value of $\alpha(0.024)$ is larger than that found numerically (0.014) and that implied in the observations discussed by Hsu (1983) – a value of 0.015 – for fetches < 50 km. Overall we believe a value of α of about 0.02 in Equation 1 should predict h to within 25–50% or so on most occasions for x between 5 and 300 km at least.

Acknowledgements

To the aircrew of the CSIRO Fokker F-27 research aircraft; to Bill Physick for participation during the flights in 1984 and to Russell O'Brien and John Wren for data analysis.

Appendix

An alternative derivation of $\partial h/\partial x$ (Equations 7 and 13 in Garratt, 1987) is given. The steady-state equation for θ, viz.,

$$u\frac{\partial \theta}{\partial x} + w\frac{\partial \theta}{\partial z} = -\frac{\partial \overline{w'\theta'}}{\partial z} \tag{A1}$$

is integrated between the surface and h assuming the following profile forms –

$$u/U = f_1(z/h); \qquad \frac{\theta - \theta_s}{\Delta\theta} = f_2(z/h); \qquad w/w_h = f_3(z/h).$$

After some manipulation, the equivalent to Equation 7 of Garratt (1987) becomes

$$\partial h/\partial x = w_h/BU + (A_0/U\Delta\theta)\overline{w'\theta'_0}.$$

Here A_0 is a profile shape factor defined as

$$A_0^{-1} = \int_0^h f_1\, \partial f_2/\partial h\, dz. \tag{A2}$$

Use of the critical layer-flux Richardson number concept (Equations 9–11 in Garratt, 1987) requires profile forms for the fluxes, so we take

$$\overline{w'\theta'} = \overline{w'\theta'_0}f_4(z/h); \qquad u_*^2 = u_{*0}^2 f_5(z/h),$$

whence (cf. Equation 12 of Garratt, 1987),

$$\overline{w'\theta'_0} = u_{*0}^2 GR_f/(g/\theta)hf(z/h),$$

with,

$$f(z/h) = \int_0^h f_4(z/h)\,\mathrm{d}(z/h)\Big/\int_0^h f_5(z/h)\partial f_1/\partial(z/h)\,\mathrm{d}(z/h).$$

Upon integration we have (cf. Equation 14a of Garratt, 1987),

$$h^2 = \left(\frac{2A_0f(z/h)R_fC_D}{\cos^3\beta}\right)U^2x(g\Delta\theta/\theta)^{-1}. \tag{A3}$$

Thus the quantity $(-A_0f(z/h))$ is equivalent to the factor A in Garratt (1987).

References

André, J. C. and Mahrt, L.: 1982, 'The Nocturnal Surface Inversion and Influence of Clear Air Radiative Cooling', *J. Atmos. Sci.* **39**, 864–878.

Caughey, S. J., Wyngaard, J. C., and Kaimal, J. C.: 1979, 'Turbulence in the Evolving Stable Boundary Layer', *J. Atmos. Sci.* **36**, 1041–1052.

Friehe, C. A., LaRue, J. C., Champagne, F. H., Gibson, C. H., and Dreyer, G. F.: 1975, 'Effects of Temperature and Humidity Fluctuations on the Optical Refractive Index in the Marine Boundary Layer', *J. Optical Soc. Amer.* **65**, 1502–1511.

Garratt, J. R.: 1977, 'Review of Drag Coefficients over Oceans and Continents', *Mon. Wea. Rev.* **105**, 915–929.

Garratt, J. R.: 1982, 'Observations in the Nocturnal Boundary Layer', *Boundary-Layer Meteorol.* **22**, 21–48.

Garratt, J. R.: 1987, 'The Stably Stratified Internal Boundary Layer for Steady and Diurnally Varying Offshore Flow', *Boundary-Layer Meteorol.* **38**, 369–394.

Garratt, J. R. and Brost, R. A.: 1981, 'Radiative Cooling Effects within and above the Nocturnal Boundary Layer', *J. Atmos. Sci.* **38**, 2730–2746.

Hsu, S. A.: 1983, 'On the Growth of a Thermally Modified Boundary Layer by Advection of Warm Air over a Cooler Sea', *J. Geophys. Res.* **88**, 771–774.

Johnson, H. D., Lenschow, D. H., and Danninger, K.: 1978, 'A New Fixed Vane for Measuring Air Motion. Preprints, Fourth Symposium on Meteorological Observations and Instrumentation', *Amer. Meteorol. Soc.* Boston, Mass., U.S.A., pp. 467–470.

Mulhearn, P. J.: 1981, 'On the Formation of a Stably Stratified Internal Boundary Layer by Advection of Warm Air over a Cooler Sea', *Boundary-Layer Meteorol* **21**, 247–254.

Pasquill, F. and Smith, F. B.: 1983, *Atmospheric Diffusion*, Third Edition, J. Wiley and Sons, New York, 437 pp.

Raynor, G. S., Sethuraman, S., and Brown, R. M.: 1979, 'Formation and Characteristics of Coastal Internal Boundary Layers during Onshore Flows', *Boundary-Layer Meteorol.* **16**, 487–514.

Ryan, B. F., Wilson, K. J., Garratt, J. R., and Smith, R. K.: 1985, 'Cold Fronts Research Programme: Progress, Future Plans and Research Directions', *Bull. Amer. Meteorol. Soc.* **66**, 1116–1122.

Venkatram, A.: 1977, 'A Model of Internal Boundary-layer Development', *Boundary-Layer Meteorol.* **11**, 419–437.

Venkatram, A.: 1986, 'An Examination of Methods to Estimate the Height of the Coastal Internal Boundary Layer', *Boundary-Layer Meteorol.* **36**, 149–156.

Wyngaard, J. C.: 1973, 'Chapter 3 in *Workshop on Micrometeorology*', D. A. Haugen (ed.), Amer. Meteorol. Soc., 392 pp.

Yamada, T.: 1979, 'Prediction of the Nocturnal Surface Inversion Height', *J. Appl. Meteorol.* **18**, 526–531.

COHERENT STRUCTURES IN THE VERY STABLE ATMOSPHERIC BOUNDARY LAYER

PAUL RUSCHER[1] and L. MAHRT

Department of Atmospheric Sciences, Oregon State University, Corvallis, Oregon 97331 U.S.A.

(Received in final form 16 May, 1988)

Abstract. This study examines the structure of horizontal modes (meandering, vortical modes or fossil turbulence) in a layer of intermittent turbulence occurring at the top of a strongly stratified nocturnal inversion layer as observed by fast response aircraft data. The spatial variation of the coefficients of the principal components identify regular coherent structures with mainly horizontal motions. Conditional sampling is formulated in terms of this spatial variation. The quasi-horizontal motions are characterized by relatively sharp edges (transition zones) where horizontal convergence or divergence, small-scale turbulence and vertical fluxes seem to be concentrated. Zones of horizontal divergence appear to be associated with ejection of cold air from the underlying surface inversion while the convergent zones might be due to random "collisions" between horizontal modes.

1. Introduction

Little is known about transporting turbulent eddies in strongly stratified flow. This lack of knowledge has inhibited progress in the formulation of turbulent transport in models of the atmospheric nocturnal boundary layer, clear-air turbulence and intermittent turbulence in stratified oceanic and lake flows. Three-dimensional information on the main eddies is largely missing and determination of fluxes from observations of turbulence in stratified flow is often plagued by severe sampling problems (Wyngaard, 1973).

The main types of motions in stratified flow seem to be gravity waves, shear-driven overturning sometimes related to Kelvin–Helmholtz instability and quasi-horizontal motions referred to as vortical motions. Shear-driven overturning has been inferred from tower measurements and various remote sensing observations and many laboratory flows (Thorpe, 1973, 1987; Koop and Browand, 1979). Quasi-horizontal modes have been inferred from spectral decomposition of time series, observations of atmospheric plume-meandering and laboratory flows using visualization techniques (e.g., Kristensen *et al.*, 1982; Lilly, 1983) but little is known about the structure or dynamics of such motions.

Eigenvector decomposition of aircraft data in strongly stratified atmospheric flow has indicated dominance of variance by quasi-horizontal modes while shear-driven overturning and gravity waves accounted for less variance (Mahrt and Frank, 1988). These modes are typically characterized by narrow zones of sharp horizontal gradients which might be the edges of the main coherent structures.

[1] Present affiliation: Department of Meteorology, The Florida State University, Tallahassee, Florida 32306, U.S.A.

Boundary-Layer Meteorology **47**: 41–54, 1989.
© 1989 *Kluwer Academic Publishers*.

The present study analyzes quasi-horizontal modes and accompanying horizontal shear zones (microfronts) in some detail from fast response aircraft data collected in intermittent turbulence at the top of a nocturnal surface inversion. Attempts will be made to isolate the structure of such modes by employing conditional sampling based on principal components of the zero-lagged correlation matrix.

2. Principal Component Analysis

Towards the goal of studying motions with very stable stratification, we shall analyze fast-response aircraft data from a synoptically quiet period during the Severe Environmental Storms and Mesoscale Experiment (SESAME). The data were collected in the early morning before surface heating altered the character of the nocturnal boundary layer. Six north-south legs were flown at the top of a strong surface inversion, where weak wind shear generated intermittent turbulence about 50 m above gently rolling terrain. The directional shear zone at 50 m marked the boundary between cool northwesterly flow in the surface inversion layer and warmer southerly flow aloft. Both flows were weak with speeds of $1-2 \text{ m s}^{-1}$. Further details may be found in Mahrt (1985) and Ruscher (1987).

a. METHODOLOGY

Coherent structures will be identified by studying the spatial variation of principal components corresponding to the eigenvectors of the multivariate zero lag correlation matrix. Conditional sampling as an initial step would have required too much advanced knowledge about the nature of the main coherent structures. The observational vector, $\phi(y)$, is given by

$$\phi(y) = \begin{Bmatrix} u \\ v \\ w \\ T \end{Bmatrix}, \tag{1}$$

where y indicates spatial position. Each of the four variables in ϕ are demeaned and high-pass filtered with a cutoff wavelength of 1 km. This filtering partially eliminates the influence of larger scale motions which are characterized by small correlation coefficients and are undoubtedly observed with significant sampling problems. The variables are then normalized by their standard deviation, and the orthonormal eigenvectors are computed form the resulting correlation matrix. The principal components are formed from the eigenvectors and normalized observational vector as

$$Z_{il} = \sum_{j=1}^{4} e_{ij}\phi_{jl}, \quad i = 1, \ldots, 4, \tag{2}$$

where ϕ_j is the observational vector consisting of L observations of the variables

u, v, w, and T and e_{ij} is the ith eigenvector element for the jth variable. The meteorological variables have at this point undergone a transformation to a new set of four uncorrelated variables (the Z_i), which are linear combinations of the original four variables and which describe progressively smaller portions of the total multivariate variance. The proportion of variance explained by the ith eigenvector is sometimes used to assess statistical significance of its associated principal component; the percent of variance explained is

$$\frac{\lambda_i}{\sum_{i=1}^{4} \lambda_i} \times 100\% \,,$$

where λ_i is the eigenvalue corresponding to the ith eigenvector.

The eigenvectors do not automatically represent any specific physical structure in the data, since the statistical technique is designed to maximize variance explained. However, the expansion coefficient for the first principal component of this study does suggest regular spatial behavior (Figure 1). Subdomain stability of the principal component analysis (e.g., Richman, 1986) is tested by splitting each leg into two parts. The principal component analysis results for each part were roughly the same as for the entire legs. Sampling errors due to eigenvectors with closely-spaced eigenvalues do not arise in the present data.

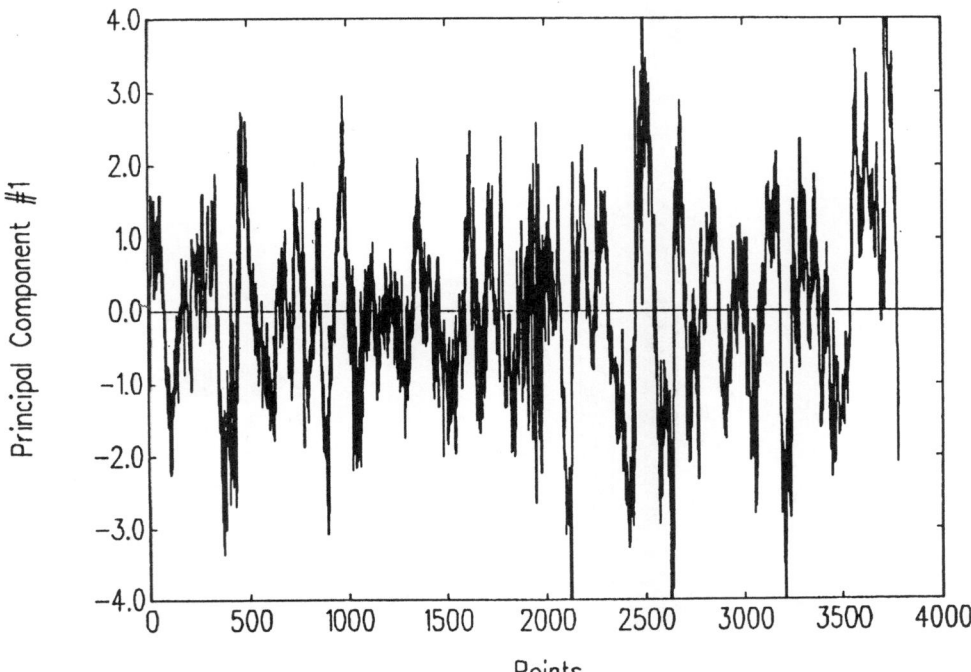

Fig. 1.   Time series of the leading principal component (including velocity and temperature) for the first flight leg from 5 May 1979.

Principal component loadings are defined as the correlation of the observations (the ϕ_j) with the principal components (the Z_i). The loadings for each variable of the first principal component versus those of the second principal component form a nearly elliptical pattern for the present data (not shown), without clustering about a diagonal line; therefore rotation of the eigenvectors (Richman, 1986) is not suggested.

b. RESULTS

The first eigenvectors for the legs on the morning of 5 May contain small coefficients for w, about a factor of ten smaller than those of the horizontal components (Figure 2). Thus the first eigenvector represents nearly horizontal flow, where vertical motion is suppressed by the strong stratification. The percentage of variance explained by the first eigenvector varies between legs

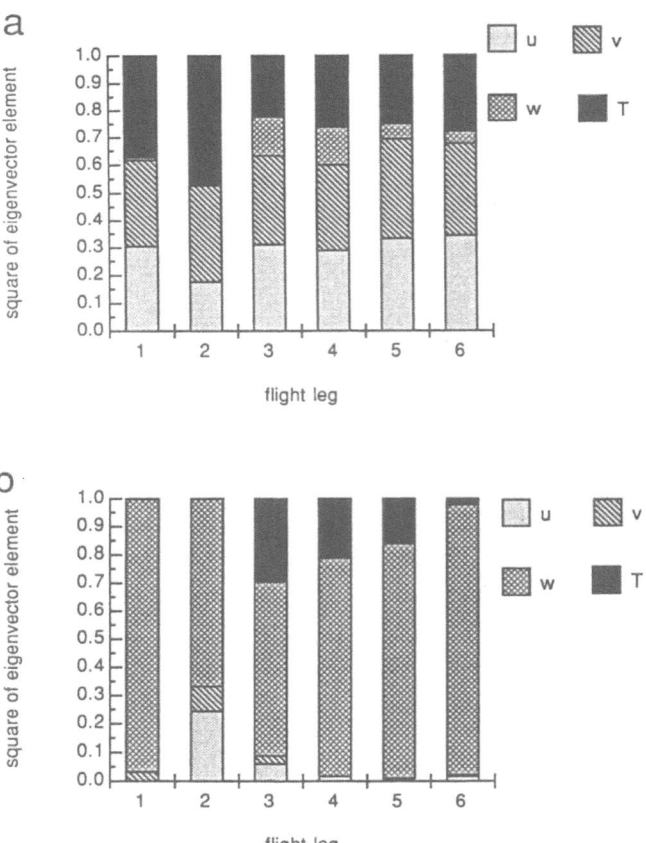

Fig. 2. Composition of the leading eigenvectors for the six horizontal flight legs of 5 May. Since the eigenvectors are unit vectors, the sums of the squares of their elements will be one: (a) the first eigenvector, (b) the second eigenvector.

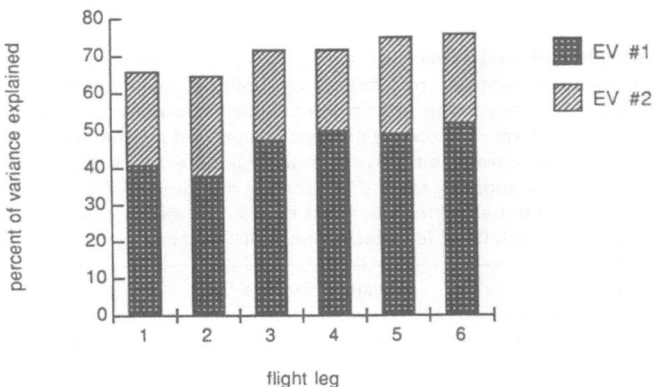

Fig. 3.   The percentage of variance explained by the first two eigenvectors for the six flight legs.

from about 35 to 50% (Figure 3), while the second eigenvector typically accounts for 20 to 25% of the variance. Vertical motion is generally the dominant variable in EV #2 (Figure 2b), probably representing shear-driven overturning. One does not expect the leading eigenvector to explain a majority of the variance, because turbulence by definition contains a variety of modes observed in different phases.

A common feature of EV #1 is that the coefficients for v and T are of the same sign, while u is of opposite sign. The physical flow pattern suggested by this first eigenvector is one of alternating cold northwesterly, characteristic of the underlying flow, and warm southeasterly, characteristic of overlying flow.

The repetitive flow structure illustrated in the leading principal component suggests the possibility of some type of regular coherent structure on the scale of roughly 400 m, a little less than half of the maximum scale permitted by the filtering. This verifies the inspection of raw times series (not shown) and results from spectral analysis (Ruscher, 1988), which also shows a peak at the same scale. It will be seen in subsequent sections that this repetitive structure does not represent simple gravity waves.

3. Conditional Sampling

In order to examine the local horizontal structure represented by the regularity of the first eigenvector, samples are conditionally selected by finding local maxima and minima of the leading principal component. To facilitate location of maxima and minima, the leading principal component is smoothed with a low pass filter with cutoff wavelength of 350 m. Samples of cycles of spatial variation between adjacent minima are then selected. Each sample is assigned ten equally-spaced points from the minima to the maxima and another ten equally-spaced points

TABLE I

Statistics for the samples obtained from each flight leg, where the conditional sampling is based on the leading principal component. Samples are defined from minima to minima in the principal component. Samples are rejected if the intervening principal component maximum is not positive or if samples result which are wider than 1 km. The scaling distance is separated into the northern (N) and southern (S) half of the samples, and represents the average distance between each point in the sample. The average width is found by multiplying each of the scaling distances by 10 (the number of points on each side of the sample), and multiplying the result by 3.5, the distance between observations. The percent of data in each flight leg which contributes to the samples is also indicated

Flight	Leg	Samples		Scaling distance (m)		Average width(m)	% of leg in samples
		Taken	Rejected	N	S		
3	1	20	1	7.3	7.8	528.5	79.9
3	2	25	2	7.1	7.3	505.7	91.5
3	3	19	5	7.1	7.8	522.6	70.4
3	4	19	0	6.3	5.9	424.8	89.8
3	5	22	5	7.2	7.5	513.4	82.9
3	6	20	6	6.3	6.2	437.3	73.1
Total		125	19				
Weighted average				6.9	7.1	490.0	81.7

from maxima to minima. Table I provides information about the samples obtained using this procedure.

This method of utilizing the shape of the leading principal component to select samples is a "natural" conditional sampling technique in that no specific *a priori* assumptions about the intensity or nature of the turbulence are necessary. Most conditional sampling techniques require some *a priori* knowledge about the main structures in order to specify sampling criteria. Such criteria are sometimes based on a threshold intermittency factor (e.g., Khalsa, 1980; Shaw and Businger, 1985), possibly in connection with the Variable Integration Time Averaging (VITA) method (e.g., Blackwelder and Kaplan, 1976; Chen and Blackwelder, 1978; Schols, 1984). Other attempts to reduce the role of the sampling criteria include the continuous weighting technique of Mahrt and Frank (1988). In the present case, information contained in the principal components seems especially useful since little is known about the main turbulent events.

Let x represent position within a sample and k the sample number. We then decompose the value of any high-pass filtered function Φ as

$$\Phi(x, k) = \phi(x, k) + \phi_0(k) , \qquad (4)$$

where

$$\phi_0(k) \equiv \frac{1}{X} \int_0^x \Phi(x, k) \, dx . \qquad (5)$$

Thus ϕ represents a variable whose sample mean (ϕ_0) has been removed.

We shall now decompose ϕ into two parts – a composite ($\phi_c(x)$) or average over all of the samples performed at each of the twenty points and the deviation ($\phi'(x, k)$) from the composited structure, so that

$$\phi_c(x) \equiv \frac{1}{K} \sum_{k=1}^{K} \phi(x, k) , \tag{6}$$

and

$$\phi''(x, k) \equiv \phi(x, k) - \phi_c(x) , \tag{7}$$

where K is the total number of samples. Note that

$$\sum_{k=1}^{K} \phi''(x, k) = 0 . \tag{8}$$

Using (5) through (7), the variance of the function ϕ may be written

$$\text{Var}(\phi) = \frac{1}{K} \sum_{k=1}^{K} \frac{1}{X} \int_{0}^{X} [\phi_c(x) + \phi''(x, k)]^2 \, dx , \tag{9}$$

where X is sample width. This variance includes the composited structure

$$\text{Var}(\phi_c) \equiv \frac{1}{X} \int_{0}^{X} \phi_c^2(x) \, dx \tag{10}$$

and the variance of deviations within a given sample structure from the composited structure

$$\text{Var}(\phi''(k)) \equiv \frac{1}{X} \int_{0}^{X} \phi''^2(x, k) \, dx . \tag{11}$$

The remaining cross term vanishes due to (8),

$$\frac{2}{X} \int_{0}^{X} \phi_c(x) \left[\frac{1}{K} \sum_{k=1}^{K} \phi''(x, k) \right] dx = 0 . \tag{12}$$

The horizontal distance within the samples is discretized to 20 points so that for any variable φ:

$$\frac{1}{X} \int_{0}^{X} \varphi(x) \, dx \Rightarrow \frac{1}{20} \sum_{j=1}^{20} \varphi(j) . \tag{13}$$

Then, using (12) and the discretized versions of (10) and (11), the total variance becomes

$$\text{Var}(\phi) = \frac{1}{20} \sum_{j=1}^{20} \phi_c^2(j) + \frac{1}{K} \sum_{k=1}^{K} \frac{1}{20} \sum_{j=1}^{20} \phi''^2(k, j). \qquad (14)$$

The first term of (14) represents the variance explained by the composite and is the discrete form of (10). The second term represents the discretized within-sample variance.

About one-fifth of the total variance of the samples (14) is explained by the composited structure (the first term on the right-hand side of Equation 14) for temperature and the two horizontal velocity components. This modest signal-to-noise ratio is characteristic of turbulence. The composited structure explains much less variance for vertical motion, compared to the other variables. The similarity of the sample structures to the composited structure varies widely between samples. An attempt to homogenize the record and improve the variance explained by the composite was attempted by eliminating samples most dissimilar to the composite and then recomputing the composite. This procedure failed to increase significantly the variance explained by the composite.

The composited fields for u, v, w, and T of the leading principal component for all 125 samples of the six legs (Figure 4) indicate the expected two-dimensional character of the main structures. Arrows just below the figure indicate the horizontal flow field in plan view (x-y plane). The lowest set of

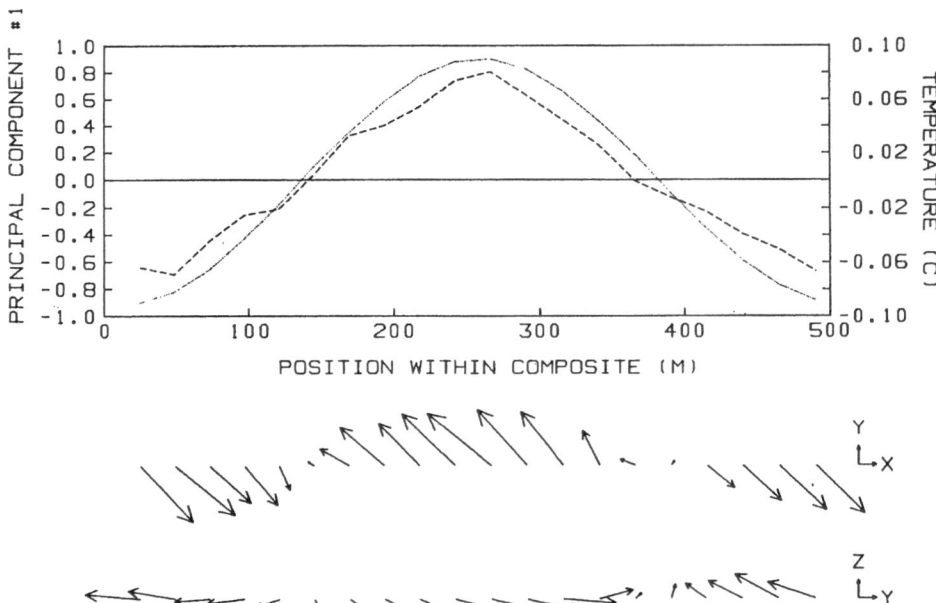

Fig. 4. Composite temperature (dashed line) and first principal component (dotted line) from 125 samples of the coherent structure found by the conditional sampling. Arrows representing the flow in two different two-dimensional planes are shown. The length of the reference vectors corresponds to a speed of 2.5 cm s^{-1}. North is to the left.

arrows indicates the flow field in a longitudinal cross section (y-z plane). The amplitude of the variation of the horizontal flow is only about $0.1 \, \mathrm{m \, s^{-1}}$ due to phase interference between the various samples. Amplitudes of many of the individual samples were much larger.

Many of the individual quasi-horizontal structures are characterized by sharp edges, although only a hint of stronger gradients survives the artificial smoothing resulting from the compositing procedure. The relative horizontal wind vector rotates gradually for some samples and for other samples shifts direction nearly 180° across narrow transition zones. These transition zones correspond to zones of convergence of the along-flight flow component (north-south) on the northern end of the composited structure and zones of divergence on the southern ends. The vertical motion is either very weak or upward in the zones of divergence and downward in the convergence zones. The flow between the transition zones is mainly horizontal.

One is tempted to relate the quasi-horizontal structures to vortical modes (Riley *et al.*, 1981; Lilly, 1983; Müller, 1984) which are considered to be remnants of the decay of three-dimensional turbulence. Temperature variations in the present observed flow are significant and strongly correlated to the horizontal flow field. This correlation agrees with the sign predicted by vertical gradient transfer even though vertical motion is apparently weak. It may be. that vertical motion had earlier produced perturbations in temperature and horizontal wind before becoming damped by the stratification, in which case the temperature perturbations would reflect the history of the motion. The sharp gradients at the transition zones and the lack of a definite phase lag between vertical motion and temperature rule out the possibility of simple linear gravity waves although the influence of nonlinear wave activity cannot be discarded.

4. Transition Zones

Since the transition zones at the edges of the structures are blurred by the compositing procedure, we now reconsider individual samples. A total of 178 well-defined transition edges were obtained objectively from the 125 samples of coherent structures. Some samples of horizontal structures from the aircraft time series yielded only rotation or meandering of the wind vector with no distinct transition zones, while a few others yielded more than the expected two zones; however most of the transition zones appeared to be at the edges of the coherent structures. The objective method involves computation of the horizontal velocity structure variance,

$$D_h \equiv [D_u + D_v], \qquad (15)$$

where D_φ represents the structure function for arbitrary variable φ, defined as

$$D_\varphi(r) \equiv \overline{(\varphi(y + r) - \varphi(y))^2} . \qquad (16)$$

and r is the separation distance, or spatial lag and the overbar represents averaging. Here we choose a value of 42 m for the separation distance. The results did not depend significantly on the choice of the value of the separation distance.

Potential transition zones of variable width are chosen, where D_h is at least four times the average value for the entire sample; the selection of this threshold is based on an observed sharp dropoff in the relative frequency of D_h occurring at about 4.3.* This requirement must be met for at least four consecutive points as the sampling window of width $r = 42$ m is moved one point at a time (3.5 m) to help eliminate more random fluctuations.

The average ratio of D_h for the resulting 62 transition zones compared to D_h for the samples as a whole is 5.74. This value is much larger than the value of unity expected for randomly-drawn zones without regard for the nature of φ. Random selection of zones with more than 100 samples did indeed yield ratio values close to unity. Application of the method to white noise yielded artificial transition zones for the horizontal flow with ratios of order unity but without correlation to vertical motion and temperature.

About half of the transition zones occur at the north end of the sampled structures with convergence of the component of motion along the flight path. Most of the transition zones at the southern ends occur with divergence of this flow component. We now examine the structure of the transition zones and their environment by compositing separately for the 21 "divergent" zones and 29 "convergent" zones. The greater number of convergent zones suggests that horizontal convergence contributes to the intensity of the gradients, as in frontogenesis.

In addition to strong velocity gradients, the composited convergence zone is characterized by significant horizontal temperature gradients (Figure 5a) but little vertical motion or heat flux (not shown). The composited amplitudes again appear weak compared to many of the individual transition zones due to some phase interference in the compositing. The weak vertical motion and large $\partial v/\partial y$ in the convergent zone, suggest that either $\partial u/\partial x$ or $\partial w/\partial z$ must also be large.

In contrast, the composited divergent zone (Figure 5b) is characterized by significant upward motion and roughly 50% stronger horizontal temperature gradient. The significant horizontal temperature structure for both the convergent and divergent transitions suggests that the flows on either side of the transition zones have originated from different levels. The narrow region of significant downward heat flux and upward flux of horizontal momentum on the south side of the divergent zone (Figure 6) indicates that much of the total flux is concentrated in a small fraction of the total area. The concentration of momentum flux in the narrow divergent transition zones implies large intermittency and sampling problems when estimating the flux for the entire record.

---

* This is a nondimensional ratio of the transition zone D_h divided by the sample D_h.

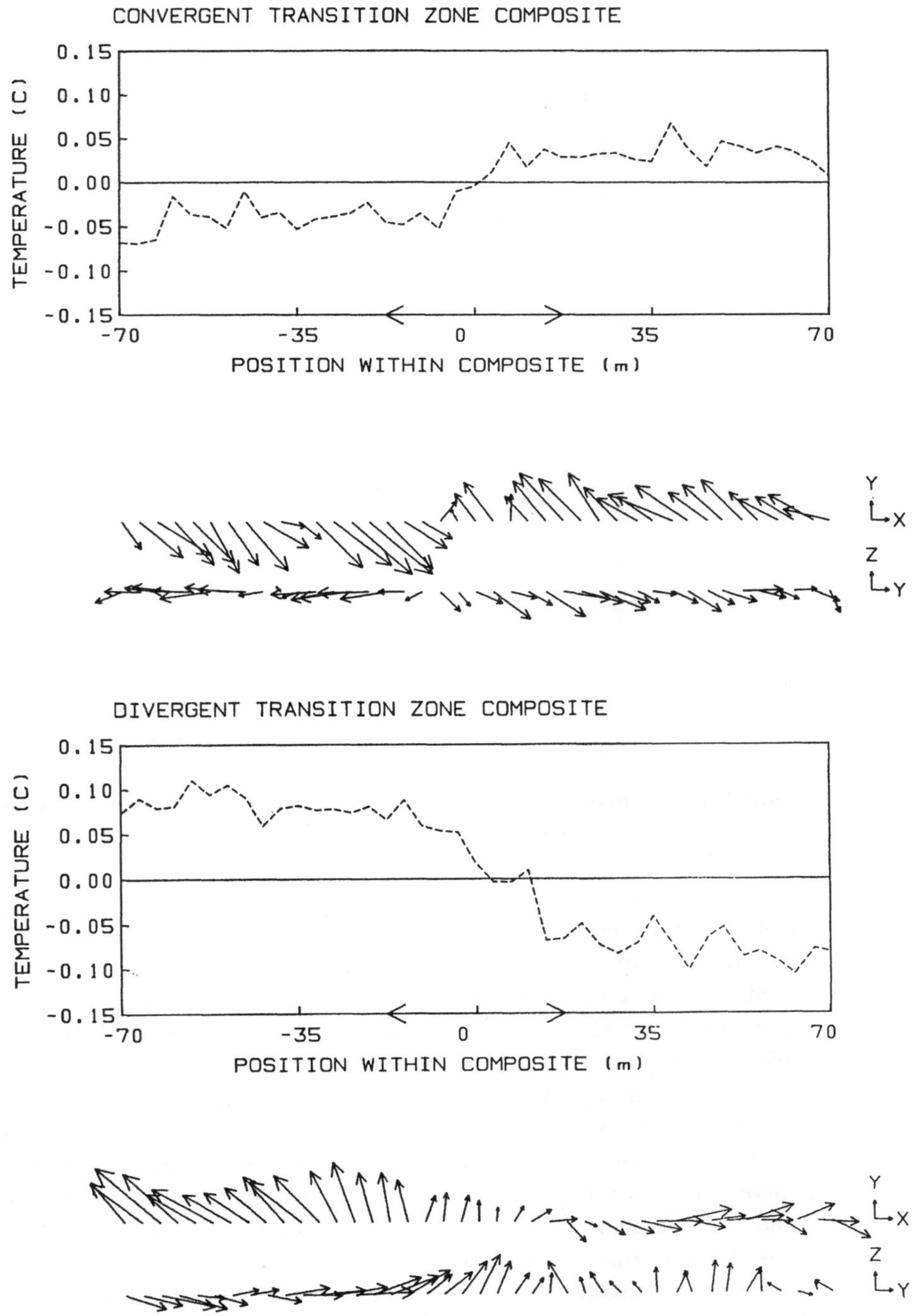

CONVERGENT TRANSITION ZONE COMPOSITE

Fig. 5.   Composite temperature pattern for the transition zones. Arrowheads on the abscissa depict the edges of the transition zone. Velocities are depicted as in Figure 4: (a) the convergent zone, based on 29 samples, (b) the divergent zone, based on 21 samples.

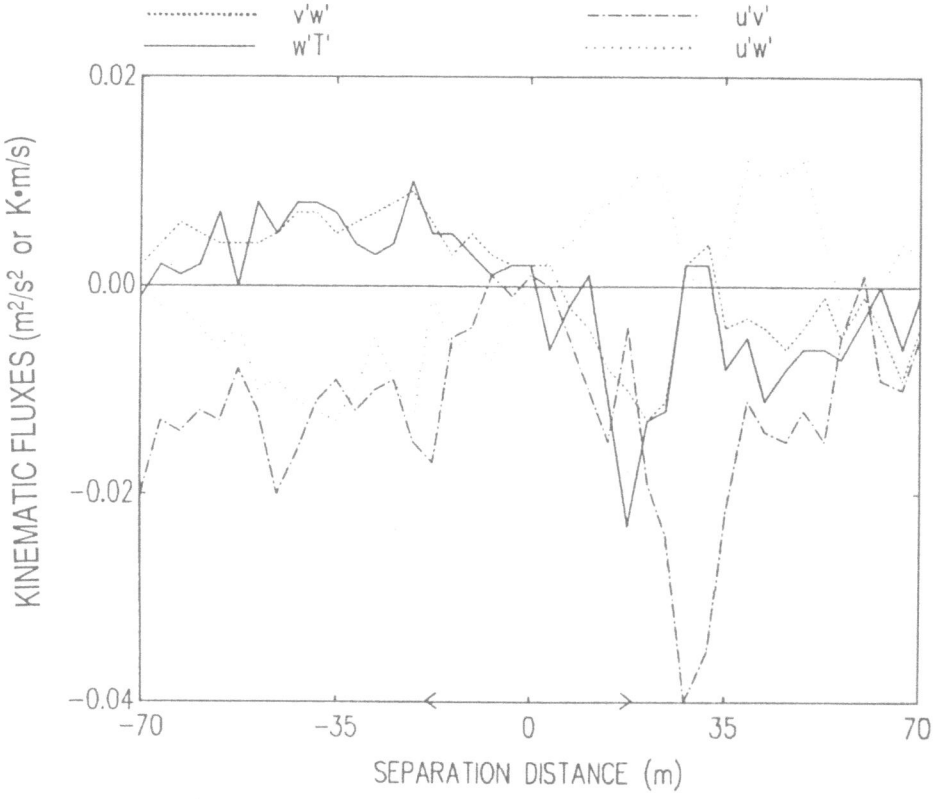

Fig. 6. Composite kinematic fluxes of horizontal momentum and vertical heat flux across the divergent transition zone.

Using this observation and the results above, an attempt is made to combine the composited features of the intense convergent and divergent zones into a conceptual model of one circulation system. Since the well-defined convergent and divergent zones do not always occur as pairs, the sketch in Figure 7 suggests only a plausible structure not necessarily describing any given sample. The sketch in Figure 7 recognizes the significant rising motion and horizontal temperature gradient in the divergent zones and the greater frequency of strong horizontal velocity gradients in the convergence zones. Apparently ejection of cold air from the surface inversion layer is responsible for the overall large fluxes in the divergent zone. We further speculate that the ejection of cold air in the divergence zone leads to the relative horizontal motion which then terminates in the convergence zone, possibly involving a "collision" with an adjacent circulation. The apparent generation of turbulence in the convergence zones may be a significant source of intermittent turbulence as well as a contributor to the loss of kinetic energy of the quasi-horizontal motions.

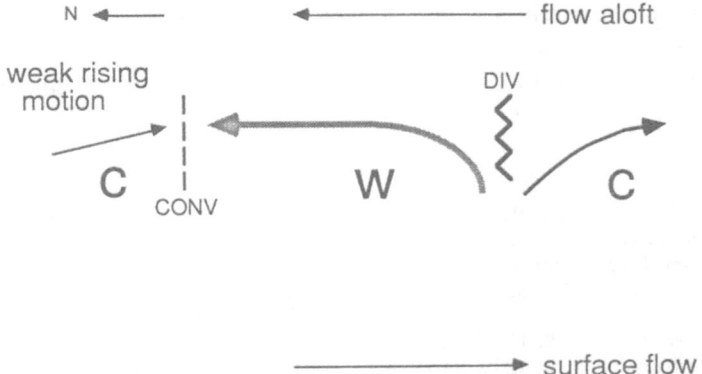

Fig. 7.   Plausible relative flow consistent with analysis of composited structures and transition zones. Maximum temperature perturbation is indicated by a 'W' and the minimum by a 'C'. The jagged line indicates a horizontal shear zone and is coincidental with the divergent zone. A microfront is indicated by the dashed line and occurs with horizontal convergence.

5. Conclusions

Regular reoccurring structures have been observed with fast response aircraft data in a layer of intermittent turbulence at the top of a strongly stratified surface inversion layer. These structures are identified from the spatial variation of the first principal component computed from the zero-lagged correlation matrix. The regular structures are often characterized by sharp edges with concentrated horizontal gradients.

Some of the transition zones occur with divergence of the along-flight wind component. The temperature structure of these zones suggests upward ejection of cold air from the underlying surface inversion layer probably associated with shear-driven overturning. This horizontal divergence leads to horizontal motions which terminate in narrow convergence zones (micro-frontogenesis). This convergence might result from random "collision" of individual horizontal motions. The convergent and divergent transition zones account for a major portion of the turbulence variance in the layer of intermittent turbulence. Significant vertical fluxes occur primarily in the divergent transition zones. The horizontal motions between the transition zones can be loosely associated with existing terminology such as "two-dimensional turbulence", "fossil turbulence" or "vortical modes".

The above results are necessarily speculative because spatial structure is inferred from aircraft observations with little previous guidance from existing studies.

Acknowledgements

This material is based upon work supported by the Meteorology Program of the National Science Foundation under Grant ATM-8521349. The National Center for Atmospheric Research is acknowledged for computer resources and use of the Queen Air research aircraft. The helpful comments of an anonymous reviewer and the computational assistance of Wayne Gibson are appreciated.

References

Blackwelder, R. F. and Kaplan R. E.: 1976, 'On the Wall Structure of the Turbulent Boundary Layer', *J. Fluid Mech.* **76**, 89–112.

Chen, C.-H. P. and Blackwelder R. F.: 1978, 'Large-Scale Motion in Turbulent Boundary Layer: A Study Using Temperature Fluctuations', *J. Fluid Mech.* **89**, 1–31.

Khalsa, S. J. S.: 1980, 'Surface Layer Intermittency Investigated with Conditional Sampling', *Boundary-Layer Meteorol.* **19**, 135–153.

Koop, C. G. and Browand F. K.: 1979, 'Instability and Turbulence in a Stratified Flow with Shear', *J. Fluid Mech.* **92**, 135–159.

Kristensen, L., Jensen N. O. and Petersen E. L.: 1982, 'Lateral Dispersion of Pollutants in a Very Stable Atmosphere – The Effect of the Meandering', *Atmos. Environ.* **15**, 837–844.

Lilly, D. K.: 1983, 'Stratified Turbulence and the Mesoscale Variability of the Atmosphere', *J. Atmos. Sci.* **40**, 749–761.

Mahrt, L.: 1985, 'Vertical Structure and Turbulence in the Very Stable Boundary Layer', *J. Atmos. Sci.* **42**, 2333–2349.

Mahrt, L. and Frank H.: 1988, 'Eigenstructure of Eddy Microfronts', *Tellus* **40A**, 107–119.

Müller, P.: 1984, 'Small Scale Vortical Motion', in P. Müller and R. Pujalet (eds.), *Proceedings, 'Aha Huliko'a Hawaiian Winter Workshop: Internal Gravity Waves and Small-Scale Turbulence*, University of Hawaii, Manoa, pp. 249–261.

Richman, M. B.: 1986, 'Rotation of Principal Components', *J. Climatol.* **6**, 293–335.

Riley, J. J., Metcalfe R. W. and Weissman M. A.: 1981, 'Direct Numerical Simulations of Homogeneous Turbulence in Density-Stratified Fluids', in B. J. West (ed.), *Nonlinear Properties of Internal Waves*, American Institute of Physics, New York, pp. 79–112.

Ruscher, P. H.: 1987, 'An Examination of Structure and Parameterization of Turbulence in the Stably-Stratified Atmospheric Boundary Layer', PhD thesis, Department of Atmospheric Sciences, Oregon State University, Corvallis, Oregon, 170 pp.

Ruscher, P. H.: 1988, 'On Rotary Spectral Analysis of Motion in the Very Stable Atmospheric Boundary Layer', to be submitted to *J. Geophys. Res.*

Schols, J. L. J.: 1984, 'The Detection and Measurement of Turbulent Structures in the Atmospheric Surface Layer', *Boundary-Layer Meteorol.* **29**, 39–58.

Shaw, W. J. and Businger J. A.: 1985, 'Intermittency and the Organization of Turbulence in the Near-Neutral Marine Atmospheric Boundary Layer', *J. Atmos. Sci.* **42**, 2563–2584.

Thorpe, S. A.: 1973, 'Turbulence in Stably Stratified Fluids: A Review of Laboratory Experiments', *Boundary-Layer Meteorol.* **5**, 95–119.

Thorpe, S. A.: 1987, 'Transitional Phenomena and the Development of Turbulence in Stratified Fluids: A Review', *J. Geophys. Res.* **92C**, 5231–5248.

Wyngaard, J. C.: 1973, 'On Surface Layer Turbulence', in D. Haughen (ed.), *Workshop on Micrometeorology*, American Meteorological Society, Boston, pp. 101–149.

AN EVALUATION OF AIRCRAFT FLUX MEASUREMENTS
OF CO₂, WATER VAPOR AND SENSIBLE HEAT

R. L. DESJARDINS[1], J. I. MACPHERSON[2], P. H. SCHUEPP[3] and F. KARANJA[3]

[1] Agrometeorology Section, Land Resource Research Centre, Research Branch, Agriculture Canada, Ottawa, Ontario K1A 0C6, Canada
[2] National Research Council, Flight Research Laboratory, Ottawa, Ontario K1A 0R6, Canada
[3] Dept. of Renewable Resources, Macdonald College of McGill University, Ste. Anne de Bellevue, Quebec H9X 1C0, Canada

(Received in final form 3 June, 1988)

Abstract. Ground-based flux measurements of carbon dioxide and water vapor integrate physiological processes taking place on a field scale. Aircraft flux measurements have recently been undertaken to attempt to widen the scope of applicability of such measurements. However, because of the intermittency of turbulent transfer, flux measurements must be averaged over long periods of time or long distances to give reproducible results. This requirement makes it difficult to relate aircraft flux measurements to local surface processes. Flux measurements of CO_2, latent and sensible heat obtained from repeated passes in four directions and at three elevations over a homogeneous wheat-growing area are compared with ground-based measurements. Averages based on four runs of 4 km in length gave results consistent with ground-based measurements. The largest percentage differences were in the sensible heat flux. Cospectral analyses showed no significant high frequency losses for the data from flight levels of 25 and 50 m, but an underestimation of approximately 10% resulted at 10 m. Flight direction with respect to wind direction was relatively unimportant at 10 and 25 m but some effects were observed at 50 m. It was also shown that at 25 m, over a relatively smooth and homogeneous surface, the means of either three or four runs 4 km in length were similar to the means of 12–16 km runs. This confirms that at this altitude, most of the flux contribution is contained at wavelengths less than 4 km and that the mean of 3 to 4 passes accounts for most of the intermittency of turbulent transfer.

1. Introduction

It is widely recognized that flux density measurements are the most effective way of integrating gas exchange and physiological processes taking place on a field scale (Anderson and Verma, 1986). However, such measurements must be averaged over long periods of time (Shaw and Businger, 1985) or long distances in order to obtain reproducible results (Wyngaard, 1986). For example, using a ground-based system, it is necessary to sample over at least 30 min to observe most of the low frequency contribution to CO_2 fluxes (Anderson and Verma, 1985; Ohtaki, 1985). Moreover, most ecosystems are not very homogeneous and point measurements are frequently not representative. For this reason, it is important to develop alternatives to ground-based measurements. In recent years, much effort has been directed towards flux measurements from aircraft (Bingham et al., 1983; Desjardins et al., 1982; Desjardins et al., 1985; Lenschow et al., 1982; Schuepp et al., 1987). This type of information is particularly promising for calculation of satellite-based algorithms to estimate vegetation and soil conditions on a regional scale (Séguin and Itier, 1983).

For aircraft systems, the requirement for long runs has frequently been discussed (Lumley and Panofsky, 1964; Wyngaard, 1986). Recently, based on theoretical considerations, Lenschow and Stankov (1986) have estimated that, at the middle of the mixed layer, scalar fluxes require a measurement length of 10^2 to 10^4 times the boundary-layer height to obtain 10% accuracy. Because of this sampling requirement, it is desirable to develop aircraft-based flux measuring techniques for flight levels close to the ground.

A comparison of flux densities between airborne and ground-based systems is complicated by the intermittency of turbulent transfer and by the variability of environmental conditions. Haugen *et al.* (1971) discussed the variability of flux densities in the first 20 m from measurements taken on a tower. Over an hour, 15-min average flux densities were constant only within 20%; even over 24 hours of carefully chosen runs, flux densities were only constant to within 5 to 10%. With aircraft flux measurements, additional factors can also contribute to run-to-run variability (Lenschow, 1986): these are usually recorded over a very short period and because of the intermittency of turbulent transport, variability is increased. Long-lived convective eddies can only strongly bias short-term observations. This is very important at higher levels. In the surface boundary layer, i.e., measuring height $h < 0.1\ Z$, where Z is the height of the mixed layer, no organized structures, such as elongated thermals and roll vortices, seem to exist (Brown, 1974).

By making repeated passes at several elevations over the site of a ground-based eddy correlation system, an aircraft-based flux system can be evaluated. Such an experiment was conducted over a wheat-growing area near Fannystelle, Manitoba during July 1985, with repeated aircraft passes at 10, 25 and 50 m. This paper demonstrates the degree of repeatability obtainable in flux densities of CO_2, latent and sensible heat from an aircraft with the available sensors. It demonstrates convergence of flux estimates from increasing sequences of passes over the same tracks by partial collection of an ensemble of samples. Documentation of differences between flux computations from along-wind and cross-wind flights is given for three elevations. Finally, the averages of three to four 4 km runs obtained in 1986 are compared with measurements over 12–16 km flight tracks.

2. Instrumentation

The Twin Otter aircraft used in this study allows steady flight trajectories at low airspeed (50–60 m s^{-1}) down to levels <10 m above the ground. The aircraft is instrumented to measure the contribution of flux densities of momentum, sensible and latent heat and CO_2 over a frequency range of 0 to 5 Hz (MacPherson *et al.*, 1981). Air motion relative to the aircraft is measured by a nose-mounted gust boom which, until 1985, carried a pitot/static probe and two orthogonal free vanes. In 1986, airspeed, altitude and the angles of attack and sideslip were sensed on the boom by a Rosemount 858 5-hole probe and associated pressure transducers. The inertial velocity of the aircraft is computed in aircraft axes using a complementary filtering technique with the high frequency contribution from integrated accelerometer and rate gyro signals, and the low frequency velocity from a 3-axis Doppler radar. The three components of the true air motion are then derived from the difference between the air velocity relative to the aircraft and the inertial velocity relative to the ground, and resolved into earth-fixed axes using the aircraft heading and attitude signals. Temperature, dew point, radio altitude, incident radiation and geographical position are also measured. Fluctuations in H_2O and CO_2 are measured by a new version of the fast response infrared gas analyzer developed by Brach et al. (1981). It has a sensitivity of 0.01 g m^{-3} for H_2O and 0.1 mg m^{-3} for CO_2 and a frequency response of 15 Hz. This instrument is installed within a duct in the rear cabin. The air intake is from a 120 cm^2 inlet on the top of the fuselage. Temperature and pressure are measured within the duct to compute flow velocity (~0.43 of aircraft true airspeed) and allow for density corrections to the CO_2 and H_2O flux measurements (Webb et al., 1980). Data are time-adjusted to correct for the separation between the gust boom and the other sensors (MacPherson et al., 1987).

The ground-based system consisted of a one-dimensional sonic anemometer, an open-path infrared gas analyzer for CO_2, a Lyman alpha for water vapor measurements and a fast response thermocouple for temperature. The data were recorded using a 21X Campbell Scientific micrologger at a frequency of 10 Hz and high pass filtered using 2-min averaging periods. These measurements were recorded at 3 m above the crop during most of the flights in 1985. Correction for density variations was applied to the CO_2 and water vapor flux (Webb et al., 1980). The averaging period used was selected to minimize the contamination of w by u (Kaimal and Haugen, 1971).

3. Observations

Flights were made around solar noon on six days between July 4 and July 11, 1985. Two tracks were flown in reciprocal directions and at three elevations with occasional repeat passes (Figure 1). At least 4 runs were flown at 50, 25 and 10 m based on the radio altimeter. The flight tracks were selected to intersect near the ground-based system. Some 77 runs were flown, each approximately 4 km in length, or 75 s duration. Each day, the twelve runs took one hour to complete. The flights were carried out during mainly sunny conditions. Flux densities measured with the ground-based system during the hour of aircraft overpass were obtained on all but the last day of aircraft observations. Mean air temperature, dew point, solar radiation, wind speed and direction and sky conditions are given in Table I for the hour of aircraft observations. Substantial rain had fallen just prior to the measuring period but no rain was observed during it. The wheat crop was just starting to head, there were no signs of moisture stress and the whole area appeared visibly homogeneous.

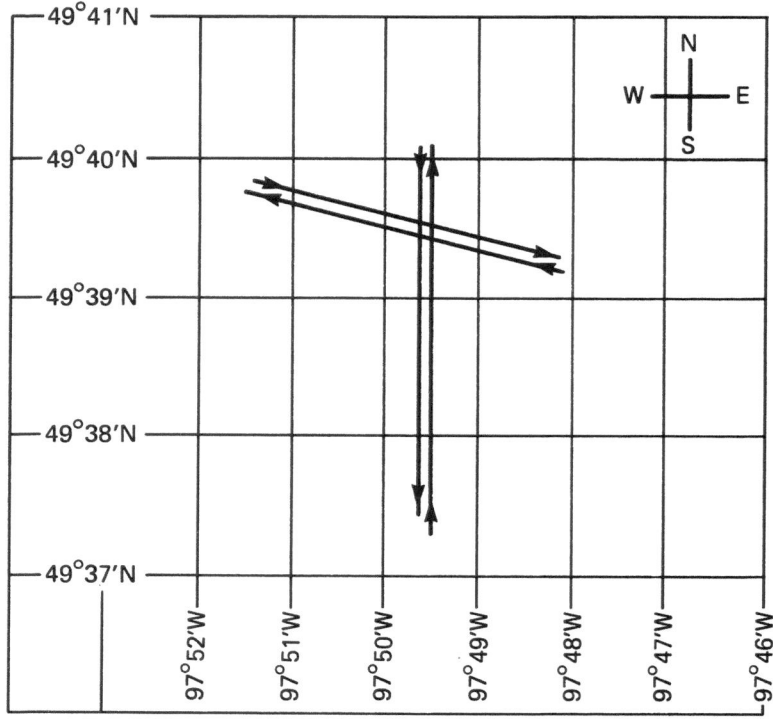

Fig. 1.   The N–S and E–W flight tracks over an homogeneous wheat growing area.

On three days during July 1986, flights were carried out to compare flux estimates from short and long runs over homogeneous crop and terrain conditions. Four 12–16 km runs were flown at 25 m above a wheat-growing area near Morris, Manitoba.

TABLE I

Summary of data over wheat field near Fannystelle in 1985

Date	July 4	July 6	July 8	July 9	July 10	July 11
Temperature (°C)	20	27	21	19	18	22
Dew Point (°C)	14	20	13	9	5	12
Solar Radiation (W m^{-2})	800	950	750	830	890	950
Winds (deg/m s^{-1})	320/9	165/7	310/5	340/6	030/4	180/4
Local Time	1030–1130	1310–1400	1020–1120	1045–1130	1130–1215	1340–1410
Sky Conditions	Cloudy periods	Clear	High broken	Hazy	Hazy	Clear

4. Results and Discussion

The results of the comparison between ground and airborne-based flux densities are shown in Table II. The ground-based data are averages of 2-min sampling corresponding to the period for the hour of aircraft overflight while the aircraft averages at each height are based on four runs taken during approximately 6 min of the hour. Aircraft-measured fluxes in Table II were computed using data high pass filtered at 0.02 Hz with a first-order filter. The variability of repeated runs demonstrated by the standard errors is considerably smaller than the variations with height. This may be due to the fact that runs for each level were flown sequentially. Intermittency is demonstrated in the 5 sec running mean flux of CO_2 (WC) sensible heat (WT) and latent heat (WQ) recorded at 25 m above vegetation during a typical run (Figure 2). Only 3 to 4 events account for most of the transfer. This is clearly not enough to obtain a representative sample. The main

TABLE II

Comparison of means and standard errors of high-pass filtered airborne and ground-based (G) flux densities above a large homogeneous area near Fannystelle, Manitoba during 1985

Date		CO_2 Flux $mg\ m^{-2}\ s^{-1}$	Latent heat $W\ m^{-2}$	Sensible heat $W\ m^{-2}$
July 4	G	-0.81 ± 0.04	213 ± 8	54 ± 3
	10 m	-1.31 ± 0.11	264 ± 24	42 ± 3
	25 m	-1.28 ± 0.08	274 ± 20	49 ± 12
	50 m	-1.03 ± 0.11	236 ± 20	36 ± 12
July 6	G	-0.86 ± 0.10	296 ± 20	36 ± 6
	10 m	-0.86 ± 0.10	313 ± 25	6 ± 4
	25 m	-0.89 ± 0.08	342 ± 37	5 ± 9
	50 m	-0.64 ± 0.14	270 ± 45	7 ± 4
July 8	G	-0.94 ± 0.08	268 ± 16	30 ± 5
	10 m	-1.06 ± 0.11	285 ± 14	30 ± 4
	25 m	-1.00 ± 0.11	254 ± 31	3 ± 5
	50 m	-0.94 ± 0.10	272 ± 29	11 ± 10
July 9	G	-0.83 ± 0.06	249 ± 14	78 ± 4
	10 m	-1.11 ± 0.06	305 ± 28	45 ± 3
	25 m	-1.17 ± 0.08	344 ± 41	43 ± 4
	50 m	-0.89 ± 0.11	285 ± 39	42 ± 10
July 10	G	-0.81 ± 0.07	352 ± 19	77 ± 4
	10 m	-0.97 ± 0.08	384 ± 17	43 ± 4
	25 m	-1.00 ± 0.08	432 ± 55	49 ± 9
	50 m	-0.89 ± 0.17	371 ± 49	43 ± 8
July 11	10 m	-0.94 ± 0.19	322 ± 58	58 ± 4
	25 m	-1.25 ± 0.08	372 ± 36	63 ± 11
	50 m	-0.89 ± 0.14	306 ± 35	40 ± 7

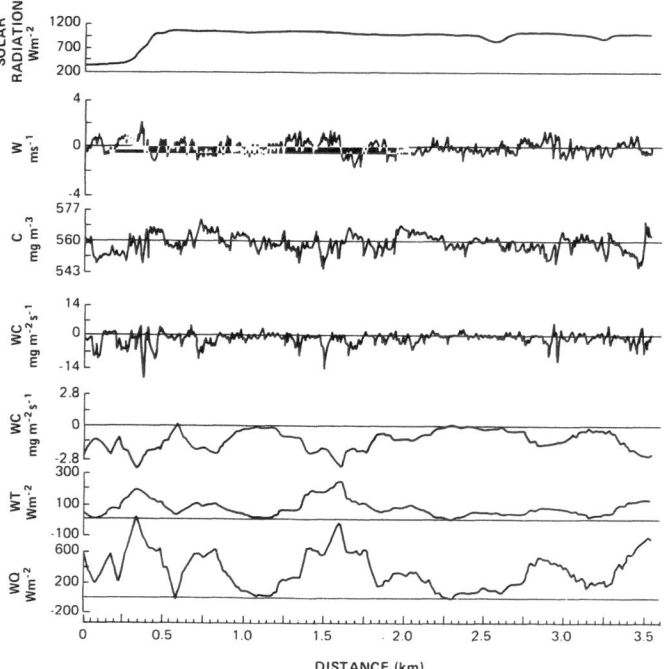

Fig. 2.   Fluctuations of solar radiation, vertical wind (W), CO_2 concentration, and CO_2 flux density (WC) for one aircraft run at 25 m above a wheat crop near Fannystelle, Manitoba. The averages of CO_2 (WC), sensible heat (WT) and latent heat (WQ) flux densities are also shown with a 5 s window centered on the plot point.

requirement is that the aircraft should sample a large number of the large eddies, nominally at least ten (Marht, 1987).

The ground-based flux data tend to be low due to the 2 min high-pass filter used. This is the case for CO_2 and latent heat flux but not for sensible heat flux where the ground-based sensible heat flux estimates are considerably larger than the aircraft-based measurements. This has repeatedly been observed and no sources of error in aircraft flux measurements of sensible heat have been detected to account for this difference. Radiative transfer and flux divergence are the two most plausible causes for the discrepancy (Deardorff, 1974; Druilhet and Durand, 1984; Lenschow and Agee, 1976).

An examination of the aircraft flux densities with height has shown that the flux densities are underestimated at the lowest level (Table II). This is substantiated by an examination of the standard deviations of vertical wind σ_w, carbon dioxide, σ_c, and air temperature, σ_T, as a function of height (Table III). The variations of vertical wind are clearly underestimated at 10 m, most likely due to the 5 Hz low pass filter applied as an anti-aliasing measure during conditioning of the raw sensor data. For temperature, a very large difference in σ_T is observed on most days between the ground and the aircraft measurements. This agrees with the

TABLE III

Standard deviations of vertical wind velocity, σ_w, carbon dioxide concentration, σ_c, and air temperature, σ_T, measured using a ground-based and an airborne system near Fannystelle, Manitoba during 1985

Date		σ_w m s^{-1}	σ_c mg m^{-3}	σ_T °C
July 6	G	0.55	7.58	0.21
	10	0.41	7.20	0.12
	25	0.56	5.28	0.13
	50	0.62	3.65	0.12
July 8	G	0.45	9.66	0.19
	10	0.41	8.22	0.19
	25	0.45	5.95	0.12
	50	0.54	4.61	0.12
July 9	G	0.55	8.38	0.31
	10	0.45	6.62	0.22
	25	0.59	4.64	0.14
	50	0.75	3.52	0.13
July 10	G	0.44	9.09	0.37
	10	0.40	7.30	0.23
	25	0.55	4.54	0.16
	50	0.65	3.26	0.12

large differences observed between ground and aircraft-based sensible heat flux measurements.

Cospectral analyses provide information on the relative contribution to flux estimates from the various frequency or wavenumber bands. Figure 3 presents the mean cospectral estimates for CO_2 (WC), latent heat (WQ) and sensible heat (WT) for the 10, 25 and 50 m levels as a function of wavenumber, K, for all 77 runs. The shape of the cospectra changes with altitude as expected: at higher altitude, the peaks shift to a lower wavenumber. At any given altitude, the co-spectrum shape is very similar for the various transport processes. The main transporting eddies are typically on the scale of hundreds to thousands of meters. In this case, no filtering except linear detrending was done on the data. Average cospectral estimates, for each altitude, are also presented as a function of $K \times Z$ (Figure 4). By taking into account the aircraft speed and the sampling height, very similar cospectra are obtained for the 25 and 50 m runs. At 10 m, an underestimation of approximately 10% is observed because of the loss of the high frequency contributions of the small eddies.

The cumulative contribution of the cospectral estimates as a function of $K \times Z$ provides information on the percentage loss in the cospectral estimates which can result by high- and low-pass filtering. Figure 5 shows that the relative con-tribution as a function of $K \times Z$ is very similar for CO_2 and water vapor at 25 m

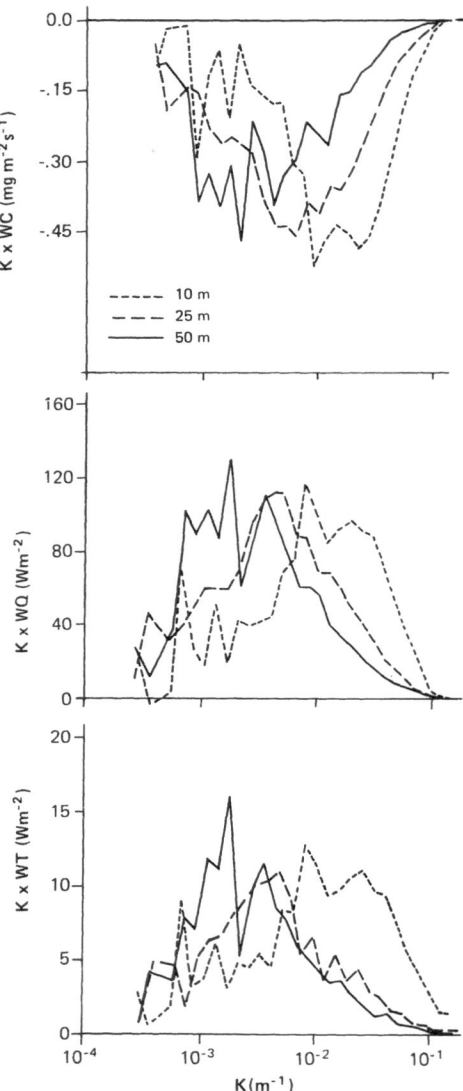

Fig. 3. Average cospectra of CO_2 (WC), latent heat (WQ) and sensible heat (WT) flux densities as a function of wavenumber K measured at three elevations near Fannystelle from July 4 to July 11, 1985.

and that sensible heat has lower frequency contributions. Figure 6 which presents the cumulative contribution to the average CO_2 flux densities as a function of $K \times Z$ for all 10, 25 and 50 m runs confirms the similarity between the transfer at 25 and 50 m levels and the underestimation at the 10 m level.

In order to quantify the directional effect in more detail, the CO_2 flux data for each altitude were grouped into upwind and downwind categories. Runs with

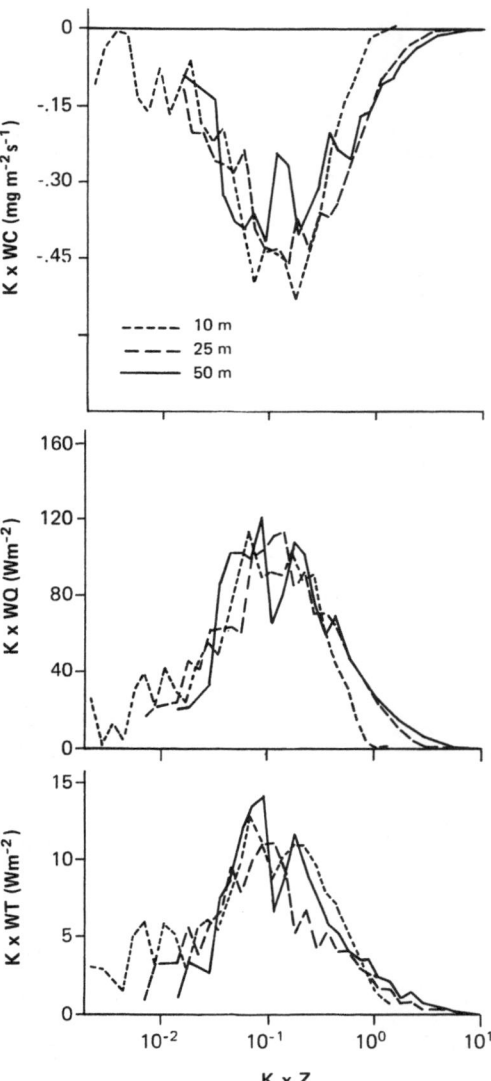

Fig. 4. Average cospectra of CO$_2$ (WC), latent (WQ) and sensible heat (WT) flux densities as a function of wavenumber (K) times the sampling height (Z) measured at three elevations near Fanny-stelle from July 4 to July 11, 1985.

head winds are defined as those in which the relative wind direction, i.e., aircraft heading minus wind direction, was less than 45 deg. Similarly, winds from within ±45 deg of the tail were designated tail winds. A good distribution of runs was obtained over the six days. Figure 7 shows little difference in the cospectra of WC at 10 and 25 m, but the downwind runs have lower contributions at smaller wavenumbers than upwind runs at 50 m. This effect is more pronounced at 50 m

TABLE IV

Means and standard errors of filtered, F, and non-filtered, NF, flux densities measured at 25 m above wheat growing area near Morris, Manitoba in 1986

Date		July 7		July 8		July 18		July 19	
Distance (km)		16.4	4.1	13.7	4.5	13.6	4.5	12.0	4.0
Solar Radiation ($W\,m^{-2}$)		884	883 ± 2	846	846 ± 2	856	856 ± 5	846	846 ± 44
Wind speed ($m\,s^{-1}$)		3.2	3.1 ± 0.4	2.1	2.1 ± 0.2	2.1	2.1 ± 0.2	5.5	5.6 ± 0.3
Heat flux ($W\,m^{-2}$)	F	36	37 ± 2	41	39 ± 4	41	43 ± 25	27	28 ± 1
	NF	45	45 ± 2	48	44 ± 1	42	44 ± 7	26	28 ± 1
Latent heat flux ($W\,m^{-2}$)	F	307	304 ± 35	308	308 ± 27	314	338 ± 17	295	298 ± 9
	NF	344	342 ± 43	428	410 ± 93	274	358 ± 4	323	312 ± 10
CO_2 flux ($mg\,m^{-2}\,s^{-1}$)	F	-0.98	-0.98 ± 0.14	-0.99	-0.97 ± 0.09	-0.99	-1.01 ± 0.06	-0.86	-0.88 ± 0.08
	NF	-1.07	-1.08 ± 0.16	-0.69	-0.76 ± 0.32	-1.33	-0.94 ± 0.12	-0.93	-0.92 ± 0.08
Momentum Flux ($N\,m^{-2}$)	F	-0.13	-0.14 ± 0.03	-0.03	-0.02 ± 0.02	-0.05	-0.06 ± 0.00	-0.22	-0.22 ± 0.02
	NF	-0.15	-0.17 ± 0.04	-0.03	-0.02 ± 0.02	-0.07	-0.04 ± 0.01	-0.23	-0.24 ± 0.03

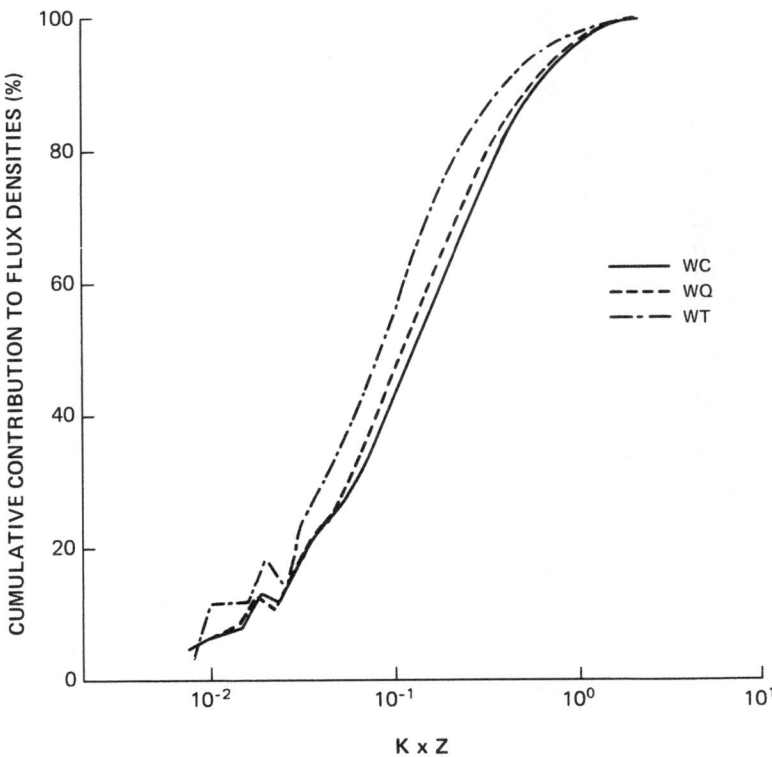

Fig. 5.   Average cumulative contribution to CO_2 (WC), latent heat (WQ) and sensible heat (WT) flux densities versus $K \times Z$ measured at 25 m above the vegetation between July 4 to July 11, 1985.

because of the general shift with height of the cospectra to lower wavenumbers. This indicates that the headwind runs at constant airspeed, therefore lower ground speed, permit longer and more complete sampling of the long-wavelength fluctuations that are still associated with the ground source and sink distributions. As expected, these observations suggest that run lengths of 4 km are marginal for adequate sampling of the long-wave components of the fluxes at altitude of 50 m or more.

Aircraft-based measurements allow spatial integration of data; however it is thought that flux variations are often on a scale comparable with the length of the aircraft runs. Table IV presents the means and standard errors of flux densities measured at 25 m above a relatively homogeneous area. Flux estimates based on high-passed filtered data at 0.02 Hz and nonfiltered data, NF, are presented to examine the repeatability of flux density measurements and the loss to the flux contribution due to the filtering used and the length of the runs. At the operating speed of the Twin Otter aircraft, high-pass filtering of 0.02 Hz corresponds to a 50% cutoff at wavelength of 4 km.

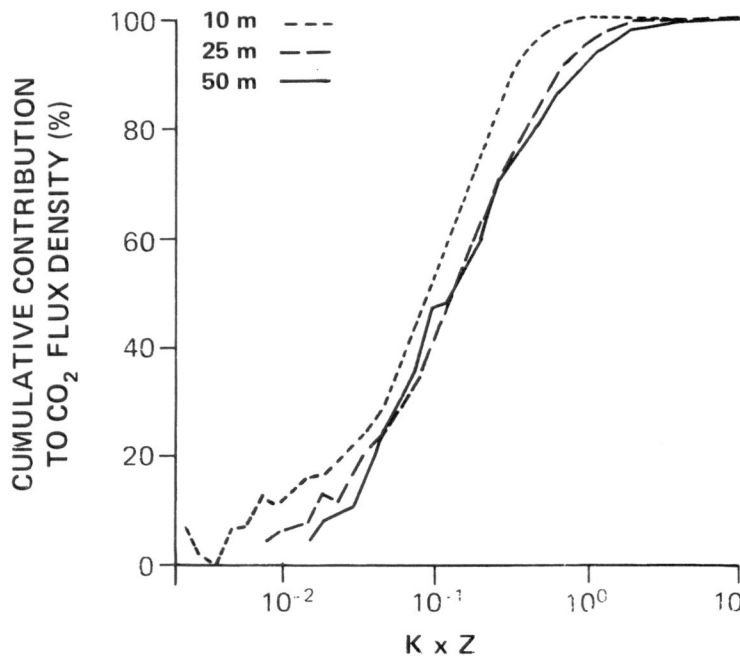

Fig. 6.   Average cumulative contribution to CO_2 flux density (WC) for 10, 25 and 50 m runs as a
function of $K \times Z$ for all runs above the wheat crop between July 4 to July 11, 1985.

The agreement between long runs on July 7th and 8th of the filtered CO_2,
latent and sensible heat flux densities is within a few percent (Table IV). The CO_2
flux values are slightly smaller on July 19th but this could be attributed to the
beginning of senescence of the wheat crop. The mean flux densities from either
three or four individual four km runs also agree within a few percent, but as
shown by the standard errors, these measurements are more variable. The
momentum flux densities are not as consistent from day to day but this is to be
expected since these values depend on wind conditions. As expected, the
nonfiltered flux estimates are also considerably more variable. However, the
means of these estimates are useful to determine the loss due to the 0.02 Hz filter
used which is of the order of 10%.

5. Conclusions

A comparison of flux measurements of CO_2, H_2O and sensible heat between
airborne and ground-based systems has shown that similar measurements can be
obtained with both techniques. Flux divergence is significant even for low
altitude runs (10 to 50 m). Hence profiles of flux measurements are necessary to
extrapolate aircraft-derived flux estimates to the surface. For a more accurate

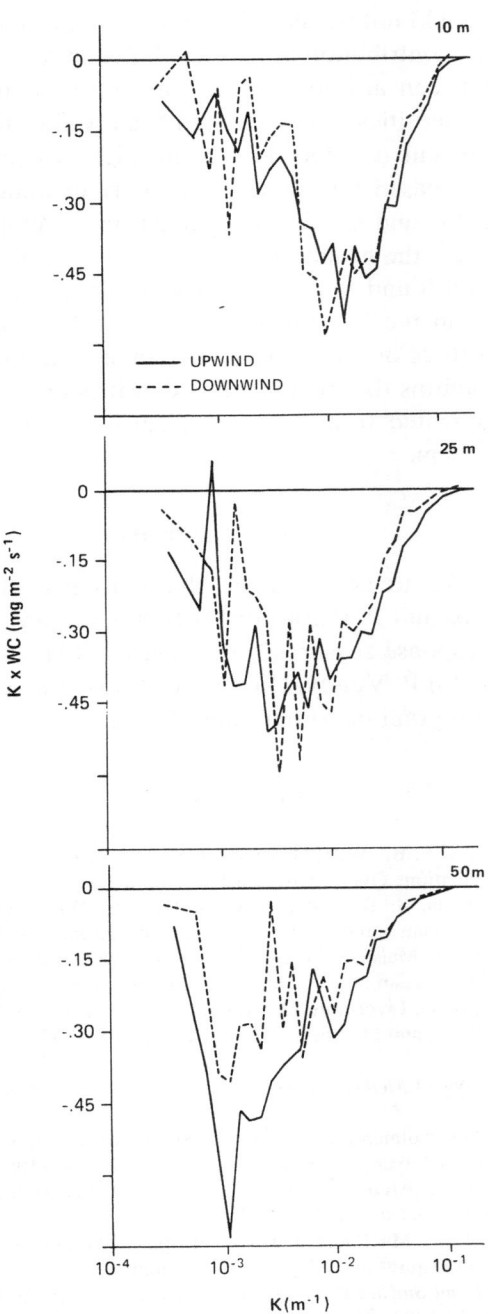

Fig. 7.   Average cospectra of CO$_2$ flux densities at 10, 25 and 50 m for upwind and downwind runs.

extrapolation to the surface with the present aircraft instrumentation, airborne flux measurements should not be recorded at 10 m because of the significant loss in the high-frequency contribution at that level. For purposes of comparison, the time of sampling between airborne and ground-based systems should coincide closely because flux densities can vary significantly due to the low-frequency intermittency of turbulent transfer. The 2-min high-pass filter sampling period used with the ground-based system was too short to obtain all the frequency contributions to the flux and hence, as expected, the aircraft CO_2 and H_2O flux measurements exceeded the ground-based values. Flight heading with respect to wind direction was not found to be significant for 10 and 25 m but resulted in a significant difference in the flux measurements at 50 m. Finally, the mean flux densities from either three or four 4 km runs calculated from 12–16 km runs were very similar. This confirms that regional flux densities of latent and sensible heat and CO_2 can be estimated from a series of runs of 4 km in length using this aircraft-based flux system.

Acknowledgements

The authors would like to acknowledge the assistance of: (1) R. Verdon of LRRC and M. Moscos and P. Poirier of ESRC for calibrating and servicing the CO_2 and H_2O fast response sensor; (2) the staff of NAE of NRC for collecting the aircraft data; and (3) P. Voisey, Director of ESRC for supporting the project through the energy program of Agriculture Canada.

References

Anderson, D. E. and Verma, S. B.: 1985, 'Turbulence Spectra of CO_2, Water Vapor, Temperature and Wind Velocity Fluctuations Over a Crop Surface', *Boundary-Layer Meteorol.* **33**, 1–14.
Anderson, D. E. and Verma, S. B.: 1986, 'Carbon Dioxide, Water Vapor and Sensible Heat Exchanges of a Grain Sorghum Canopy', *Boundary-Layer Meteorol.* **34**, 317–331.
Bingham, G. E., Gilmer, R. O., Maish, S. M. and Hansen, K.: 1983, 'Area CO_2 Flux Studies: Aircraft Measurements Over Open Ocean and Sea Ice', *Annual Report to the Department of Energy, CO_2 and Climate Office*, Lawrence Livermore Laboratory, Livermore, CA 94550, 75 pp.
Brach, E. J., Desjardins, R. L., and St. Amour, G.: 1981, 'Open-path CO_2 Analyzer', *J. Phys. E.* **14**, 1415–1419.
Brown, R. A.: 1974, *Analytical Methods in Planetary Boundary Layer Modelling*', John Wiley, New York, 150 pp.
Deardorff, J. W.: 1974, 'Three-dimensional Numerical Study of the Height and Mean Structure of a Heated Planetary Boundary Layer', *Boundary-Layer Meteorol.* **7**, 81–106.
Desjardins, R. L., Brach, E. J., Alvo, P., and Schuepp, P. H.: 1982, Aircraft Monitoring of Surface Carbon Dioxide Exchange', *Science*, **216**, 733–735.
Desjardins, R. L., Brach, E. J., MacPherson, J. I., Schuepp, P. H. and Austin, L.: 1985, 'Regional Measurements of Evapotranspiration Using Aircraft Mounted Sensors', *Proceedings of Conference on Parameterization of Land Surface Characteristics in Rome*, Italy pp. 381–385.
Druilhet, A. and Durand, P.: 1984, 'Étude de la Couche Limite Convective Sahélienne en Présence de Brumes Sèches (Expérience ECLATS)', *Boundary-Layer Meteorol.* **28**, 51–77.
Kaimal, J. C. and Haugen, D. A.: 1971, 'Comments on "Minimizing the Levelling Error in shearing Stress Measurements by Filtering", *J. Appl. Meteorol.* **10**, 337–339.

Kaimal, J. C., Wyngaard, J. C., Izumi, Y., and Cote, O. R.: 1972, 'Spectral Characteristics of Surface-Layer Turbulence', *Quart. J. Roy. Meteorol. Soc.* **98**, 563–589.

Lenschow, D. H. and Agee, E. M.: 1976, *Preliminary Results from the Air Mass Transformation Experiment (AMTEX)*, GARP Topics 57, pp. 1346–1355.

Lenschow, D. H., Pearson, R. Jr., and Stankov, B. B.: 1982, 'Measurements of Ozone Vertical Flux to Ocean and Forest', *J. Geophys. Res.* **87**, 8833–8837.

Lenschow, D. H.: 1986, 'Aircraft Measurements in the Boundary Layer', in D. H. Lenschow, (ed.), *Probing the Atmospheric Boundary Layer*, Amer. Meteorol. Soc., pp. 39–55.

Lenschow, D. H. and Stankov, B. B.: 1986, 'Length Scales in the Convective Boundary Layer', *J. Atmos. Sci.* **43**, 1198–1209.

Lumley, J. L., and Panofsky, H. A.: 1964, *The Structure of Atmospheric Turbulence*, Interscience, New York, 239 pp.

MacPherson, J. I., Morgan, J. M., and Lum, K.: 1981, *The NAE Twin Otter Atmospheric Research Aircraft*, National Research Council of Canada Report LTR-FR-80, 21 pp.

MacPherson, J. I., Desjardins, R. L., and Schuepp, P. H.: 1987, 'Gaseous Exchange Measurements Using Aircraft-Mounted Sensors', *Sixth Symp. on Meteor. Observ. and Instrum.*, Am. Meteorol. Soc., Boston, Mass., pp. 128–131.

Marht, L.: 1987, *Boundary-Layer Eddies and Surface Inhomogeneity and Their Impact in Measurement Strategy and Data Analysis for FIFE*, Report from Oregon State Univ. for NASA-Goddard Space Flight Center, 94 pp.

Ohtaki, E.: 1985, 'On the Similarity in Atmospheric Fluctuations of Carbon Dioxide, Water Vapour and Temperature over Vegetated Fields', *Boundary-Layer Meteorol.* **32**, 25–37.

Schuepp, P. H., Desjardins, R. L., MacPherson, J. I., Boisvert, J., and Austin, L. B.: 1987, 'Airborne Determination of Regional Water Use Efficiency and Evapotranspiration: Present Capabilities and Initial Field Tests', *Agr. and Forest Meteorol. J.* **41**, 1–9.

Séguin, B. and Itier, B.: 1983, 'Using Midday Surface Temperature to Estimate Daily Evaporation from Satellite Thermal IR Data', *Int. Journ. Rem. Sens.* **4**, 371–383.

Shaw, W. J. and Businger, J. A.: 1985, 'Intermittency and the Organization of Turbulence in the Near-Neutral Marine Atmospheric Boundary Layer', *J. Atmos. Sci.* **42**, 2563–2584.

Webb, E. K., Pearman, G. I. and Leuning, R.: 1980, 'Correction of Flux Measurements for Density Effects Due to Heat and Vapor Transfer', *Quart. J. Roy. Meteorol. Soc.* **106**, 85–100.

Wyngaard, J. C.: 1986, 'Observational Strategies', in D. H. Lenschow (ed.), *Probing the Atmospheric Boundary Layer*, AMS, Boston, 269 pp.

SPECTRA OF SURFACE WIND SPEED AND AIR TEMPERATURE OVER THE OCEAN IN THE MESOSCALE FREQUENCY RANGE IN JASIN-1978

HIROSHI ISHIDA

Department of Nautical Sciences, Kobe University of Mercantile Marine, Kobe, Hyogo 658, Japan

(Received in final form 27 June, 1988)

Abstract. Based on the data from an array of buoys during the JASIN-1978 field experiment made in an area northwest of Scotland, power spectra of surface wind speed and air temperature over the ocean in the mesoscale frequency range were studied. The averaged composite spectrum of wind speed for the whole period shows the existence of a spectral gap in the frequency range from 10^{-4} to 5×10^{-3} Hz. However, significant peaks in this range are often seen in particular spectra under certain weather conditions. Mesoscale spectral peaks of wind speed occur in 14 segments of the data record, approximately 10% of the total duration of the observations. In 4 of these segments, the mesoscale spectral peaks of both wind speed and air temperature occurred simultaneously. Several wave patterns of mesoscale atmospheric disturbances when mesoscale spectral peaks were seen are derived from phase differences between buoys. Significant mesoscale peaks in spectra appear in relatively strong winds and unstable or near-neutral atmospheric conditions, and none in stable atmospheric conditions. A criterion of wind speed and atmospheric stability is found for the mesoscale spectral peak appearance.

1. Introduction

Van der Hoven (1957) first showed the existence of a low kinetic energy region, the so-called spectral gap, in a wind speed power spectrum over land, in the mesoscale frequency range centered at a period of about 1 hr between synoptic and micro-scale frequency ranges. Oort and Taylor (1969) and Hess and Clarke (1973) showed the existence of a gap in wind speed spectra in the planetary boundary layer over land based on data from weather stations and balloon ascents, respectively. Panofsky (1969) and Fiedler and Panofsky (1970) reviewed various types of wind speed spectra in the lowest few hundred meters over land, and concluded that the spectral gap was a normal occurrence. Later observations and theoretical arguments in support of the spectral gap are found in Smedman-Högström and Högström (1975), Thomson (1979) and Weinstock (1980) in moderate and averaged weather conditions. On the other hand, LeMone (1973) observed fluctuating wind speeds and air temperatures with a period of about 30 min associated with mesoscale convection from tower observations.

Over the ocean, Millard (1968) and Mitsuta and Fujitani (1974) showed a similar spectral gap in the long averaged spectra of wind speed observed on a buoy and the mast of a ship, respectively, in fair weather conditions. Burt *et al.* (1974, 1975) and Ishida *et al.* (1984), however, reported from buoy observations that there were significant spectral peaks of wind speed at periods of 30–60 min in the mesoscale frequency range associated with the presence of mesoscale convective cells.

As mentioned above, the existence and generality of the spectral gap in the mesoscale frequency range of power spectra of meteorological entities have been interesting problems. Existence of the spectral gap is important in the discussion of atmospheric motions because these motions can usually be divided into two categories, synoptic and micro-scale, by this spectral gap. Therefore, in general numerical models, the grid size is chosen in the dimension corresponding to the scale of the spectral gap to filter out turbulent noise. However, if there exist significant mesoscale atmospheric motions, large errors in numerical prediction over a long period will be induced from aliasing. Pierson (1983) mentioned that mesoscale motions were dominantly a nuisance factor both for correct synoptic-scale analysis and for the interpretation of microturbulent fluxes. Overland and Wilson (1984) observed some effects on microturbulent fluxes by mesoscale wind field variations under mesoscale atmospheric structures from aircraft measurements.

In fact, several researchers, listed above, have pointed out that there may be an energy peak in the spectral range corresponding to the energy gap when mesoscale convective motions are present, both over land and sea. Such peaks are hidden when a long-period average is made. Mesoscale atmospheric activity over the ocean appeared to exist during particular periods during the field experiment of the Joint Air–Sea Interaction (JASIN) in 1978 (Ishida, 1986). More detailed studies are presented here.

The JASIN-1978 field experiment took place in an area northwest of Scotland for about two months in the summer of 1978. It was proposed and conducted by the British Royal Meteorological Society as a contribution of the United Kingdom to the GARP program. A summary of scientific and operational plans for this experiment is given by Pollard *et al.* (1983). As a part of the JASIN-1978 field experiment, observations from an array of anchored buoys were made by Professor W. V. Burt of Oregon State University. This paper shows spectral analyses of surface wind speed and air temperature, focusing on the mesoscale frequency range between the periods of about 1.5 hr and 10 min of the spectral gap, and characteristics of mesoscale atmospheric disturbances are discussed, based on the data from the buoys in the JASIN-1978 experiment.

2. Observations

In the JASIN-1978 field experiment, toroidal buoys of Oregon State University were moored at the 4 locations, B1, B2, B3 and B4 (see Figure 1 of Ishida (1986)) in an area about 400 km northwest of Scotland. Observations were made for about 40 days from July 28 to September 6, 1978. These buoys were similar to those used in the JASIN-1972 field experiment (Burt *et al.*, 1974). Measurements were briefly interrupted on August 12 and 30 to change the data tape at each buoy. The details concerning these buoy observations are described in the reports of Weller *et al.* (1983) and Ishida (1986).

Wind speed and direction, air temperature and solar radiation were measured at a height of 2.5 m above mean sea level, and sea temperatures were measured at depths of 0.5 and 2 m. Sampling intervals for the data were 3.5 min in the period from July 28 to August 30 and 1.75 min from August 30 to September 6. Wind speed was averaged over the sampling interval, while the other variables were sampled instantaneously. These data were recorded digitally on magnetic tape on the buoy. The sensors and recorders on the buoys worked satisfactorily throughout most of the observational period as reported by Ishida (1986).

3. Analyses in the Mesoscale Frequency Range

3.1. SPECTRA

The averaged composite spectrum of wind speed for the whole observational period, computed from the data of buoys B1 and B3, is shown in Figure 1 (Ishida, 1986). The spectrum was computed by FFT. The composite spectrum is patched

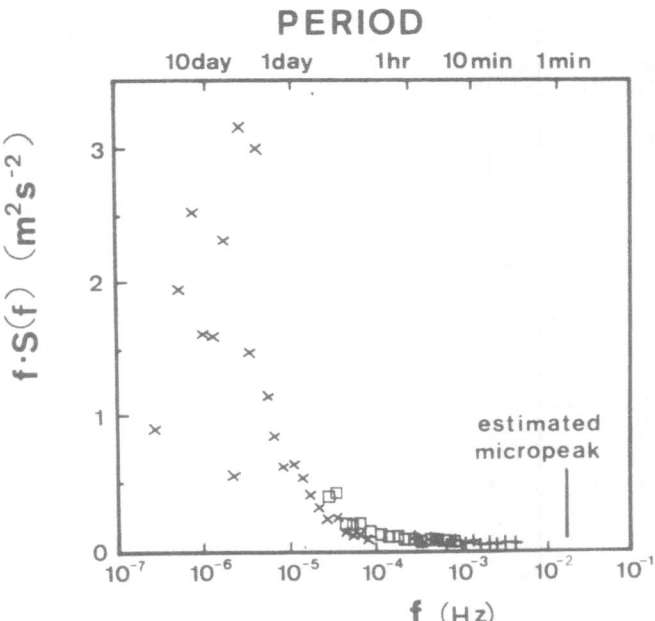

Fig. 1. Averaged composite spectrum of wind speed from buoys B1 and B3, after Ishida (1986). Symbols ×, □ and + are spectral estimates from hourly, 3.5-min and 1.75-min averages, respectively. The position of a micropeak is estimated from the similarity spectrum of Pond *et al.* (1971).

by 3 spectra computed from 3 time series of 1-hr, 3.5-min and 1.75-min
averages. The frequency ranges of the spectra overlap. The composite spectrum
does not extend to sufficiently high frequency ranges to demonstrate the existence of the microturbulence peak, which is estimated to be at about 0.02 Hz
from the similarity spectrum of Pond *et al.* (1971).

The composite spectrum suggests a flat and low kinetic energy region between
10^{-4} and 5×10^{-3} Hz in the mesoscale frequency range. There is no significant
spectral peak at the diurnal period, unlike wind speed spectra from near-shore
locations (e.g., Halpern, 1974; O'Brien and Pillsbury, 1974). The general features of the composite spectrum are in relatively good agreement with previous
spectra obtained by other observers.

Mesoscale atmospheric activity was suggested to exist in a particular period in
the data of JASIN-1978 (Ishida, 1986). To see the detailed features of the
spectrum in the mesoscale frequency range, mean spectra of wind speed and air
temperature were estimated from every 8-hr spectrum of buoy B3 data as shown
in Figure 2. Each 8-hr spectrum was computed from the data for 7.5 hr

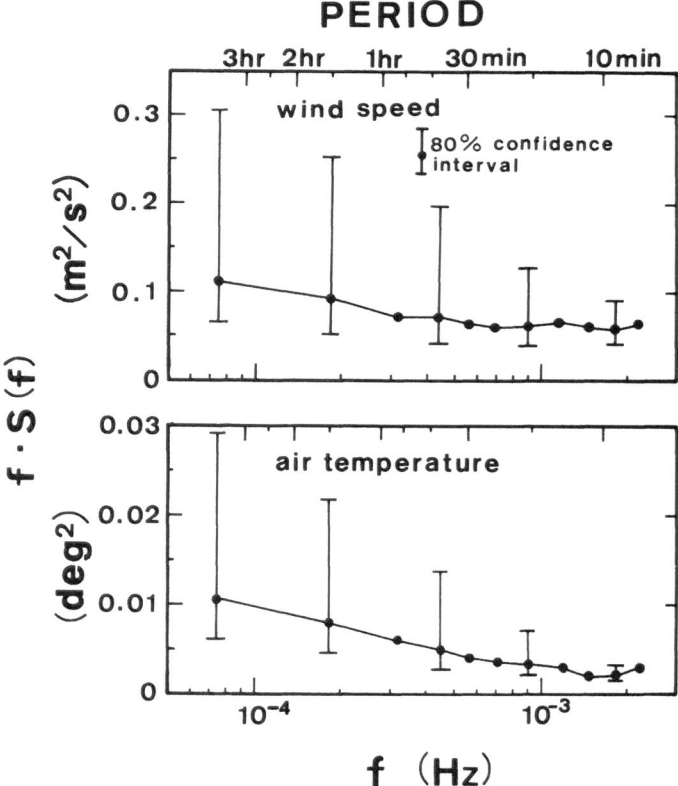

Fig. 2.   Mean spectra of wind speed and air temperature in the mesoscale frequency range obtained
from buoy B3 data for all the observations.

(3.5 min × 128) by FFT. Vertical bars show the 80% confidence interval assuming the chi-square distribution (Båth, 1974). The data were high-pass filtered at a period of 3.5 hr to remove the effect of long-period fluctuations. Spectral analyses hereafter are made in the same manner.

The mean spectra of wind speed and air temperature are flat and in magnitude approximately $0.05-0.1\ m^2/s^2$ and $0.003-0.01\ deg^2$, respectively. These are in good agreement with previous results of the spectral gap over land (e.g., Van der Hoven, 1957; Smedman–Högström and Högström, 1975). Pierson (1983), however, pointed out that the spectral values in the mesoscale gap increased with wind speed, and that sampling variability effects were very large.

Spectra under particular weather conditions do not always show the existence of a spectral gap as shown in Figures 1 and 2. Figure 3 is a typical example of the

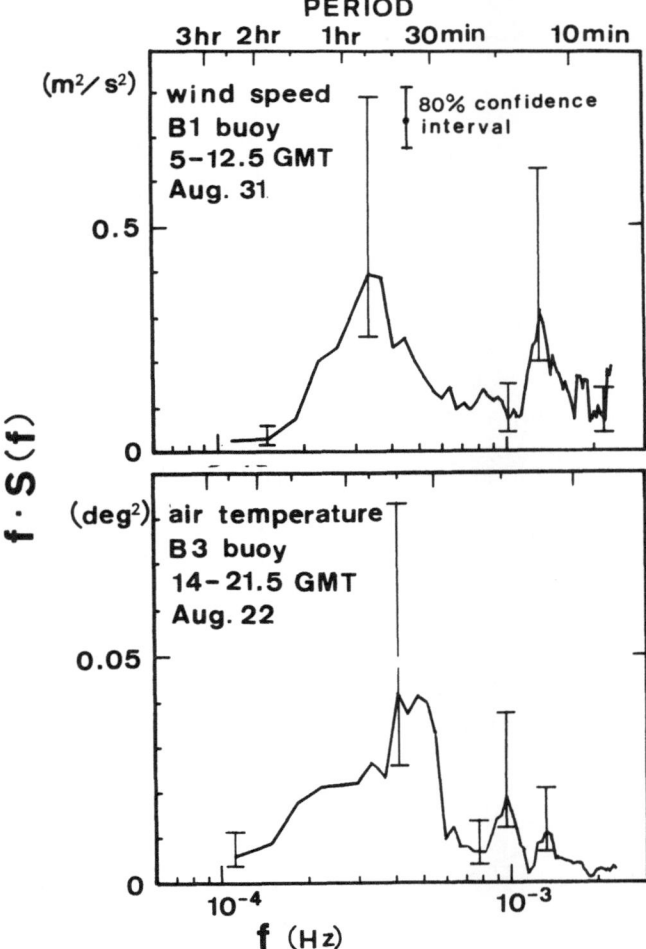

Fig. 3. Spectra of wind speed (the upper figure) and air temperature (the lower figure) computed from the B1 data from 5 to 12.5 GMT on August 31 and from the B3 data from 14 to 21.5 GMT on August 22, respectively.

results of spectral analyses when large peaks in the spectra of wind speed and air temperature were seen. Large spectral peaks occur at periods of 13 and 50 min in wind speed and at a period of 40 min in air temperature. They are greater in magnitude than the upper limits of the 80% confidence interval of the mean

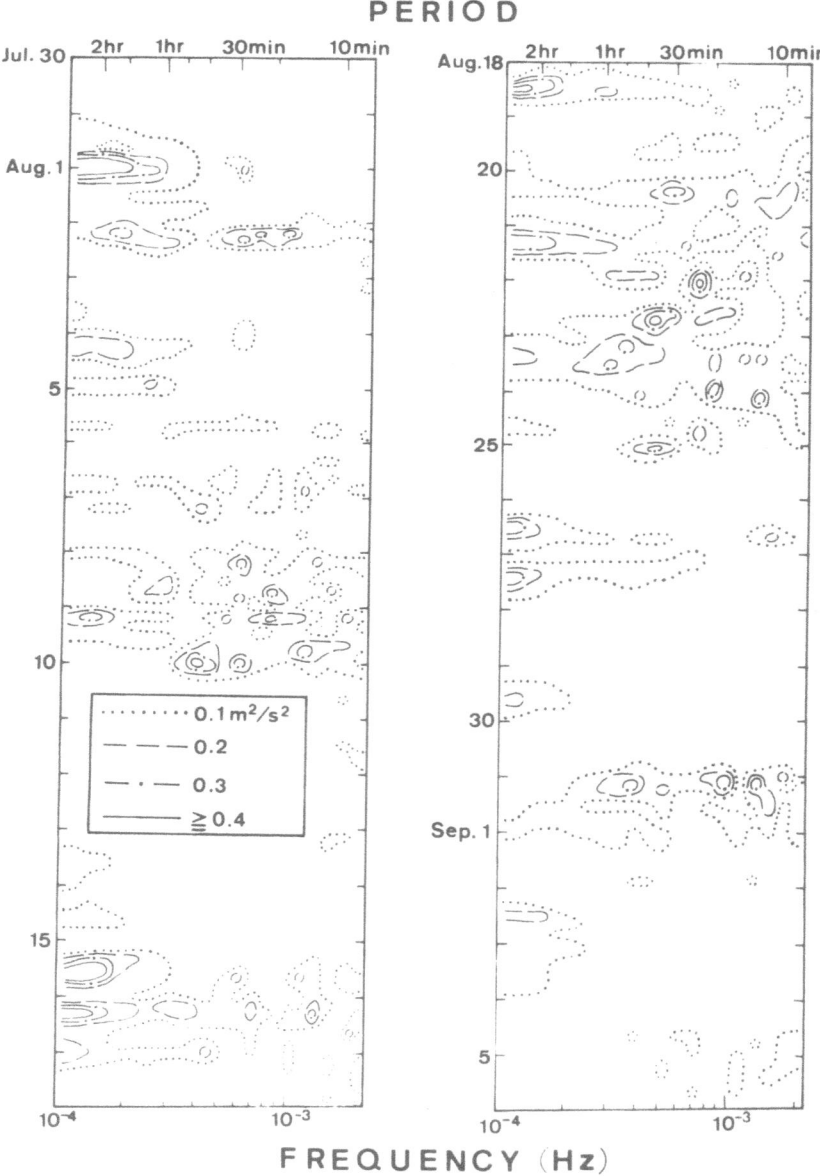

Fig. 4.   Spectrograph of wind speed, time series of isopleths of spectra $(f \cdot S(f) \ vs. \ f)$ produced from spectra at 4-hr intervals from buoy B3 data from July 30 to September 6.

spectra shown in Figure 2, which were about $0.2 \, \mathrm{m^2/s^2}$ in wind speed and $0.02 \, \mathrm{deg^2}$ in air temperature, respectively. They are also significant at the 80% confidence level. Thus, spectral features in the mesoscale frequency range are not always consistent with a spectral gap. Significant peaks show up in the spectral gap frequency range from time to time.

Figures 4 and 5 show spectrographs of wind speed and air temperature,

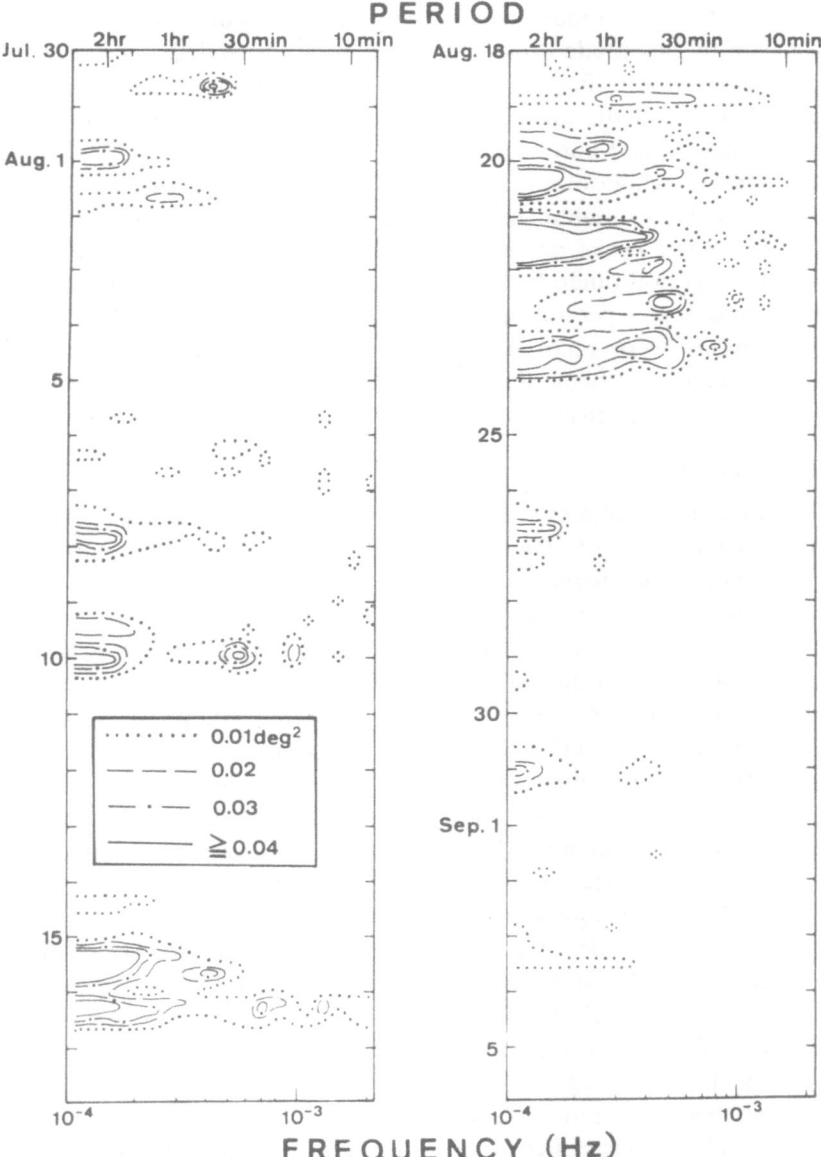

Fig. 5. Spectrograph of air temperature from July 30 to September 6, as in Figure 4.

respectively, time series of isopleths of spectra $(f \cdot S(f)$ vs. $f)$ produced from spectra at 4-hr intervals of buoy B3 data from July 30 to September 6. Large spectral densities in low frequency ranges can be attributed to synoptic weather changes such as passing low-pressure systems and fronts. It is clear from these figures that the spectrum in the mesoscale frequency range is not uniform and low. It is characterized by small peaks that show the existence of short-lived mesoscale atmospheric activity with narrow frequency ranges.

In order to distinguish mesoscale activity, the following criteria are used: (1) The peaks are in magnitude greater than $0.2 \text{ m}^2/\text{s}^2$ in wind speed or 0.02 deg^2 in air temperature (these limits correspond to the upper limits of the 80% confidence interval of the mean spectra); (2) the peaks are seen at least in the data from 3 out of the 4 buoys.

Throughout the whole observational period, there were 4 cases when mesoscale peaks occurred simultaneously in wind speed and air temperature spectra and 10 cases when peaks occurred only in wind speed spectra. These peaks occurred at the period range from 15 to 60 min. The total time period in the 14 cases when mesoscale spectral peaks were found is about 10% of the total time that observations were recorded. Low-level clouds such as cumulus and/or somewhat broken stratocumulus were observed from satallite photographs whenever mesoscale spectral peaks appeared.

3.2. CROSS-SPECTRA

Figure 6 shows an example of cross-spectral analyses between wind speed and air temperature from buoy B3, when significant peaks in the spectra of wind speed and air temperature occurred simultaneously in the mesoscale frequency range. Mesoscale peaks in the spectra of both components are seen at a period of 30–40 min, coherence is as high as 0.9, and the phase difference is about 170°. This shows that wind speed and air temperature fluctuated in opposite phase coherently with a period of 30–40 min. The coherence and phase shift may have been caused by the convection associated with some form of mesoscale atmospheric structures. These fluctuations would have an important effect on the air-sea exchanges of momentum and heat energy in the surface boundary layer.

Similar results were obtained in cross-spectral analyses in 2 out of 4 other cases when mesoscale spectral peaks of both components were seen simultaneously. In the one exceptional case from 20 GMT on August 21 to 3.5 GMT on Auguest 22, mesoscale spectral peaks of air temperature were found at a period of about 40 min while peaks in wind speed were at a period of about 20 min except for buoy B3 data, and coherence was low at the periods of those spectral peaks. The data from buoy B3 in the exceptional case showed that the mesoscale spectral peaks of both components were at mostly the same period, about 40 min, while coherence was low.

Cross-spectral analyses between the wind speed data from buoy pairs (B1-B2,

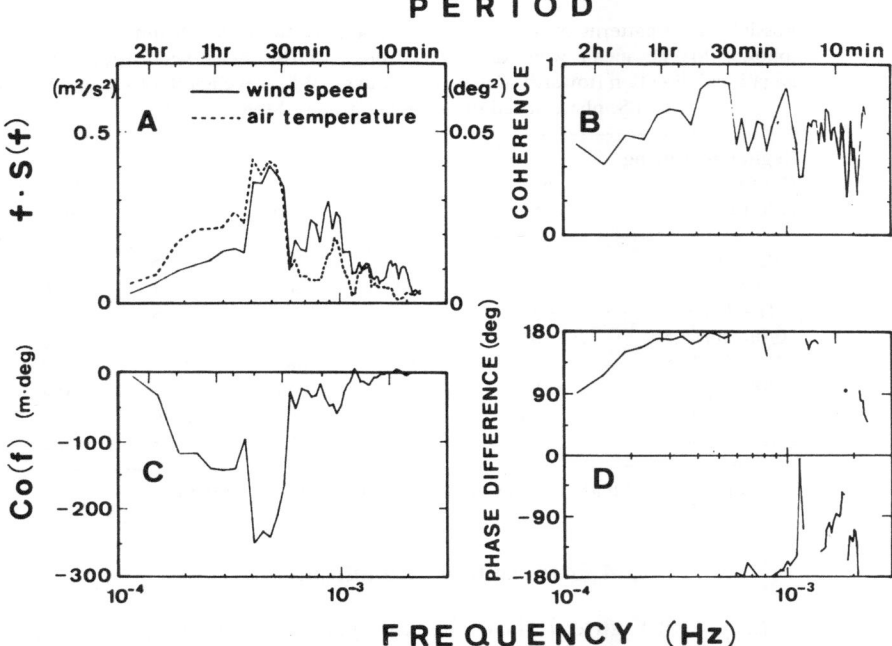

Fig. 6. Cross-spectral analysis between wind speed and air temperature from buoy B3 for 14–21.5 GMT on August 22. (A) auto-spectrum; (B) coherence; (C) co-spectrum; (D) phase difference. In (D), plus means that air temperature leads wind speed.

B1-B3 and B2-B3) were also made to investigate the possible wave patterns of mesoscale atmospheric disturbances in 14 cases when mesoscale spectral peaks of wind speed were seen. Wavelength, phase speed and propagational direction (toward) of mesoscale atmospheric waves were derived from phase differences between the data of buoy pairs at the periods of mesoscale spectral peaks, assuming that mesoscale atmospheric waves have a wavelength longer than or approximately equal to the dimension of the buoy triangle (B1-B1-B3) in the propagational direction of mesoscale atmospheric waves.

Possible solutions for mesoscale wave patterns in 8 among the 14 cases show that the propagational direction differs by no more than about 25° with the mean downwind direction, and that the phase speed is approximately equal to the mean wind speed (within about 30% of the mean wind speed). These results are listed in Table I. The 2 cases of high wind speed and air temperature correlation in the mesoscale period range are indicated in Table I with an asterisk. In these 8 cases, the mean difference of the propagational direction minus the mean downwind direction is −14°, and the standard deviation is 16°. The mean ratio of the phase speed to the mean wind speed is 0.97, and the standard deviation is 0.18.

TABLE I

Possible wave patterns of mesoscale atmospheric disturbances. Symbols are defined as follows: MWD, mean wind direction (from); MWS, mean wind speed; PD, propagational direction (toward) of mesoscale waves; WL, wavelength of mesoscale waves; PS, phase speed of mesoscale waves. Month is August.

Beginning–Ending Date/time (GMT)	MWD (deg)	MWS (m/s)	PD (deg)	WL (km)	PS (m/s)
2/05–2/12.5	057	4.3	231	18.5	4.6
6/16–6/23.5	358	6.9	204	10.1	5.6
8/21–9/04.5	038	4.9	198	4.8	3.9
20/08–20/15.5	181	11.7	345	28.3	15.6
22/14–22/21.5*	285	8.2	080	16.1	8.0
23/06–23/13.5*	254	9.0	051	22.5	10.4
23/22–24/05.5	264	9.8	061	20.9	8.4
24/23–25/06.5	287	5.8	084	9.3	4.2

* denotes cases of high wind-speed and air temperature correlation.

4. Atmospheric Conditions

The upper 2 graphs in Figure 7 show the time series at 4-hr intervals of 7.5-hr averaged air-sea temperature difference $(AT - ST)$ and wind speed (WS), respectively, from buoy B3. The lower 2 graphs show the occurrence of mesoscale spectral peaks and cloudiness estimated from the solar radiation data and/or satellite photographs. Atmospheric conditions when significant spectral peaks of wind speed occurred are $WS > 4.5$ m/s and $AT - ST < -1$ °C or $WS > 8$ m/s and $AT - ST < 0.3$ °C, except for one case on August 25 when $WS = 5.8$ m/s and $AT - ST = -0.2$ °C. Conditions when significant spectral peaks of air temperature appeared are $WS > 8$ m/s and $AT - ST < 0.1$ °C, except for one case on August 10 when $WS = 6.7$ m/s and $AT - ST = -0.9$ °C. No significant spectral peaks are found in stable atmospheric conditions when $AT - ST > 0.3$ °C.

Figure 8 is a plot of 7.5-hr averaged wind speed versus $AT - ST$ to show relationships between atmospheric conditions for time periods when mesoscale spectral peaks were found (circles) and those when mesoscale spectral peaks were not seen (crosses). These variables are averages for 7.5 hr from all available buoys so that they differ slightly from the values in Figure 7. There appears to be a distinct boundary, denoted by the broken line in Figure 8, between atmospheric conditions, wind speed and $AT - ST$, when mesoscale spectral peaks occurred and when no peaks occurred. It also appears that there is an excluded region, at least for the JASIN area and season, where unstable temperature differences do not occur in high winds. The mesoscale peaks occur along the edge of this excluded area.

These results qualitatively conform with the results of a numerical study by Agee and Chen (1973) and radar observations by Kelly (1984) on mesoscale

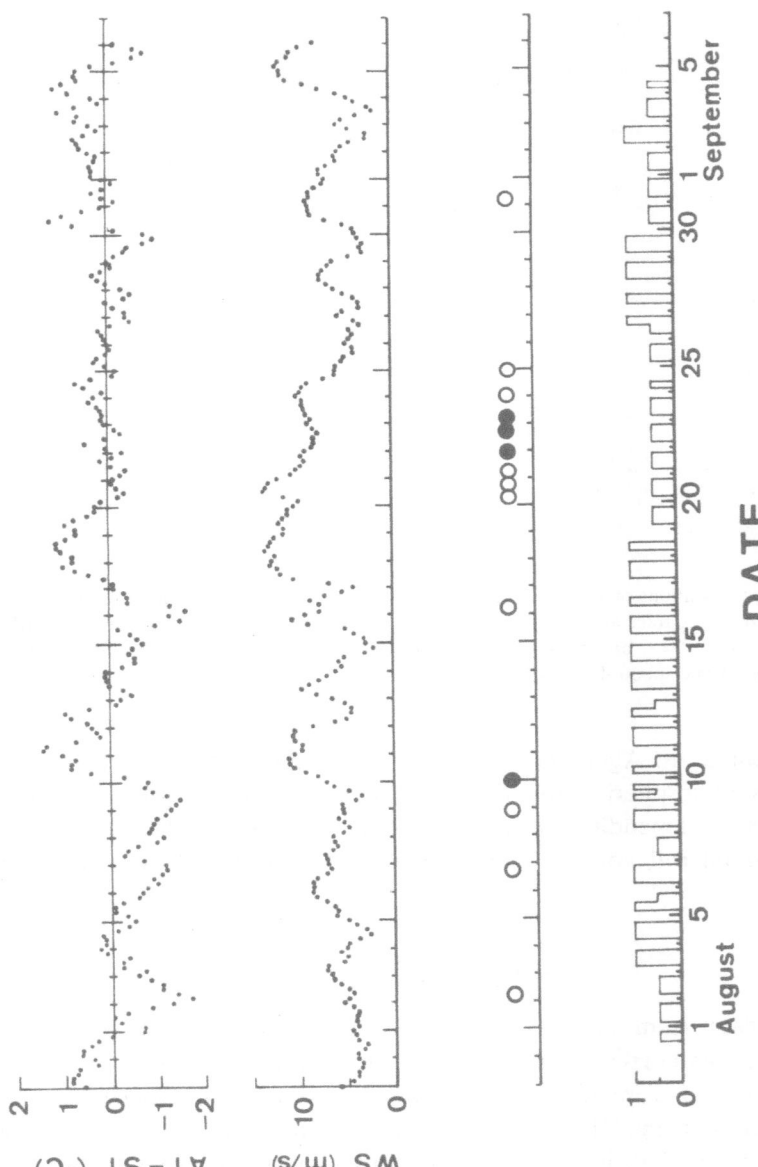

Fig. 7. Time series at 4-hr intervals of 7.5-hr averages of air-sea temperature difference (AT − ST) and wind speed (WS) from buoy B3. In the third graph from the top, open circles indicate cases when mesoscale peaks only in wind speed spectra occurred, and solid ones when mesoscale spectral peaks of both wind speed and air temperature occurred simultaneously. The lowest graph shows cloudiness estimated from the radiation data and/or satellite photographs. Cloudiness 1 means mostly overcast, 0.5 is cloudy, and 0 is mostly clear.

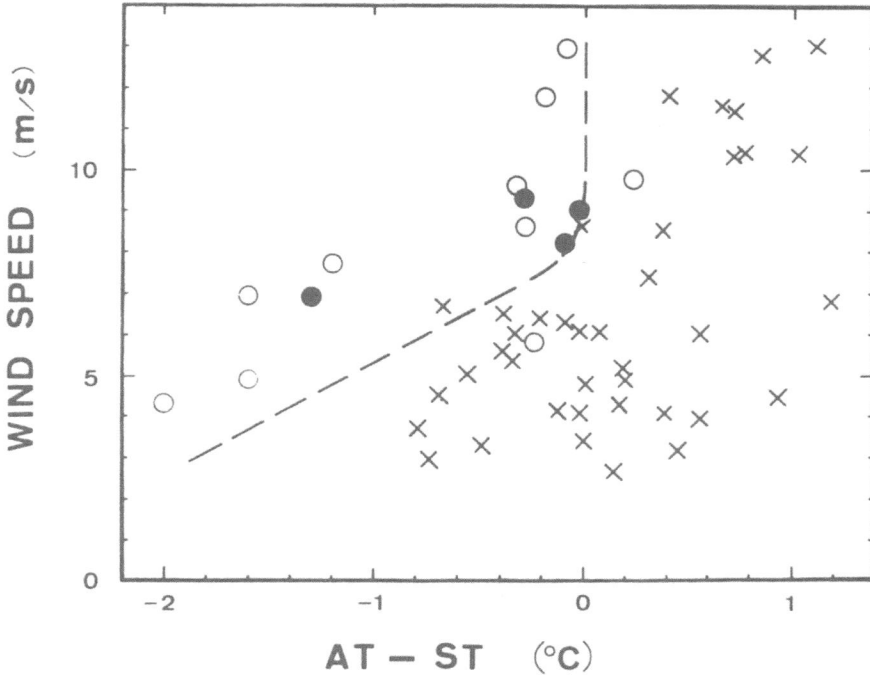

Fig. 8.   Atmospheric conditions of 7.5-hr averaged wind speed versus air-sea temperature difference (AT – ST). Open and solid circles are defined in Figure 7. The symbol × is the atmospheric condition when no peaks appeared. The broken line is a boundary between atmospheric conditions when mesoscale spectral peaks occurred and when they did not occur.

atmospheric circulations. Agee and Chen (1973) find that mesoscale convective cells develop in atmospheric conditions of Rayleigh number greater than a certain critical value depending on the sizes of the cells. Kelly (1984) observed mesoscale horizontal roll vortices in certain ranges of Reynolds and Rayleigh numbers.

5. Conclusion

Based on the data from an array of buoys during the JASIN-1978 field experiment, in an area northwest of Scotland, a study has been undertaken of the spectra of surface wind speed and air temperature over the ocean in the mesoscale frequency range. The averaged composite spectrum of wind speed for all the observational periods is generally consistent with the spectral gap reported first by Van der Hoven (1957). However, large peaks in the spectra of wind speed and air temperature sometimes appear for short periods in the mesoscale frequency range under particular weather conditions.

During the observational period, there were 4 cases when mesoscale spectral peaks of both wind speed and air temperature occurred and 10 cases with mesoscale peaks only in the spectra of wind speed. The total length of record for the 14 cases is approximately 10% of the total duration of the observations. The peaks occurred at periods ranging from 15 to 60 min. In 3 cases when significant mesoscale spectral peaks of wind speed and air temperature were seen simultaneously at a period of 30–40 min, both components fluctuated in opposite phase with high coherence. These fluctuations would have an important effect on air-sea exchanges of turbulent transfers in the surface boundary layer. These may be caused by convection in some form of mesoscale atmospheric structures.

The wave patterns of mesoscale atmospheric disturbances in the 14 cases when mesoscale spectral peaks of wind speed were seen were derived from phase differences between buoys at the periods of mesoscale spectral peaks. In 8 of the 14 cases, the possible mesoscale wave patterns showed that the propagational direction (toward) differed by no more than about 25° from the mean downwind direction, and that the phase speed was almost equal to the mean wind speed (within 30% of the mean wind speed). In the 8 cases, the mean difference of the propagational direction minus the mean downwind direction was $-14°$, and the standard deviation was 16°. The mean ratio of the phase speed to the mean wind speed was 0.97, and the standard deviation was 0.18.

Significant mesoscale spectral peaks occurred when wind speed and air-sea temperature difference were in the range above the boundary denoted by the broken line in Figure 8. Mesoscale peaks occurred in relatively strong winds and unstable or near-neutral atmospheric conditions. No peaks appeared in stable atmospheric conditions.

Acknowledgements

The author wishes to express heartfelt appreciation to Professors Y. Mitsuta of Kyoto University and W. V. Burt and S. J. Neshyba of Oregon State University for their helpful suggestions and criticisms. He is grateful to the members of the Buoy Groups of Oregon State University, who planned and operated the buoy observations, to the Woods Hole Oceanographic Institute, who assisted in installation and retrieval of the moorings and to the officers and crew of the R/V Atlantis for their cooperation. This research was supported by the Office of Naval Research through contract N00014-76-C-0067 and N00014-79-C-0004 under project NR083-102 with Oregon State University.

References

Agee, E. M. and Chen, T. S.: 1973, 'A Model for Investigating Eddy Viscosity Effects on Mesoscale Cellular Convection', J. Atmos. Sci. 30, 180–189.
Båth, M.: 1974, Spectral Analysis in Geophysics, Elsevier, Amsterdam, 563 pp.
Burt, W. V., Cummings, T., and Paulson, C. A.: 1974, 'The Mesoscale Wind Field Over the Ocean', J. Geophys. Res. 79, 5625–5632.

Burt, W. V., Crew, H., and Poole, S. L.: 1975, 'Evidence for Roll Vortices Associated with a Land Breeze', *J. Marine Res.* **33**, 61–68.

Fiedler, F. and Panofsky, H. A.: 1970, 'Atmospheric Scales and Spectral Gaps', *Bull. Amer. Meteorol. Soc.* **51**, 1114–1119.

Halpern, D.: 1974, 'Summertime Surface Diurnal Period Winds Measured Over an Upwelling Region Near the Oregon Coast', *J. Geophys. Res.* **79**, 2223–2230.

Hess, G. D. and Clarke, R. H.: 1973, 'Time Spectra and Cross-Spectra of Kinetic Energy in the Planetary Boundary Layer', *Quart. J. Roy. Meteorol. Soc.* **99**, 130–153.

Ishida, H., Burt, W. V., and Mitsuta, Y.: 1984, 'The Effects of Mesoscale Convective Cells on the Surface Wind Field Over the Ocean', *Boundary-Layer Meteorol.* **29**, 75–84.

Ishida, H.: 1986, 'Mesoscale Spatial and Temporal Variability of Meteorological Observations From an Array of Buoys in JASIN-1978', *Boundary-Layer Meteorol.* **37**, 149–165.

Kelly, R. D.: 1984, 'Horizontal Roll and Boundary-Layer Interrelationships Observed Over Lake Michigan', *J. Atmos. Sci.* **41**, 1816–1826.

LeMone, M. A.: 1973, 'The Structure and Dynamics of Horizontal Roll Vortices in the Planetary Boundary Layer', *J. Atmos. Sci.* **30**, 1077–1091.

Millard, R. C.: 1968, 'Wind Measurements From Buoys: A Sampling Scheme', Ref. 68-68, 34 pp., Woods Hole Oceanogr. Inst., Woods Hole, Massachusetts.

Mitsuta, Y. and Fujitani, T.: 1974, 'Direct Measurement of Turbulent Fluxes on a Cruising Ship', *Boundary-Layer Meteorol.* **6**, 203–217.

O'Brien, J. J. and Pillsbury, R. D.: 1974, 'A Note on Rotary Wind Spectrum in a Sea Breeze Regime', *J. Atmos. Sci.* **13**, 820–825.

Oort, A. H. and Taylor, A.: 1969, 'On the Kinetic Energy Spectrum Near the Ground', *Mon. Wea. Rev.* **97**, 623–636.

Overland, J. E. and Wilson, J. G.: 1984, 'Mesoscale Variability in Marine Winds at Mid-Latitude', *J. Geophys. Res.* **89**, 10,599–10,614.

Panofsky, H. A.: 1969, 'Spectra of Atmospheric Variables in the Boundary Layer', *Radio Sci.* **4**, 1101–1109.

Pierson, W. J.: 1983, 'The Measurement of the Synoptic Scale Wind Over the Ocean', *J. Geophys. Res.* **88**, 1683–1708.

Pollard, R. T., Guymer, T. H., and Taylor, R. K.: 1983, 'Summary of the JASIN-1978 Field Experiment', *Phil. Trans. Roy. Soc. London* **A308**, 221–230.

Pond, S., Phelps, G. T., Paquin, J. E., McBean, G., and Stewart, R. W.: 1971, 'Measurements of Turbulent Fluxes of Momentum, Moisture and Sensible Heat Over the Ocean', *J. Atmos. Sci.* **28**, 901–917.

Smedman-Högström, A. and Högström, U.: 1975, 'Spectral Gap in Surface-Layer Measurements', *J. Atmos. Sci.* **32**, 340–350.

Thomson, M. R.: 1979, 'The Wind Speed Spectrum in an Industrial Environment or Does the Spectral Gap Exist', Rept. No. ARL/STRUC Note 457, Aeronautical Research Laboratories, Melbourne, Australia.

Van der Hovan, I.: 1957, 'Power Spectrum of Horizontal Wind Speed in the Frequency Range From 0.0007 to 900 Cycles Per Hour', *J. Meteorol.* **14**, 160–164.

Weinstock, J.: 1980, 'A Theory of Gaps in the Turbulence Spectra of Stably Stratified Shear Flow', *J. Atmos. Sci.* **37**, 1542–1549.

Weller, R. A., Payne, P. E., Large, W. G., and Zenk, W.: 1983, 'Wind Measurements From an Array of Ocenaic Moorings and From F/S Meteor During JASIN-1978', *J. Geophys. Res.* **88**, 9689–9705.

SHAPES OF ANNUAL FREQUENCY DISTRIBUTIONS
OF WIND SPEED OBSERVED ON HIGH
METEOROLOGICAL MASTS

JON WIERINGA

Royal Netherlands Meteorological Institute, De Bilt*
The Netherlands

(Received 30 August, 1988)

Abstract. Annual wind distributions from six masts are used to investigate the annual variability of hourly-average wind. A variability extremum occurs at the height where the average diurnal cycle of wind reverses its phase from a nocturnal minimum to a nocturnal maximum. A simple profile model shows that in non-complex terrain, this reversal height varies approximately between 50 m at coasts and 90 m inland. The Weibull distribution shape factor k has a maximum at the reversal height, and with decreasing height, k decreases approximately as a linear function of height. Therefore estimation of wind distribution shape from surface data is possible in the surface layer. In the upper PBL, however, such estimation is not very feasible, because no simple methods are available to estimate the reversal height from routine surface observations.

1. Introduction

The annual spectrum of boundary-layer wind speeds shows a gap around the period of one hour (see e.g., Fiedler and Panofsky, 1970). Investigations of boundary-layer meteorology have almost exclusively dealt with the high-frequency side of this gap. The reason is obvious: that part of the spectrum contains the turbulent wind variations, and turbulence is a defining property of the atmospheric boundary layer. Variability of wind and temperature at the low-frequency side of the spectral gap is of minor interest in boundary-layer research, hardly worth a single article per year in *Boundary-Layer Meteorology*.

A distinct reason for the lack of research into the average behavior of the planetary boundary layer (PBL) is a scarcity of useful data. Routine radiosonde observations are not suitable because of the low-sampling frequency, 6 hours at best, and too-fast inaccurate sampling of the lowest kilometer. Special projects for observing the PBL seldom last more than a couple of months because of limitations in manpower and costs. Evaluations then tend to focus on the structure of interesting cases, rather than on the average situation. A clear illustration is the Wangara project, for which there are more than a hundred analyses of day 33, while only a solitary study gives an in-depth analysis of the combined 40 Wangara days (Clarke, 1974).

This study deals with the variability of hourly-averaged wind speed around the annual average. For low-frequency variability description, the shape of wind frequency distributions proves a useful alternative to the variance. The occur-

---

* Also: Dept. of Applied Physics, Technical University, Delft, Netherlands.

Boundary-Layer Meteorology **47**: 85–110, 1989.
© 1989 *Kluwer Academic Publishers*.

rence of a variability extremum at heights of 50 to 100 m is investigated, using long time series from several good meteorological masts. The necessary quality requirements for such masts are listed, and the results are discussed in relation to the engineering requirement of estimating average high-level wind behavior from very simple surface observations.

2. Selection of Masts

Meteorological masts offer the opportunity of nearly continuous PBL observation, and do so at relatively low staff requirements and modest costs after the initial effort of mast establishment. This makes it possible to describe average PBL behavior under specified boundary conditions, or to calibrate the average reliability of PBL models towards operational use; also masts offer the opportunity to observe infrequent but interesting PBL situations.

However, the siting of existing masts generally has been chosen to fit operational purposes; in particular, the requirements for diffusion climatologies for atomic reactors (e.g., Brookhaven, Hanford, Karlsruhe, Mol, München, Obninsk, Risø, Savannah River) are different from those of meteorological research. As a result, there are extremely few meteorological masts from which the observations are analyzable over the full mast height range. Moreover, even if all instruments worked regularly, much tedious work must be undertaken to obtain a well-checked mast data set over a reasonably complete representative period, say a year. It is easier, and usually more rewarding, to pick out the plums from the observations.

Even if long data sets are generated from meteorological masts, they are hardly fitting material for journal publication and generally are hidden in limited-distribution in-house reports. Finding out about existing mast data sets for comparison with observations at the Cabauw meteorological mast was an object of a USA roundtrip which I made in 1978. My quest was structured during several discussions with Hans Panofsky, and we agreed that the following information would be required before dealing with data from a mast or tower:

(a) **Terrain specifications**: Specification of roughness and of terrain elevation changes within several km radius, azimuth-dependent in sectors $\leq 45°$. Photos of mast and surroundings in several directions. Maps showing existence of large-scale terrain features (coasts, cities, mountains) and location of weather stations within a radius of 200 mast heights.

(b) **Observation**: (1) Boom height (relative to tower base), mast width, boom length, boom direction; (2) Instrument locations and characteristics, including calibration procedures and dynamical response; (3) Recording facilities, data sampling and checking procedures; (4) Archival procedures; (5) Period, quality and status of long data records.

(c) **Organization**: (1) Listing of directly involved technical and scientific staff; (2) Full address and telephone of manager responsible for mast experiments; (3) Information on degree of interest of the controlling and funding agencies in boundary-layer research; (4) Publications.

For most masts it is not much of a problem to get a useful amount of information about observation and organization – but it is surprisingly difficult, even with the active cooperation of colleagues like Panofsky, to unearth a sufficient description of terrain features around the masts. Often a visit to the mast site is the only way to obtain details.

For this study, it is necessary that the lower and upper part of the mast profiles can be analyzed together, i.e., that the surface layer is not too rough and the distant terrain is reasonably level: not mountainous, no cliffs or major obstacles nearby, at most a simple coastline. This ruled out masts with nearby elevation variations of the order of 50 m (e.g., Boulder, Savannah River), or masts located in a wood or in an urbanized area (e.g., Brookhaven, Obninsk, Tsukuba), or masts in complicated surroundings (Risø – see e.g., Peterson, 1975). It also ruled out masts for which there was insufficient information on mast surroundings (and on observation procedures), and for which I did not have personal acquaintance with the local situation.

Additionally, for this study of annual wind distributions, it is necessary that wind distribution statistics over a full year with $\geq 80\%$ completeness be available in classes of ≤ 2 m/s width as tables or neat histograms, at least for 3 heights – the lowest at ≈ 10 m, the highest at least at 70 m.

Six masts proved to be accessible and useful within these limits, these masts are listed in Table I. In addition, for the purpose of independent checking, three single-height observation series from the Netherlands were available: Hazerswoude (65 m), Appingedam (80 m) and Oirschot (80 m); they were compared with nearby surface stations. In Appendix A, references are listed giving the necessary background information on the masts.

The actual annual wind frequency distributions used are listed in Appendix B. Either they have been published in nearly inaccessible reports (R in Table I), or they have not yet been published (N in Table I). To make further progress,

TABLE I

Used meteorological masts (height = highest observation level).

Mast	Height (m)	Location	Geography	Data status
Cabauw	200	52.0° N, 4.9° E	Inland	N
Cape Kennedy	150	28.6° N, 80.6° W	Near sea coast	R
NSSL Oklahoma	444	35.6° N, 97.5° W	Inland	R
Quickborn	70	53.7° N, 9.9° E	Coastal region	N
Vlaardingen	80	51.9° N, 4.3° E	Coastal region	N
Wallops Island	77	37.8° N, 75.5° W	At sea coast	R

N – data not yet published; R – data available in nearly inaccessible reports.

readers should be able to make checks on the data and develop alternatives. For instance, the fitting of Weibull parameters to a distribution can be done by many different methods, and it should be possible to ascertain whether the choice of method is appropriate.

3. Low-Frequency Spectra and the Behavior of the Diurnal Cycle

The distribution of wind variance over time scales from seconds to years is best summarized in spectral form. Figure 1 shows two published long-term spectra, the pioneering effort by Van der Hoven (1957) and a second one made by Gomes and Vickery (1977). Both spectra are quite similar with respect to the location of a kinetic energy minimum at periods of ≈ 1 hr, and the existence of a variability peak for periods between 2 and 7 days. The latter peak is due to synoptic pressure system variations. There are minor differences in the sizes of the annual peaks (unobtainable from Van der Hoven's one-year record) and the high-frequency turbulence peaks (which Van der Hoven obtained from a hurricane). However, the major difference is the absence of a 24-hour peak in Van der Hoven's spectrum, while this peak is a major feature for Gomes and Vickery. This is a consequence of the fact that Van der Hoven's data were observed at 100 m height on the Brookhaven mast, while Gomes and Vickery took their data from surface stations.

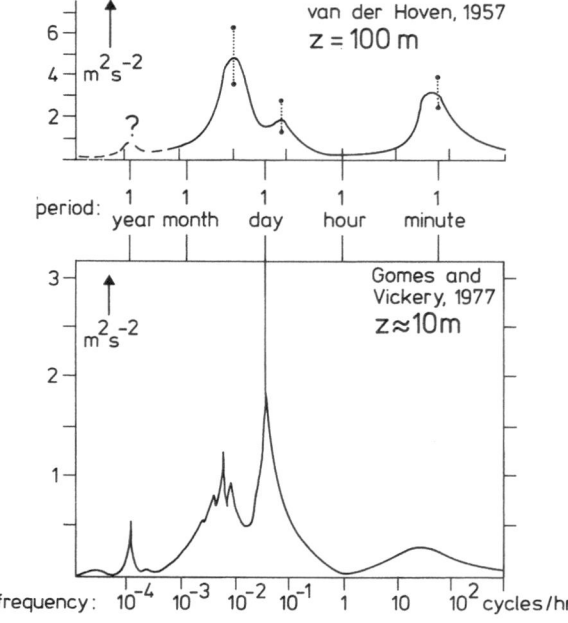

Fig. 1.   Comparison of broad-range spectra, published respectively by Van der Hoven (1957) and
Gomes and Vickery (1977) from data observed at different heights.

It has been known for seventy years (Hellmann, 1917; Peppler, 1921) that on clear days, the diurnal wind cycle changes phase with height. In daytime, efficient boundary-layer mixing causes near-equality in wind speeds near the surface and at higher levels. On stable nights, the transfer of air down to the surface is much less, resulting in low surface wind speeds – while simultaneously the wind speeds above the surface inversion increase, because these upper layers lose less momentum to lower layers than in the daytime. For a given level of geostrophic flow, the diurnal cycle of wind therefore has a nighttime minimum near the surface, while it has a nighttime maximum in the upper PBL. In other words, the diurnal cycle changes its phase halfway in the PBL, quite distinctively on clear days, less markedly on overcast days. The average height at which this phase reversal occurs will be called here the reversal height z_r.

Figure 2 shows that at Cabauw, the seasonal-average reversal height is around 80 m. Dependence of z_r on terrain conditions and on stability climate has hardly been investigated, maybe because accurate determination of z_r is difficult. A tall mast is needed to observe reversal, but the number of observation levels will be limited, and the vertical distance between two heights with clearly opposite phases in their diurnal courses will typically be ≥ 50 m. Published graphs of diurnal cycles indicate that reversal heights may vary between 40 and 100 m. In Oklahoma, the annual-average reversal height lies between the 45 m and 177 m levels on the mast, somewhere around the 90 m level (Crawford and Hudson, 1973). At Nauen in Germany, Hellmann (1917) estimated z_r at 70 m, somewhere between the 32 and 123 m levels.

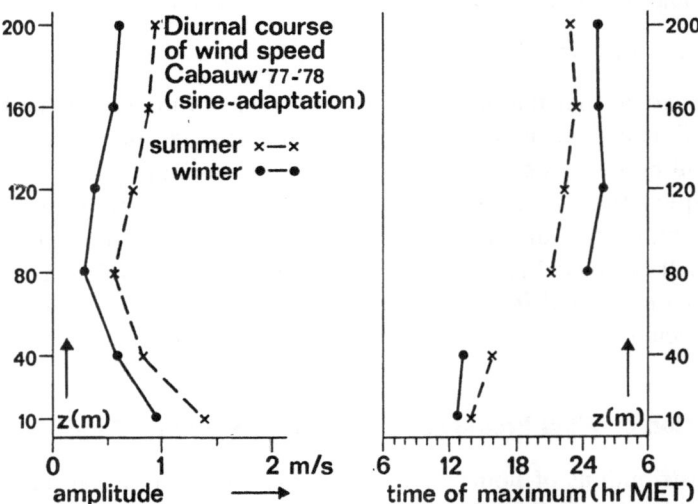

Fig. 2. Change with height of the half-year-average amplitude and phase of the diurnal wind cycle, observed at the Cabauw 200 m mast in the period March 1977 – February 1978 (Wieringa, 1987). Amplitude and phase are those of a sinusoid, least-square fitted to 24 hourly wind speed season-averages.

In any case, the disappearance of the diurnal peak in Van der Hoven's spectrum can be explained from the fact that his observation height (≈ 100 m) was close to the local reversal height. For the 136 m mast at Millstone Point, near Brookhaven, an estimate of $z_r = 85$ m was obtained from some data published by Verholek (1977), using methods discussed below.

From Figure 2 it can be concluded that reappearance of a diurnal cycle peak in the spectrum can be expected well above 100 m. However, it is not obvious what value the average diurnal cycle amplitude, A_u, will have at the top of the PBL. Figure 2 shows that at Cabauw, A_u still increases at 200 m. Crawford and Hudson (1973) show that in Oklahoma, the amplitude of the diurnal cycle continues to increase with height at least to 440 m. Vinnichenko (1970) shows that in the free troposphere, the spectrum of horizontal wind speed has not only a large synoptic peak, but also a lesser diurnal peak with a gap in between. Manier and Weingärtner (1979) have shown that in central Europe, the geostrophic wind has a diurnal cycle with an early morning peak, and that its existence is not related to terrain slope or to orography. Macroscale model studies by e.g., Krishna (1968), Paegle and Rasch (1973), Hoxit (1975) and Roth (1981) show that such phenomena can be explained from the interaction between the diurnal cycle in the PBL sensible heat flux and corresponding variations in cross-isobar mass flux.

Anyhow, it is a sensible working hypothesis to separate the spectrum of hourly-averaged wind into two parts, a diurnal peak and a synoptic "weekly" peak, separated by a secondary spectral gap at periods around 36 hrs. In other words, the long-term variance σ_u^2 can be split into two partial variances, a diurnal variance σ_d^2 and a synoptic variance σ_w^2, related to different time scales:

$$\sigma_u^2 \approx \sigma_d^2 + \sigma_w^2 . \tag{1}$$

It is to be expected that the seasonal value of σ_w^2 is a function of the large-scale circulation climate (e.g., location relative to the average polar front, continentality) and will be the same at all levels of the local boundary layer. On the other hand, we expect that σ_d^2 is height-dependent, with a minimum at the average reversal height of the diurnal cycle of wind speed. In addition, σ_d^2 will be relatively small at coastal locations, because there the diurnal variations will be small in the case of wind from the sea, so the seasonally averaged amplitude of the diurnal cycle of wind will be small compared to nearby inland locations.

4. Relation of Wind Frequency Distribution Shape to Wind Variance

Frequency distributions of hourly-averaged wind speed often can be reasonably well described by the Weibull distribution function:

$$F(U) = 1 - \exp\left[-\left(\frac{U}{a}\right)^k \right] \tag{2}$$

where $F(U)$ is the cumulative probability of occurrence of wind speeds $< U$. The scale parameter a has dimensions of wind speed, is proportional to the average wind speed calculated from the entire distribution, and increases monotonously with increasing height.

The shape parameter $k(\geq 1)$ describes the skewness of the distribution function. For $k \approx 3.5$, the function is nearly symmetrical, not unlike Gaussian distributions; for decreasing values of k, the distribution modus shifts to lower wind speeds and the probability of high wind speeds increases simultaneously. For typical wind speed distributions, the k-parameter has values around 2.

The Weibull distribution will give a fully adequate representation of a circular normal frequency distribution of wind (Tuller and Brett, 1984). This requires that the orthogonal components of the wind velocity, transformed by raising to the power $k/2$, have the following properties: (a) they are normally distributed; (b) they have equal variances; (c) they have zero means; (d) they are uncorrelated. These conditions will be approximately met for wind climates, where the frequency distribution does not vary strongly with wind direction and has a low frequency of calms. In other words, the Weibull distribution should be a good variability descriptor for hourly-averaged wind speeds at rather windy locations without strong topographical influences. In practice, the Weibull formula is used to describe wind distributions as long as a reasonable fit can be obtained, because its parameters are rather simply related to statistical moments and to the average nth power of wind speed.

In particular, if the Weibull parameters a and k of some wind distribution are known, the standard deviation of the distribution can be calculated:

$$\sigma_u = a\left[\Gamma\left(1 + \frac{2}{k}\right) - \Gamma^2\left(1 + \frac{1}{k}\right)\right]^{1/2} \tag{3}$$

where Γ is the well-known gamma-function. It is obvious that k and σ_u are closely and inversely related; in fact, a simple calculation shows that

$$k = (0.94 \pm 0.02)(a/\sigma_u) \pm 0.06 \qquad \text{for } 1.5 \leq k \leq 2.7 \,.$$

This means that we can investigate the behaviour of the diurnal cycle with height by way of the height changes in the frequency distribution. Because σ_d has a minimum at the reversal height z_r and σ_w is locally constant, σ_u will have a minimum at z_r. Since the scale parameter is a monotonous function of height, it follows that the shape parameter k must have a maximum near the reversal height of the diurnal cycle. This hypothesis needs checking.

From the available distributions, Weibull parameters were obtained by least-square fitting of the Weibull function to the cumulative distribution over the range 4–16 m/s. This method is straightforward, economic and sufficiently reliable (Stevens and Smulders, 1979; Tuller and Brett, 1984). In particular, it gives better fits than the method of moments because it allows elimination of the low wind speeds, which in the surface layer can deviate from the Weibull function due to stability effects (Rijkoort, 1972). Use of an alternative Weibull evaluation method could give an additive k-bias of ≈ 0.1.

Summarized input data obtained from Appendix B are listed in Table II. One should realize that the accuracy of k-determination is about ± 0.05; for that reason, k-values from very low levels (e.g., 20 m at Cabauw) are risky to interpret, because their difference with k_s (the surface-level value of k) is very inaccurate.

5. Shape Evaluation of Mast Wind Distributions

For a preliminary analysis of the vertical variation of wind variability, we need a workable profile function to fit the change of k with height. We know that at the surface, A_u is maximal and k therefore minimal, that then k is expected to increase with height to a maximum at height z_r, and then k should decrease again to some asymptotic value related to variability in the troposphere. A simple function which has the appropriate behavior is:

$$k_z - k_s \equiv \Delta k_z = c_k(z - z_s) \exp\left[-\frac{z - z_s}{z_r - z_s} \right] \qquad (4)$$

where k_s is the value of k at the surface observation level z_s, typically 10 m. This function has a k-maximum at height z_r. In the upper PBL, it will revert asymptotically to the value k_s; in advance, this is not certain to be the right k-value, but according to our limited data, it has the right order of magnitude. In Oklahoma, $k_s = 1.72$, and we find $k_z = 1.73$ at $z = 444$ m. In Cabauw, we have $k_s = 1.78$, and the local geostrophic wind distribution has $k = 1.82$.

In order to evaluate the value of z_r and of c_k for all levels of a mast together, we rewrite Equation (4), using the notation $\Delta z \equiv z - z_s$:

$$\ln(\Delta k/\Delta z) = \left(-\frac{1}{z_r - z_s} \right) \Delta z + \ln c_k \, . \qquad (5)$$

This is a linear equation in Δz, and if it represents adequately the variation of k with height, the points calculated for the levels of any particular mast should lie on a straight line with its slope inversely proportional to $(z_r - z_s)$.

Results are shown graphically in Figure 3. The first impression from this figure is, that for the three tall masts, the linearity of model Equation (4) is quite good. NSSL Oklahoma and Cabauw are both inland masts in rather homogeneous flat terrain; nevertheless, the fact that their data points are nearly collinear should not be overrated. The line through their combined data points above 50 m has a least-square averaged slope corresponding to a reversal height of 85 m. This agrees well with the z_r-estimates of 90 and 80 m from their published diurnal course graphs (see Section 3). In this respect, the arguments of Section 4 relating z_r to k seem to be vindicated.

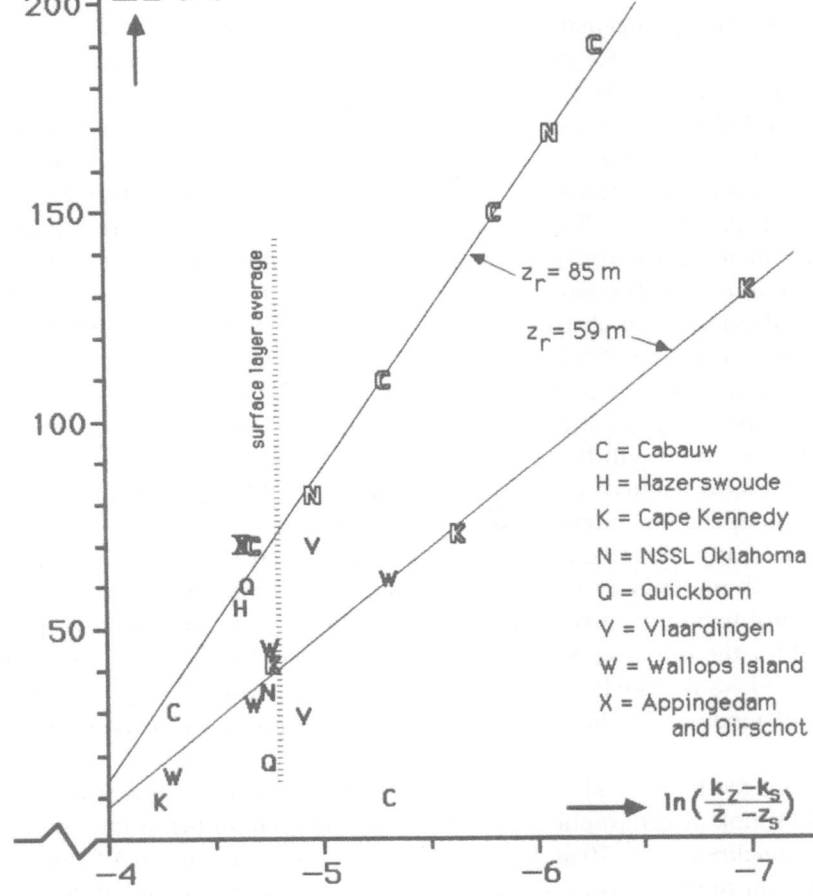

Fig. 3.   Height-variation on various masts of wind distribution shape, normalized according to Equation (5). Open letters are data points used to determine the z_r-isolines. The "surface-layer average" line corresponds to Equation (7).

The third high mast, Cape Kennedy, is located near a coast and its data points above 50 m give $z_r = 59$ m. This could be related to the low diurnal variability of the shore climate, maybe also to the possibility of low advective inversions with winds from the sea – one should not forget that we are discussing an annual average of many different situations. The separately calculated z_r-lines prove to have similar constants: $c_k = 0.023$ for Cape Kennedy, and $c_k = 0.022$ for Cabauw and NSSL Oklahoma combined.

In the lower 80 m, data points from the other masts give extra checks of the model description. In the layer between 50 and 80 m, the position of their data points tallies with the continentality situation. Inland locations at Appingedam and Oirschot give points near the upper line, the coastline mast at Wallops Island agrees well with Cape Kennedy, and inland coast-influenced locations (Quickborn, Vlaardingen, and Hazerswoude) are halfway between.

However, below z_r most data points deviate from the model lines. Below 50 m, increasing scatter is expected due to the lesser accuracy of Δk at decreasing height, but there is a systematic bias towards lower k-values than would correspond to Equation (4). This implies that in the surface layer, there is additional variability in σ_u^2 apart from the components accounted for by (4). An obvious extra source is the decrease of average wind speed in just a few selected directions due to terrain obstructions or to high roughness. This would agree with the fact that e.g., the Wallops Island location, which is rather open in all directions, still conforms to Equation (4) at relatively low levels. On the other hand, Vlaardingen had variable mesoscale surroundings, a suburb to the south and very open country to the west and north (Appendix A).

A full k-model extension for the surface layer would require explicit handling of azimuth-dependent roughness and of stability. This falls outside the scope of the present study, since nearly all mast distributions used were not available in such detail. A preliminary very simple equation describing the available surface layer k-information is given in Section 6, where also additional support is presented for the use of $c_k = 0.022$ at masts which are lower than Cabauw, Cape Kennedy and NSSL Oklahoma. This allows estimation of the local reversal height at masts of medium height, say 60 to 100 m, by application of model Equation (4) or (5).

Using this method, we shall briefly investigate the range of possible z_r-values. First, let us look at geographical variations. Application of (4) to the other masts gave for Quickborn $z_r = 70$ m, for Vlaardingen $z_r = 63$ m, and for Wallops Island $z_r = 52$ m. On small islands, z_r-values might be less than 50 m, and at inland sites, large z_r-values could occur if the local diurnal variation of heat flux is large. Experimental information on upper z_r-limits may be difficult to get for lack of very tall masts in tropical regions. However, it was found at Tsumeb (19° S, 17° E, in arid savannah country) that in both dry and wet seasons, the diurnal course of wind speed already had reversed its phase at 120 m (Mayer, 1974; Walk and Wieringa, 1988). In complex terrain, the reversal height can only be defined in a very local sense and may be suppressed by orographic effects (e.g., at Boulder – see Mikhail, 1985). Martner and Marwitz (1982) found that $z_r \geqslant 107$ m in the Wyoming mountains.

Second, the seasonal variation of z_r was preliminarily investigated with the data from the single year of Cabauw data used here. It was found that z_r decreased to 50 m in the winter and rose to 110 m in the summer. Statistical inaccuracies, caused by restriction to two-month data sets, make it inappropriate to report those results here in more detail.

6. Shape Factor Profile Estimation in the Surface Layer

Practical engineering problems require wind distribution estimation for the upper part of the surface layer, if an observed distribution is available at a height of about 10 m. Correct solution of this modelling problem becomes difficult, if in addition it is assumed that only wind data are available, because realistic modelling of wind above 30 m requires some information on atmospheric stability. Anyhow, in the present context we are restricted to indivisible annual wind distributions. For their shape parameters, it will be shown that a stability-independent vertical extrapolation appears feasible in the surface layer, but not at higher levels.

There are two previous studies of the vertical variation of k in the surface layer. The first is by Rijkoort (1972), who analyzed Vlaardingen 80 m mast data and found that daytime wind distributions conformed well to the two-parameter Weibull formula (2), but that modelling of nighttime distributions required an additional stability parameter. This approach has merits and was applied with success to model the probability of extreme winds with long return periods (Rijkoort, 1983; Rijkoort and Wieringa, 1983), but it does not fit the two-parameter Weibull analysis discussed here.

More generally known is the study by Justus and Mikhail (1976). They summarized data from three of the masts mentioned above – Cape Kennedy, Wallops Island and NSSL Oklahoma – and a fourth 122 m mast at Hanford (47° N, 120° W), located in rather complex terrain (Verholek, 1977). Justus and Mikhail noticed the occurrence of maxima in the observed k-profiles and assumed their level to be 100 m for all these masts. They then fitted a k-ratio power law to all observations below 100 m, as follows:

$$\frac{k_z}{k_{10}} = \frac{1}{1 - 0.088 \ln[z/10]} (\approx 1 + 0.11 \ln[z/10]) . \tag{6}$$

Doran and Verholek (1978) tried to validate this relation with data from more than 40 nuclear reactor masts (Verholek, 1977) by comparing k-values estimated with Justus and Mikhail's formula (k_{JM}) with values observed at the same level (k_z). They observed a ±18% uncertainty and also noticed that k_{JM}/k_z showed a significant increase with increasing k_{10}-values. The mast data in our study are given in Figure 4 in the form of Justus and Mikhail's ratio k_z/k_{10}. We notice the same uncomfortable degree of scatter reported by Doran and Verholek and also the same k_{10}-trend, since for inland masts (with rather low k_{10}-values) k_{80} is underestimated by about 15% from Equation (6).

Justus and Mikhail's trend may result from the simple arithmetic fact that in a multiplicative equation like (6) with fixed constant (0.088), the rate of height increase in k depends on the value of k_{10}. The value of k_{10} is co-determined by the synoptic variance σ_w^2, which has no direct relation to the local stability climate or to other factors influencing the increase of k with height. Therefore (6) introduces scatter for unrelated causes. If we really do want to retain the empirical multiplicative formulation (6), we might have to introduce different slopes for low-k and high-k locations. For our masts, Cabauw and NSSL Oklahoma (with for both $k_{10} \approx 1.75$) give jointly $k_z/k_{10} = 1 + 0.16 \ln(z/10)$, and from the Cape Kennedy mast (assuming $k_{10} \approx k_{18} = 2.16$), we get $k_z/k_{10} = 1 + 0.08 \ln(z/10)$. These examples are shown in Figure 4.

From Figure 4, we may also find an explanation for the rather low value of the constant, 0.088, in the original Justus–Mikhail formula (6). When these authors averaged all data from their four masts, of which two are coastal and have a low reversal height (Cape Kennedy and Wallops Island), the average was biased by use of high-level observations which are not in the surface layer. Changes at the reversal height appear to be a risk factor in indiscriminate analysis of climatological data from high masts.

Concluding, modelling the k-change in the surface layer by use of k_z/k_{10}-ratios according to Justus and Mikhail (1976) gives a very scattered picture, not really promising even for an empirical summary.

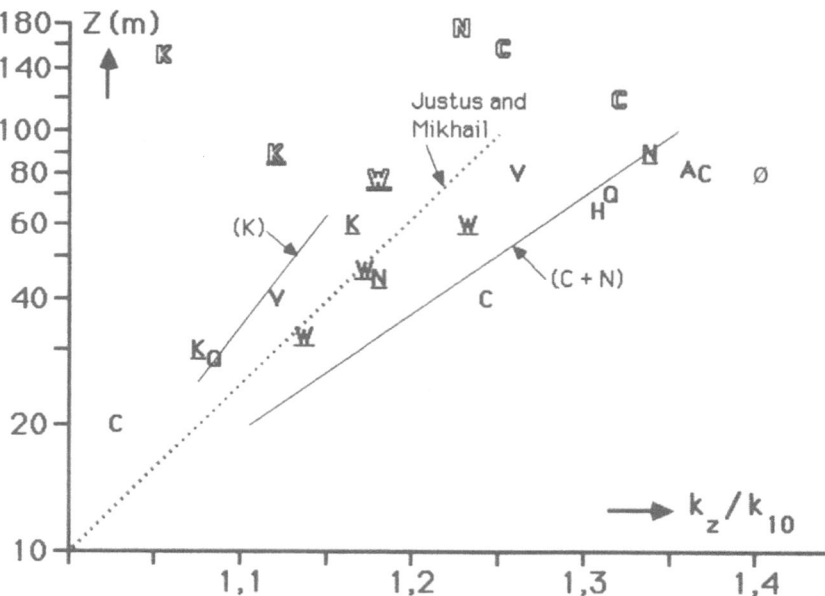

Fig. 4.   Height variation of the annual wind distribution shape parameter k, as described by the ratio to the value at 10 m. Masts are indicated with the same letters as in Figure 3 (also: A = Appingedam, Ø = Oirschot). Lengths of the averaging lines indicate the height range over which they were computed. Underlined datapoints (and also Hanford, not given here) were used by Justus and Mikhail (1976). Open-lettered datapoints are significantly above the reversal height z_r of their own mast as estimated from Equation (4).

If one wishes to separate the large-scale wind climate from the local boundary-layer situation, it is consistent and preferable to use an additive formulation analogous to Equation (4), i.e., to formulate the k-slope equation by modelling $(k_z - k_{10})$ instead of (k_z/k_{10}).

A plot of the available $(k_z - k_{10})$-data against z is shown in Figure 5. The surface-layer points with $z_s = 10$ m (C, N, Q, V) are rather collinear up to the reversal height, with a horizontal k-scatter <0.1. Least-square averaging for all heights below and around the reversal height gives:

$$k_z = k_{10} + 0.0084(z - 10) . \tag{7}$$

This "surface-layer average"-line is also shown in Figure 5, and it is simple to deduce that it crosses the lines modelled by Equation (4) at $z = z_s + 0.96 (z_r - z_s)$. Reversing the argument, if we want this line (7) derived from all surface-layer data (below z_r) to cross all lines modelled by Equation (4) at their respective reversal heights, then $c_k = 0.0228$. This is in satisfactory agreement with the value $c_k = 0.022$ which we derived earlier from the observations *above* z_r.

For the K- and W-masts, with surface observation heights of 18 and 16 m, respectively, Equation (7) now is used to estimate the k_{10}-value. In Figure 5, the K- and W-data points are related to this k_{10}-estimate, with acceptable results. This solves the problem of analyzing masts with $z_s \neq 10$ m.

It may be that the observed near-linearity of the $\Delta k - \Delta z$-relation (7) in the surface layer is an accidental result of the bias towards low k-values, which was noted in Figure 3 and tentatively ascribed to variability in meso-scale roughness with azimuth. The absolute shift to the left due to this cause should be small both at low levels (for arithmetical reasons) and near z_r because of decreased roughness influences at higher levels. Halfway into the surface layer, say around 40 m, the linearizing shift would be largest. An additional practical advantage of the additive approach seems to be that no significant difference is noticeable between coastal and inland locations.

The development of Equations (4) and (7) has been done with the use of the data from all six masts in Table I. Practical application of Equation (4) requires knowledge of the reversal height, which we cannot determine without a mast. (Against better knowledge, the data from the six available masts were used to find out if z_r could be empirically related to k_s, but not the slightest correlation was found.) On the other hand, Equation (7) only needs knowledge of k_s and therefore seems useful. However, operational use requires first a reliability check of Equation (7) in realistic circumstances on data not used before.

For control purposes, independent data are available from Appingedam, Hazerswoude and Oirschot, single high-level observation series of good quality which can serve as check points. At nearby representative surface stations, we have 30-year distributions of potential wind speed, corrected to 10 m height above open terrain (Wieringa, 1986); their Weibull parameters are listed in Table II along with those from the check points. We assume that their reversal height is

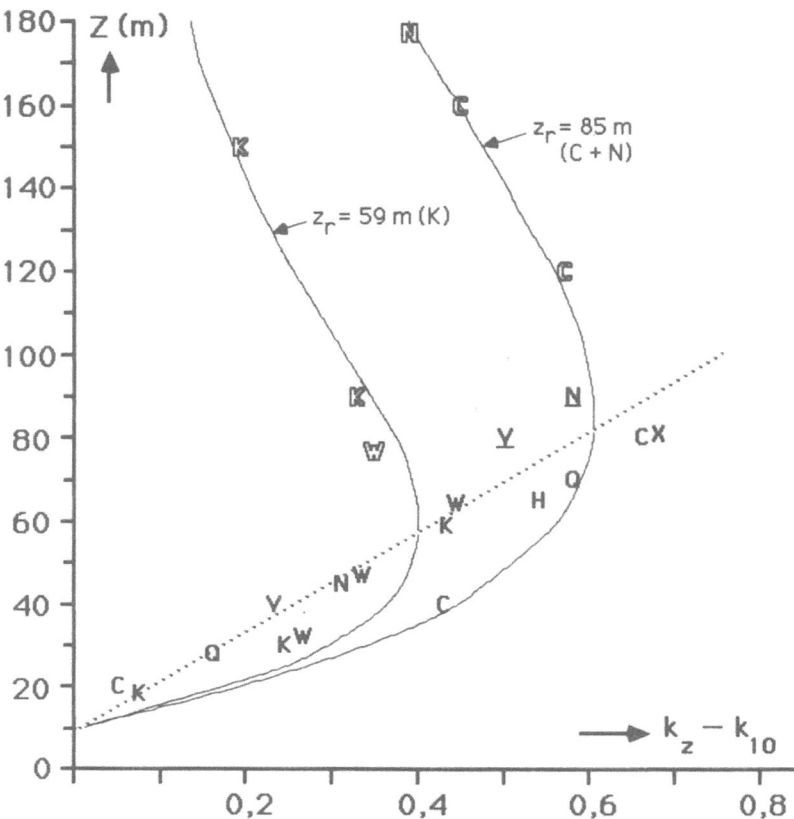

Fig. 5. Additive height variation of the annual wind distribution shape parameter k relative to its surface value at 10 m. Masts are indicated with the same letters as in Figure 3. Underlined datapoints are less than 10 m above the reversal height z_r of their own mast. Open-lettered datapoints (significantly above z_r), the K- and W-datapoints ($z_s > 15$ m), and checkpoints H and X were not used to determine the (dotted) averaging line Equation (7). The continuous lines are drawn according to Equation (4) for the indicated masts.

similar to that of Cabauw, 85 m. The results of comparing the shape factors of the check points with those of the surface stations are shown in Figure 5 and agree with Equation (7) within 0.1. The accidental equality of Δk-estimates for Appingedam and Oirschot (they had to be plotted as a single point, X) is quite interesting, because Appingedam (and Leeuwarden) are about 10 km south of the coast, while Oirschot (and Eindhoven) are more than 100 km inland. More checks are desirable, but these three make Equation (7) look moderately acceptable for application purposes.

Concluding, it appears from Figure 5 that the k-profile function Equation (4), using $c_k = 0.022$, gives satisfactory representations for homogeneous terrain. Effects of azimuthal non-homogeneity deform this profile in the surface layer to the linear function Equation (7) for level terrain, with the two functions matching at the reversal height z_r. The outcome is a simple two-layer PBL model to describe the profile of the long-term wind variability parameter k.

TABLE II

Summarized annual distribution parameters of the masts used

Cabauw
(March '78/February '79)

	$z = 10$	$z = 10p$	$z = 20$	$z = 40$	$z = 80$	$z = 120$	$z = 160$	$z = 200$	$z = $ geo
$k =$	1.78	1.80	1.83	2.21	2.44	2.35	2.23	2.15	1.82
$a =$	4.70	4.80	5.36	6.73	7.97	8.90	9.45	9.98	11.39
$U =$	4.18	4.27	4.77	5.96	7.07	7.89	8.37	8.84	10.12

NSSL Oklahoma (June '66/May '67)
(Crawford and Hudson, 1970)

Quickborn
(1957–1967)

	$z \approx 10$	$z = 45$	$z = 90$	$z = 177$	$z = 266$	$z = 355$	$z = 444$	$z = 10$	$z = 28$	$z = 70$
$k =$	1.72	2.03	2.30	2.11	1.81	1.81	1.73	1.86	2.02	2.44
$a =$	4.73	6.37	7.72	8.79	9.84	10.54	10.62	4.21	5.31	7.16
$U =$	4.22	5.65	6.84	7.78	8.75	9.38	9.46	3.74	4.70	6.35

Cape Kennedy (Dec. '65/Nov. '66)
(McVehil and Camnitz, 1969)

Wallops Island (1961–1965)
(Cochrane and Henry, 1968)

	$z = 18$	$z = 30$	$z = 60$	$z = 90$	$z = 150$	$z = 16$	$z = 32$	$z = 47$	$z = 62$	$z = 77$
$k =$	2.16	2.33	2.52	2.42	2.28	1.67	1.89	1.96	2.06	1.97
$a =$	9.91	11.17	13.81	15.33	15.57	5.11	6.22	7.04	7.68	7.58
$U =$	8.78	9.90	12.26	13.59	13.79	4.56	5.52	6.24	6.80	6.72

	Hazerswoude (May '85/April '86) 06240 ↓		Appingedam (1985) 06270 ↓		Oirschot (1985) 06370 ↓		Vlaardingen (April '67/March '68)		
	$z = 10p$	$z = 65$	$z = 10p$	$z = 80$	$z = 10p$	$z = 80$	$z = 10$	$z = 40$	$z = 80$
$k =$	1.76	2.30	1.88	2.55	1.71	2.40	1.93	2.16	2.43
a	6.08	8.68	5.91	8.74	5.14	7.29	5.58	7.21	8.89
$U =$	5.42	7.69	5.25	7.76	4.58	6.46	4.95	6.39	7.88

N.B.: Hazerswoude, Appingedam and Oirschot are compared with 30-year average distributions of potential wind ($10p$) at nearby representative meteostations: 06240 = Schiphol, 06270 = Leeuwarden, 06370 = Eindhoven. The values of average wind speed (U) are not independent, but have been derived from the corresponding Weibull-parameters a and k.

7. Discussion

The reliable mast data analyzed here make it obvious that in describing average long-term wind variability in the boundary layer, it is necessary to distinguish two layers, separated by the reversal height of the diurnal cycle. This fact does not depend on the particular way in which I have attempted to model it. It is a rather unfortunate complication that at present we know too little about the diurnal cycle to estimate the height of its reversal in a simple manner with some confidence. The erection of a mast, or brute-force calculation of thousands of diabatic profiles, is not simple.

The mast data, with help from model Equation (4), have provided some rough climatological estimates of the reversal height. In temperate latitudes and level terrain without orographic effects, z_r is typically found at 50–60 m near the coast and 80–90 m at inland sites. We don't know systematically, to what extent this is modified with slight terrain slopes (Holton, 1967) or in tropical climates, where the diurnal course has its maximum in the early morning instead of the early afternoon (Abu Bakr, 1988). Also, we still have to quantify the interaction in the surface layer between azimuth-dependency of major terrain features and the wind distribution shape. Follow-up studies would have to look for data of sufficient quality (see Section 1) from additional masts with heights in the range of 60 to 100 m, either from historical masts mentioned in reviews by Wallace (1967), Brook (1974) and Verholek (1977), or from still active mast projects.

To determine z_r if we have no mast, the only present way is by brute force, i.e., hour-by-hour stability-dependent application of similarity modelling (Van Ulden and Holtslag, 1985) for a full year of surface data. In this way, one can obtain new upper-PBL distributions and diurnal cycles, if sufficient computer time and information on the surface heat balance and roughness is available. There are indications that this similarity modelling gives less reliable climatological results at heights above 80 m (Holtslag, 1984). Anyhow, a more elegant alternative to brute force seems desirable.

For generalization of these estimates, we should understand the climate-, terrain- and height-dependent behavior of the diurnal cycle. This will require statistical dynamic stability-dependent analysis, which is able to deal e.g., with occasional (but not necessarily regular) occurrence of low-level wind maxima. Relevant earlier research was undertaken by Buajitti and Blackadar (1957), Haltiner (1961) and Matveev (1967), and more recently e.g., by Clarke (1974) and Heald and Mahrt (1981). Unless we understand these phenomena quantitatively, we cannot link surface wind climate to macroscale climate.

Assuming that in the surface layer, up to 60 to 80 m, we can estimate the profile of the shape parameter by way of Equation (7), we still need the profile of the scale parameter to obtain at such heights a complete wind distribution. In many regions, the average annual surface-layer stability climate is more or less stable. The consequence of this is that for heights as low as 40 m, the long-term average ratio U_z/U_{10} exceeds by $\geq 10\%$ the ratio value calculated by some appropriate adiabatic profile. This means that existing reliable methods to estimate annual average wind at heights above 30 m all have to use some form of stability input: either Pasquill classes (e.g., Smedman–Högström and Högström, 1978; Sedefian, 1980), or station-estimated Obukhov lengths (Petersen et al., 1985; Van Wijk et al., 1985; Wieringa, 1988), or direct mast data (Petersen et al., 1981).

If now we have only wind data, as in this study, one may well ask if it is possible to get from those simple data, sufficient stability information to estimate a reliable average wind profile. In particular, since the standard deviation of short turbulent gusts around an hourly average is strongly correlated to stability, one might hope that the annual standard deviation of hourly wind σ_u could be a similarly useful parameter. Unfortunately this is not the case because, as was shown in Section 4, σ_u has such a large synoptic component that the diurnal σ_d – which *is* stability-dependent – cannot be distinguished without extra analysis. In other words, the σ_u-value which can be calculated from the Weibull parameters, by way of Equation (3), is not useful as a stability parameter. It is possible to obtain stability information by special analysis of wind data; for example, Rijkoort (1972, 1983) handled stability by evaluating separately the wind distributions for day (unstable) and night (stable). Also A_u is an indicator of stability, but again it requires separate evaluation. In other words, if stability information is needed for estimating a U-profile along with a k-profile from Equation (7), then the data must be extensively re-analyzed; it cannot be done with the data in Table II alone.

8. Conclusions

This study belongs to the class which Tennekes (1973) calls data representation: how should the available observations be combined in order to demonstrate most clearly, that there is "law and order" in nature? A preliminary answer to this question with respect to the shape factors k of wind distributions is summarized in Figure 5. It shows that this long-term variability measure can be usefully modelled as the inverted sum of diurnal, synoptic and terrain variations. The effect of terrain variations is not explicitly accounted for in model curve (4), but seems responsible for the systematic bias of surface-layer observations towards lower k-values.

If in practice we have a surface-observed or -estimated value of k, the k-value at higher levels can be roughly estimated by increasing the surface value by 0.8% of the height difference in meters. In most local conditions, one can get away with this simple estimate up to the reversal height z_r of the diurnal cycle – over land typically to heights of 60 to 80 m.

If we want to extend estimation of k to greater heights, the problem is that in the upper PBL the value of k depends heavily on knowledge of the local average value of the reversal height. If k_s is given and z_r varies realistically from 50 to 100 m, k at a height of 200 m varies by ≈ 0.5 and the corresponding wind energy by $\approx 20\%$. Determination of the reversal height from the local surface climate requires more knowledge about the behavior of the diurnal cycle of wind in the PBL than we have now. Therefore at present it is sensible to restrict practical k-estimation to the surface layer.

Acknowledgements

In writing this study I felt in debt to Peter Rijkoort and his persistent efforts to apply distribution statistics to boundary-layer meteorology. With respect to U.S. mast information, I thank the hosts of my 1978 journey: Chandran Kaimal, Edwin Kessler, George Fichtl, Bruce Hicks, and especially Hans Panofsky. Part of that trip was made possible by a travel grant from the Netherlands Organisation for the Advancement of Pure Research (Z.W.O.).

Moreover I am indebted to the Seewetteramt of the Deutscher Wetterdienst at Hamburg for making the Quickborn data available for publication. Similarly I thank the S.E.P. (Cooperating Electricity Producers, Netherlands) for availability and publication permission of the data from Hazerswoude, Appingedam and Oirschot.

The existence of reliable mast data is based on patient technical and organizational work of those who actually ran the masts, and who are seldom mentioned. The Cabauw observations were organized by J. H. Van Ameijde and W. Hovius, the Vlaardingen observations by F. J. B. Smulders, the SEP-observations by D. J. Kleyn, and Quickborn was run by E. Frankenberger. In the U.S.A., the mast supervision was done by Leonard Johnson and John Carter in Oklahoma, and by Dennis Camp at Cape Kennedy. The raw Cabauw data were processed by J. G. Van der Vliet, H. R. A. Wessels, J. M. Koopstra, C. A. Engeldal and many others.

Appendix A: References on Setup and Surroundings of the masts

Cabauw:

Terrain maps, photos, roughness:	Driedonks (1981), Holtslag (1984), Monna and Van der Vliet (1987)
Instruments and data handling:	Driedonks et al. (1978), Monna and Van der Vliet (1987)
Mast influence on observation:	Wessels (1984b)
Data source, review:	Van Ulden et al. (1976), Wessels (1984a)

Cape Kennedy:

Terrain maps, photos, roughness:	Fichtl and McVehil (1970), Pielke and Panofsky (1970), Sinclair et al. (1973)
Instruments and data handling:	McVehil and Camnitz (1969)
Mast influence on observation:	Camp and Kaufman (1970)
Data source, review:	McVehil and Camnitz (1969)

NSSL Oklahoma:

Terrain maps, photos, roughness:	Sanders and Weber (1970), Hanafusa and Mitsuta (1971), Sinclair et al. (1973)
Instruments and data handling:	Carter (1970)
Mast influence on observation:	Carter (1970), Angell and Bernstein (1976)
Data source, review:	Crawford and Hudson (1970), Goff and Hudson (1972), Cormier (1975)

Quickborn:

Terrain maps, photos, roughness:	Frankenberger (1955), Franken (1962)
Instruments and data handling:	Frankenberger (1955), Franken (1962)
Mast influence on observation:	Tops of three separate masts
Data source, review:	Frankenberger (1968)

Vlaardingen:

Terrain maps, photos, roughness:	Rijkoort et al. (1970), Raaff (1975)
Instruments and data handling:	Rijkoort et al. (1970)
Mast influence on observation	2 opposed 4.5 m booms, 1.8 m open mast
Data source, review:	Rijkoort (1972), Raaff (1975)

Wallops Island:

Terrain maps, photos, roughness:	Cochrane and Henry (1968), Tieleman (1980)
Instruments and data handling:	Cochrane and Henry (1968), Tieleman (1980)
Mast Influence on observation:	2 opposed 1.8 m booms, broad open mast
Data source, review:	Cochrane and Henry (1968), Tieleman (1980)

Appingedam, Hazerswoude and Oirschot:

Terrain maps, photos, roughness:	Kleyn et al. (1985), Kleyn (1985)
Instruments and data handling:	Kleyn et al. (1985), Kleyn (1985)
Mast influence on observation:	Top of mast
Data source, review:	—

Appendix B: Detailed Wind Frequency Distributions from the Used Masts

The frequency distributions given below are all based on hourly averages. The distributions of the Cape Kennedy and NSSL Oklahoma masts were extracted from reported histograms. The Wallops Island mast distributions were derived from reported cumulative distributions. The other distributions given here have not been published before.

All the data series have been checked by the originators for instrument malfunctions etc., which accounts for the incomplete coverage (see bottom line of each table for percentage). However, for all of them it appears that hiatuses in the series were rather randomly distributed over the seasons. The very high coverage at the NSSL Oklahoma mast is plausible because of the very simple setup: heavy but very reliable Aerovanes, and registration designed for continuous operation rather than for high original data compression.

The Cabauw mast data were corrected for boom and mast interference effects (Wessels, 1984b). Other data are as measured. At Quickborn and on the SEP-masts, the instrument positions were very free indeed. The masts at Cape Kennedy, Vlaardingen and Wallops Island each had two booms, allowing automatic selection of the upwind side of the tower. NSSL Oklahoma has only a single boom on an open mast, but winds from the tower-shaded azimuth sector have a very low occurrence frequency (McVehil and Camnitz, 1969).

Concluding, it is estimated that for some masts, interference effects could have caused systematic shifts of maybe 5% or 10% in the annual all-azimuth distributions of wind speed, and therefore also in the a-factors and U-values in Table II. However, such an overall shift should have a negligible influence on the distribution shape and the computed k-factors.

APPENDIX B 1: ANNUAL WIND SPEED FREQUENCY (DISTRIBUTIVE PERCENTAGES)

Cabauw (March '78/February '79)

Wind (m/s)	$z = 10$	$z = 10p$	$z = 20$	$z = 40$	$z = 80$	$z = 120$	$z = 160$	$z = 200$	$z = $ geo
0–1	1.85	1.74	0.85	0.42	0.46	0.37	0.33	0.14	0.97
1–2	11.34	10.78	7.53	4.14	3.47	3.03	2.83	2.58	2.77
2–3	21.18	19.63	14.83	7.86	6.07	4.90	4.67	4.14	4.49
3–4	18.30	19.12	19.10	11.76	7.74	7.24	6.90	6.58	5.65
4–5	14.90	15.08	17.17	15.79	8.80	6.95	6.82	6.81	6.47
5–6	11.08	11.11	11.92	16.69	11.65	8.67	7.86	7.68	6.84
6–7	7.89	8.29	9.51	12.95	12.83	10.49	9.30	8.28	6.13
7–8	5.68	5.92	6.78	9.25	12.82	11.39	10.30	9.61	6.37
8–9	3.91	4.10	4.73	7.21	11.14	10.34	9.16	9.44	7.25
9–10	1.82	2.05	3.65	4.97	8.10	9.48	9.60	8.46	7.41
10–11	1.03	1.11	1.68	3.90	6.00	7.19	7.28	7.59	6.52
11–12	0.45	0.48	1.03	2.32	4.52	6.48	6.43	6.00	5.72
12–13	0.33	0.32	0.52	1.30	2.74	5.10	5.50	5.53	6.02
13–14	0.13	0.17	0.30	0.68	1.72	3.46	4.41	4.62	4.69
14–15	0.07	–	0.25	0.30	0.94	2.10	3.67	3.65	3.80
15–16	0	0.07	0.05	0.25	0.56	1.40	2.16	3.32	3.47
16–17	–	0	0.04	0.12	0.22	0.69	1.60	2.33	2.95
17–18	0	–	0	0	0.12	0.37	0.67	1.75	2.24
18–19	–	0	–	0	0.06	0.18	0.30	0.89	2.11
19–20	–	–	0	–	0	0.09	0.09	0.37	1.56
20–21	–	–	–	–	0	0	0.05	0.10	1.58
21–22	–	–	–	0	–	0	0.04	0	1.26
22–23	–	–	–	–	0	0	0	0	1.10
23–24	–	–	–	–	0	0	0	0	0.66
24–25	–	–	–	–	–	0	0	0	0.70
>25	–	–	–	–	–	–	0	0	1.29
Complete	99%	99%	91%	92%	92%	89%	91%	88%	99%

N.B.: All distribution parameter comparisons were done with respect to the original measured wind at 10 m. The distribution of potential surface wind (10p) is added for comparison purposes. The geostrophic wind at Cabauw is computed from 19 synoptic stations (Cats, 1977).

Appendix B 2: annual wind speed frequency (distributive percentages)

NSSL Oklahoma (June '66/May '67) (Crawford and Hudson, 1970) Wind (m/s)							Quickborn (1957–1967)		
	$z \approx 10$	$z = 45$	$z = 90$	$z = 177$	$z = 266$	$z = 355$	$z = 444$	$z = 10$	$z = 28$	$z = 70$
0–1	9.0	3.2	2.0	2.6	2.7	1.9	2.0	12.4	2.5	↓
1–2	14.0	5.6	4.0	3.5	3.6	3.1	3.8	12.1	6.6	4.2
2–3	16.6	9.2	5.7	5.5	5.0	4.5	5.1	17.7	14.6	↓
3–4	15.8	14.1	8.9	7.4	6.6	6.1	5.8	17.3	17.7	15.1
4–5	10.5	13.2	9.9	7.5	7.5	6.7	6.5	14.7	17.3	↓
5–6	10.1	14.6	11.4	9.4	8.3	8.5	8.1	11.2	14.2	29.5
6–7	8.3	11.7	12.8	9.8	8.9	8.3	7.9	7.2	10.7	↓
7–8	6.8	8.0	12.0	8.5	7.4	7.6	7.5	3.9	7.0	28.1
8–9	4.2	6.3	9.2	7.7	6.7	6.7	6.6	2.0	4.3	↓
9–10	2.2	4.9	7.3	8.0	6.3	5.9	5.5	0.9	2.5	15.1
10–11	1.3	3.7	5.6	8.1	6.1	6.1	6.0	0.3	1.4	↓
11–12	0.6	2.6	4.7	6.2	6.3	5.7	5.4	0.2	0.6	5.4
12–13	0.2	1.6	3.1	5.9	5.6	5.2	5.7	0.1	0.3	↓
13–14	0.1	0.7	1.7	4.2	4.8	4.9	4.6	0	0.2	1.9
14–15	–	0.3	0.7	2.9	3.3	3.7	3.2	–	0.1	↓
15–16	–	0.2	0.5	1.6	3.0	3.1	3.1	–	0	0.5
16–17	–	0.1	0.3	0.8	2.9	2.7	2.5	–	–	↓
17–18	–	0	0.1	0.3	2.1	2.3	2.3	–	–	0.1
18–19	–	–	0	0.1	1.6	2.1	2.0	–	–	↓
19–20	–	–	–	0	0.7	1.5	1.6	–	–	0.1
20–21	–	–	–	0	0.3	1.4	1.6	–	–	–
21–22	–	–	–	–	0.2	1.0	1.2	–	–	–
22–23	–	–	–	0	0.1	0.5	0.9	–	–	–
23–24	–	–	–	–	0	0.3	0.6	–	–	–
24–25	–	–	–	–	0	0.1	0.2	–	–	–
>25	–	–	–	–	–	0.1	0.3	–	–	–
Complete	99%	99%	99%	99%	99%	99%	99%	89%	86%	86%

N.B.: Due to small-scale terrain inhomogeneities, the height of the surface level at NSSL Oklahoma is uncertain, $z = 9$ m ± 2 m. At Quickborn, the data series consists of 7 years, since in 1959, 1960 and 1961 no measurements were made. The Quickborn 70 m data were in 2 m/s-classes (see comment Wallops Island).

APPENDIX B 3: ANNUAL WIND SPEED FREQUENCY (DISTRIBUTIVE PERCENTAGES)

Cape Kennedy (Dec. '65/Nov. '66) (McVehil and Camnitz, 1969)					Wallops Island (1961–1965) (Cochrane and Henry, 1968)				
Wind (m/s) $z=18$	$z=30$	$z=60$	$z=90$	$z=150$	$z=16$	$z=32$	$z=47$	$z=62$	$z=77$
0–1 0.92	0.46	0.15	0.31	0.23	1.8	1.2	1.4	1.1	1.6
1–2 2.22	1.53	0.77	0.69	0.92	12.7	7.8	6.2	4.6	5.7
2–3 3.67	2.52	1.45	1.22	1.38	↓	↓	↓	↓	↓
3–4 5.51	3.90	2.30	1.99	1.91	33.2	25.4	19.6	16.8	17.3
4–5 7.19	5.43	2.91	1.91	2.52	↓	↓	↓	↓	↓
5–6 10.10	7.11	3.67	3.29	3.98	24.1	26.0	23.9	21.2	20.2
6–7 9.10	7.42	4.36	3.83	3.75	↓	↓	↓	↓	↓
7–8 10.33	10.10	5.97	5.20	5.51	17.5	21.5	23.4	25.4	24.6
8–9 8.11	8.03	5.81	5.36	4.36	↓	↓	↓	↓	↓
9–10 8.80	9.03	7.88	6.04	5.97	6.7	10.0	13.0	14.4	14.1
10–11 6.27	7.88	7.50	6.12	5.89	↓	↓	↓	↓	↓
11–12 5.58	6.96	8.87	7.19	6.12	2.6	4.9	7.0	8.4	8.0
12–13 4.13	4.59	6.73	6.50	5.89	↓	↓	↓	↓	↓
13–14 4.82	5.13	7.50	6.27	5.89	1.0	2.2	3.5	5.1	5.2
14–15 3.44	4.44	5.89	5.66	5.43	↓	↓	↓	↓	↓
15–16 2.75	3.67	5.97	6.04	4.44	0.2	0.7	1.0	1.5	1.7
16–17 1.84	2.45	4.13	4.90	4.36	↓	↓	↓	↓	↓
17–18 1.38	2.30	4.36	4.82	4.51	0.1	0.1	0.6	0.9	1.0
18–19 0.99	1.30	2.52	3.21	3.29	↓	↓	↓	↓	|||
19–20 0.99	1.30	3.29	4.28	3.52	–	0.1	0.2	0.3	0.4
20–21 0.46	0.99	1.68	3.06	3.06					
21–22 0.46	0.46	1.53	2.68	3.29					
22–23 0.23	0.61	0.99	1.38	2.07					
23–24 0.23	0.46	0.99	1.61	1.76					
24–25 0.23	0.23	0.46	0.92	1.45					
>25 0.24	1.70	2.31	5.52	8.51					
Complete 87%	86%	86%	80%	77%	95%	95%	93%	93%	93%

N.B.: The data from the 120 m-level at Cape Kennedy were considered less reliable by McVehil and Camnitz. Arrows in the Wallops Island distribution (in 2 m/s classes) indicate that e.g., at $z=16$ m, 33.2% is in class $2 < U \leq 4$ m/s.

APPENDIX B 4: ANNUAL WIND SPEED FREQUENCY (DISTRIBUTIVE PERCENTAGES)

Wind (m/s)	Hazerswoude (May '85/April'86) 06240		Appingedam (1985) 06270		Oirschot (1985) 06370		Vlaardingen (April '67/March '68)		
	$z = 10p$	$z = 65$	$z = 10p$	$z = 80$	$z = 10p$	$z = 80$	$z = 10$	$z = 40$	$z = 80$
0–1	4.95	↑	2.39	↑	7.29	↑	1.09	0.72	0.31
1–2	6.97	15.31	7.57	13.26	9.74	22.08	8.00	3.14	2.13
2–3	11.27	↓	12.01	↓	12.98	↓	13.87	6.44	3.42
3–4	13.84		14.85		16.40		17.44	12.31	7.41
4–5	13.58	8.64	14.41	7.48	14.96	10.03	14.22	13.13	7.98
5–6	11.57	9.77	13.00	8.96	11.23	12.11	13.59	15.60	9.64
6–7	10.70	12.12	12.22	11.69	9.76	13.43	9.84	10.76	12.91
7–8	8.22	12.33	7.75	13.67	6.65	14.20	9.50	10.35	12.13
8–9	5.83	10.39	5.55	13.05	4.20	11.26	4.72	8.29	9.90
9–10	4.31	8.51	3.80	10.84	2.81	6.75	3.57	6.64	8.04
10–11	3.12	6.14	2.46	6.86	1.67	4.04	2.13	4.94	7.47
11–12	1.94	4.42	1.66	4.53	0.99	2.74	0.86	2.88	5.86
12–13	1.45	3.80	1.03	3.05	0.59	1.61	0.52	2.16	4.82
13–14	0.86	2.87	0.55	2.17	0.29	0.77	0.23	0.82	2.59
14–15	0.50	2.23	0.35	1.95	0.20	0.47	0.29	0.67	2.23
15–16	0.30	1.39	0.17	·1.12	0.11	0.39	0	0.51	0.78
16–17	0.19	0.78	0.11	6.5	0.07	0.20	0.06	0.10	0.67
17–18	0.17	0.39	0.06	3.7	0.04	0	0.06	0	0.73
18–19	0.12	–	0.02	–	0.02	–	0	0	0.47
19–20	0.05	0.84	0.02	(2.8)	0.01	–	0	0.15	0.16
20–21	0.03	–	0.01	–	0.01	–	0	0.25	0.21
21–22	0.02	–	–	–	–	–	0	0	0.05
22–23	0.01	–	–	–	–	–	0	0	0.05
>23	0.01	–	–	–	–	–	0	0	0.05
Complete	99%	95%	99%	94%	99%	88%	79%	89%	88%

N.B.: Hazerswoude, Appingedam and Oirschot are compared with 30-year average distributions of potential wind $(10p)$ at nearby meteostations: 06240 = Schiphol; 06270 = Leeuwarden; 06370 = Eindhoven. The mast observations below 4 m/s were combined by SEP into one single class. The Vlaardingen data are hourly averages at 4-hourly intervals.

References

Abu Bakr, E. H.: 1988, 'The Boundary Layer Wind Regime of a Representative Tropical African Region, Central Sudan', Ph.D. Thesis, Techn. Univ. Eindhoven, The Netherlands.

Angell, J. K. and Bernstein, A. B.: 1976, 'Evidence for a Reduction in Wind Speed on the Upwind Side of a Tower', *J. Appl. Meteorol.* **15**, 186–188.

Brook, R. R.: 1974, 'A Study of Wind Structure in an Urban Environment', Meteor. Study 27, Austr. Dept. Science.

Buajitti, K. and Blackadar, A. K.: 1957, 'Theoretical Studies of Diurnal Wind Structure Variations in the Planetary Boundary Layer', *Quart. J. Roy. Meteorol. Soc.* **83**, 486–500.

Camp, D. W. and Kaufman, J. W.: 1970, 'Comparison of Tower Influence on Wind Velocity for NASA's 150 m Meteorological Tower and a Windtunnel Model of the Tower', *J. Geophys. Res.* **75**, 1117–1121.

Carter, J. K.: 1970, 'The Meteorologically Instrumented WKY-TV Tower Facility', ESSA-ERLTM-NSSL-50.

Cats, G. J.: 1977, 'Berekening van de Geowind', Roy. Neth. Meteor. Inst. Sc. Rep. 77-2.

Clarke, R. H.: 1974, 'Attempts to Simulate the Diurnal Course of Meteorological Variables in the Boundary Layer', *Atm. Oc. Phys.* **10**, 360–374.

Cochrane, J. A. and Henry, R. M.: 1968, 'Wind Data from the 250-foot (76-meter) Tower at Wallops Island, Virginia', NASA-TN-D-4395.

Cormier, R. V.: 1975, 'The Behavior of Vertically Integrated Boundary-Layer Winds', *Boundary-Layer Meteorol.* **9**, 315–324.

Crawford, K. C. and Hudson, H. R.: 1970, 'Behavior of Winds in the Lowest 1500 ft in Central Oklahoma: June 1966–May 1967', ESSA-ERLTM-NSSL-48.

Crawford, K. C. and Hudson, H. R.: 1973, 'The Diurnal Wind Variation in the Lowest 1500 ft in Central Oklahoma: June 1966–May 1967', *J. Appl. Meteorol.* **12**, 127–132.

Doran, J. C. and Verholek, M. G.: 1978, 'A Note on Vertical Extrapolation Formulas for Weibull Velocity Distributions', *J. Appl. Meteorol.* **17**, 410–412.

Driedonks, A. G. M., Van Dop, H., and Kohsiek, W. H.: 1978, 'Meteorological Observations on the 213 m Mast at Cabauw, in the Netherlands', Prepr. 4th AMS Symp. Meteor. Obs. Instr. (Denver) 41–46.

Driedonks, A. G. M.: 1981, 'Dynamics of the Well-Mixed Atmospheric Boundary Layer', Roy. Neth. Meteor. Inst. Sc. Rep., 81-2.

Fichtl, G. H. and McVehil, G. E.: 1970, 'Longitudinal and Lateral Spectra of Turbulence in the Atmospheric Boundary Layer at the Kennedy Space Center', *J. Appl. Meteorol.* **9**, 51–63.

Fiedler, F. and Panofsky, H. A.: 1970, 'Atmospheric Scales and Spectral Gaps', *Bull. Am. Meteorol. Soc.* **51**, 1114–1119.

Franken, E.: 1962, 'Über den Geländeeinfluß auf Windrichtung und Windgeschwindigkeit im Raum Hamburg', DWD Seewetteramt Rep. Nr. 34.

Frankenberger, E.: 1955, 'Über vertikale Temperatur-, Feuchte- und Wind-gradienten in den untersten 7 Dekametern der Atmosphäre, den Vertikalaustausch und der Wärmehaushalt an Wiesenboden bei Quickborn/Holstein 1953/1954', Ber. Deutschen Wetterdienstes 20–3.

Frankenberger, E.: 1968, 'Untersuchungen über Intensität, Häufigkeit und Struktur von Starkwinden über Quickborn in Holstein', *Meteorol. Rundschau* **21**, 65–69.

Goff, R. C. and Hudson, H. R.: 1972, 'The Thermal Structure of the Lowest Half Kilometer in Central Oklahoma: Dec. 9, 1966–May 31, 1967', NOAA-ERL-NSSL-58.

Gomes, L. and Vickery, B. J.: 1977, 'On the Prediction of Extreme Wind Speeds from the Parent Distribution', *J. Industr. Aerodyn.* **2**, 21–36.

Haltiner, G. J.: 1959, 'The Diurnal Variation of the Wind', *Tellus* **11**, 452–458.

Hanafusa, T. and Mitsuta, Y.: 1971, 'Structure of the Planetary Boundary Layer in High Winds Observed from a 0.5 km TV Tower', Proc. 3rd Internat. Conf. Wind Effects on Buildings and Structures (Tokyo) 69–78.

Heald, R. C. and Mahrt, L.: 1981, 'The Dependence of Boundary-Layer Shear on Diurnal Variation of Stability', *J. Appl. Meteorol.* **20**, 859–867.

Hellmann, G.: 1917, 'Über die Bewegung der Luft in den unterste Schichten der Atmosphäre', *Meteorol. Zeitschr.* **34**, 273–285.

Holton, J. R.: 1967, 'The Diurnal Boundary Layer Wind Oscillation above Sloping Terrain', *Tellus* **19**, 199–205.

Holtslag, A. A. M.: 1984, 'Estimates of Diabatic Wind Speed Profiles from Near-Surface Weather Observations', *Boundary-Layer Meteorol.* **29**, 225–250.

Hoxit, L. R.: 1975, 'Diurnal Variations in Planetary Boundary Layer Winds Over Land', *Boundary-Layer Meteorol.* **8**, 21–38.

Justus, C. G. and Mikhail, A. S.: 1976, 'Height Variation of Wind Speed and Wind Distribution Statistics', *Geophys. Res. Lett.* **3**, 261–264.

Kleyn, D. J., Kuipers, J. A., and Muiser, C. F.: 1985, 'The SEP Wind Measurement Network', Proc. EWEC Conf., Hamburg 1984, W. Palz (ed.), Stephens, U.K. 588–593.

Kleyn, D. J.: 1985, 'Het Landelijk Windmeetnet van de SEP', *Energiespectrum* **9**, 176–183.

Khrishna, K.: 1968, 'A Numerical Study of the Diurnal Variation of Meteorological Parameters in the Planetary Boundary Layer – Diurnal Variation of Winds', *Monthly Weath. Rev.* **96**, 269–276.

Manier, G., and Weingärtner, H.: 1979, 'Gibt es ein Tagesgang des Geostrophischen Windes?', *Meteorol. Rundschau* **32**, 51–53.

Martner, B. E. and Marwitz, J. D.: 1982, 'Wind Characteristics in Southern Wyoming', *J. Appl. Meteorol.* **21**, 1815–1827.

Matveev, L. T.: 1967, 'Diurnal March of the Wind Velocity in the Boundary Layer', *Physics of the Atmosphere* (Gidrometeoizdat/Isr. Progr. Sc. Transl.) 339–345.

Mayer, H.: 1974, 'Statistische Bearbeitung der Windgeschwindigkeits- und Windrichtungsdaten (Tsumeb/SWA Mitt. 6)', *Meteorol. Rundschau* **27**, 181–187.

McVehil, G. E. and Camnitz, H. G.: 1969, 'Ground Wind Characteristics at Kennedy Space Center', NASA-CR-1418.

Mikhail, A. S.: 1985, 'Height Extrapolation of Wind Data', *J. Solar Energy Engin.* **107**, 10–14.

Monna, W. A. A. and Van der Vliet, J. G.: 1987, 'Facilities for Research and Weather Observations on the 213 m Tower at Cabauw and at Remote Locations', Roy. Neth. Meteor. Inst. Sc. Rep. 87-5.

Paegle, J. and Rasch, G. E.: 1973, 'Three-Dimensional Characteristics of Diurnally Varying Boundary-Layer Flows', *Monthly Weath. Rev.* **101**, 746–756.

Peppler, A.: 1921, 'Windmessungen auf dem Eilveser Funkturm', *Beitr. Phys. fr. Atm.* **9**, 114–129.

Petersen, E. L., Troen, I., Frandsen, S., and Hedegaard, K.: 1981, 'Windatlas for Denmark', Risø-R-428.

Petersen, E. L., Troen, I., and Wieringa, J.: 1985, 'Development of a Method for Wind Climate Analysis for Non-mountainous Terrain in Europe', Proc. EWEC Conf., Hamburg 1984, W. Palz (ed.), Stephens, U.K., 6–12.

Peterson, E. W.: 1975, 'The Risø Profiles: a Study of Wind and Temperature Data from the 123 m-Tower at Risø, Denmark', *Quart. J. Roy. Meteorol. Soc.* **101**, 107–117.

Pielke, R. A. and Panofsky, H. A.: 1970, 'Turbulence Characteristics Along Several Towers', *Boundary-Layer Meteorol.* **1**, 115–130.

Raaff, W. R.: 1975, 'Windmetingen aan de 80 m-Mast te Vlaardingen in 1967/68', Roy. Neth. Meteor. Inst. Sc. Rep., 75-3.

Roth, R.: 1981, 'The Daily Variation of Geostrophic Wind Speed', *Beitr. Phys. Atm.* **54**, 101–106.

Rijkoort, P. J., Schmidt, F. H., Velds, C. A., and Wieringa, J.: 1970, 'A Meteorological 80 m Tower Near Rotterdam', *Boundary-Layer Meteorol.* **1**, 5–17.

Rijkoort, P. J.: 1972, 'The Variation of the Wind Distribution from Observations at 10, 40 and 80 m Height on the Meteorological Mast at Vlaardingen', Roy. Neth. Meteor. Inst.-Sc. Rep., 72–4.

Rijkoort, P. J.: 1983, 'A Compound Weibull Model for the Description of Surface Wind Velocity Distributions', Roy. Neth. Meteor. Inst. Sc. Rep., 83–13.

Rijkoort, P. J. and Wieringa, J.: 1983, 'Extreme Wind Speeds by Compound Weibull Analysis of Exposure-Corrected Data', *J. Wind Eng. Ind. Aerod.* **13**, 93–104.

Sanders, L. D. and Weber, A. H.: 1970, 'Evaluation of Roughness Lengths at the NSSL-WKY Meteorological Tower', ESSA-ERLTM-NSSL-47.

Sedefian, L.: 1980, 'On the Vertical Extrapolation of Mean Wind Power Density', *J. Appl. Meteorol.* **19**, 488–493.

Sinclair, R. W., Anthes, R. A., and Panofsky, H. A.: 1973, 'Variation of Low Level Winds During the Passage of a Thunderstorm Gust Front', NASA-CR-2289.

Smedman-Högström, A. S. and Högström, U.: 1978, 'A Practical Method for Determining Wind

Frequency Distributions for the Lowest 200 m from Routine Meteorological Data', *J. Appl. Meteorol.* **17**, 942–954.

Stevens, M. J. M. and Smulders, P. T.: 1979, 'The Estimation of the Parameters of the Weibull Wind Speed Distribution for Wind Energy Utilization Purposes', *Wind Engineering* **3**, 132–145.

Tennekes, H.: 1973, 'Similarity Laws and Scale Relations in Planetary Boundary Layers', *Workshop on Micrometeorology* D. A. Haugen (ed.), Am. Meteor. Soc., 177–216.

Tieleman, H. W.: 1980, 'Planetary Boundary Layer Wind Model Evaluation at a Mid-Atlantic Coastal Site', Rep. US-DOE/ET/23007.

Tuller, S. E., and Brett, A. C.: 1984, 'The Characteristics of Wind Velocity that Favor the Fitting of a Weibull Distribution in Wind Speed Analysis', *J. Clim. Appl. Meteorol.* **23**, 124–134.

Van der Hoven, I.: 1957, 'Power Spectrum of Horizontal Wind Speed in the Frequency Range from 0.0007 to 900 Cycles per Hour', *J. Meteorol.* **14**, 160–164.

Van Ulden, A. P., Van der Vliet, J. G., and Wieringa, J.: 1976, 'Temperature and Wind Observations at Heights from 2 to 200 m at Cabauw in 1973', Roy. Neth. Meteor. Inst. Sc. Rep. 76–7.

Van Ulden, A. P. and Holtslag, A. A. M.: 1985, 'Estimation of Atmospheric Boundary Layer Parameters for Diffusion Applications', *J. Clim. Appl. Meteorol.* **24**, 1196–1207.

Van Wijk, A. J. M., Holtslag, A. A. M., and Turkenburg, W. C.: 1985, 'Wind Profile Stability Corrections: Their Influence on Wind Energy Assessment Studies', Proc. EWEC Conf., Hamburg 1984 W. Palz, (ed.), Stephens, U.K., 96–101.

Verholek, M. G.: 1977, 'Summary of Wind Data from Nuclear Power Plant Sites', Contr. Rep. Battelle-NWL-2220-WIND-4.

Vinnichenko, N. K.: 1970, 'The Kinetic Energy Spectrum in the Free Atmosphere – 1 Second to 5 Years', *Tellus* **22**, 158–166.

Walk, O. and Wieringa, J.: 1988, 'Tsumeb Studies of the Tropical Boundary-Layer Climate – Synopsis of Publications on a Savannah Observation Project of Karlsruhe University', Wiss. Ber. Meteor. Inst. Univ. Karlsruhe **11**.

Wallace, J. A.: 1967, 'An Annotated Bibliography of Meteorological Tower and Mast Studies', USDC-ESSA-WB/BS-5.

Wessels, H. R. A.: 1984a, 'Cabauw Meteorological Data Tapes 1973–1984: Description of Instrumentation and Data Processing for the Continuous Measurements', Roy. Neth. Meteor. Inst. Sc. Rep. 84–6.

Wessels, H. R. A.: 1984b, 'Distortion of the Wind Field by the Cabauw Meteorological Tower', W.M.O. Instr. Obs. Meth. Rep. 15 (TECEMO), 251–255.

Wieringa, J.: 1986, 'Roughness-Dependent Geographical Interpolation of Surface Wind Speed Averages', *Quart. J. Roy. Meteorol. Soc.* **112**, 867–889.

Wieringa, J.: 1987, 'On the Spectral Gap Between Wind Engineers and Meteorologists', Roy. Meteor. Inst. Belgium Publ. A 119/1.

Wieringa, J.: 1988, 'Kartering van Nederland's Windklimaat Boven 40 m Hoogte', Proc. Nation. Windenergie Conf. (publ. Energie Anders, Rotterdam) 102–106.

VERTICAL PROFILES OF THE STRUCTURE PARAMETER
OF TEMPERATURE IN THE STABLE, NOCTURNAL
BOUNDARY LAYER

J. W. M. C U I J P E R S and W. KOHSIEK

Royal Netherlands Meteorological Institute, De Bilt, The Netherlands

(Received in final form 2 September, 1988)

Abstract. Vertical profiles of the structure parameter of temperature C_T^2 in the stable, nocturnal boundary layer (NBL) have been obtained with the analytic models described by Nieuwstadt (1984, 1985) and Sorbjan (1986) and the numerical model of Duynkerke and Driedonks (1987). These theoretical profiles are compared with observed profiles from the meteorological mast at Cabauw, The Netherlands. From the observations, it is found that C_T^2 is large in the surface layer and small at the top of the NBL. Observations during nights with moderate geostrophic winds or during the first few hours of nights with a high geostrophic wind show a continuous decrease of C_T^2 from the surface layer to the top of the NBL. Observations made later on nights with a high geostrophic wind show the development of a maximum of C_T^2 at about three quarters of the NBL. From the comparison with the models, we conclude that the observed profiles are most satisfactorily described by the model of Duynkerke and Driedonks.

1. Introduction

The temperature structure parameter, which is a measure of temperature fluctuations in a turbulent atmosphere, is of interest for several reasons. It can be used as a diagnostic tool, such as sensing the height of the turbulent layer with an acoustic radar, or inferring the surface heat flux either acoustically (Coulter and Wesely, 1980) or from *in-situ* observations of C_T^2 (Champagne *et al.*, 1977). It may also be applied to study the influence of atmospheric turbulence on the propagation of electromagnetic and acoustic waves, such as the effect on stellar imaging (Coulman, 1985) and on satellite-earth communication links (Herben, 1983). The atmospheric phenomenon, known as scintillation, is exploited in so-called scintillometers, instruments by which line-averaged surface fluxes of heat, moisture and momentum may be measured by means of the influence of the atmosphere on the propagation of a usually horizontal beam of light or radiation (Hill and Ochs, 1983; Ochs and Hill, 1985). In this paper we present observations of vertical profiles of the structure parameter of temperature obtained from the meteorological mast at Cabauw, The Netherlands, during clear nights and we compare them with the analytic models described by Nieuwstadt (1984, 1985) and Sorbjan (1986) and the numerical model of Duynkerke and Driedonks (1987).

During clear nights, the atmospheric boundary layer usually is stably stratified due to radiative cooling at the land surface. Vertical motion is restricted and turbulence is weak. Moreover the dynamics of the stable boundary layer may be influenced by more processes than continuous turbulence alone. Internal gravity waves (Finnigan and Einaudi, 1981), longwave atmospheric radiation (Garratt and Brost, 1981) and intermittent turbulence (Kondo *et al.*, 1978) may be important and can complicate the structure of the stable boundary layer. We selected nights that are characterized by continuous turbulence and that did not show the presence of significant gravity wave activity.

Observations of turbulence variables in general and of the structure parameter of temperature in particular in a stable boundary layer are not as abundant as in convective circumstances, e.g., Wyngaard and LeMone (1980) and Fairall (1987). Except for the data shown by Caughey *et al.* (1979), which are based on runs during early evening, we are not aware of other observations of vertical profiles of the temperature structure parameter.

2. Observations

Between September 1977 and February 1979 and in August 1983, experiments were conducted on the meteorological mast at Cabauw during clear stable nights (Nieuwstadt, 1984). The observations were begun about 2–3 hours after sunset to avoid the transition period during which turbulence is dominated by non-stationary effects.

The mast at Cabauw is 213 m high and the surrounding terrain is flat and homogeneous on a scale of approximately 20 km (Monna and van der Vliet, 1987). At height intervals of 20 m, booms are installed in three directions. The instruments and measuring heights used during the experiments are given in Table I.

Turbulent wind fluctuations were measured with a trivane (Wieringa, 1967 and 1972). A trivane consists of a propeller attached to one end of a rod, that can turn around two axes and is kept in the wind direction by means of an annular fin at the other end. The azimuth and elevation angle of the rod are determined by potentiometers.

Turbulent temperature fluctuations were measured from September 1977 to February 1979 with a pair of 100 μm copper-constantan thermocouples, mounted

TABLE I

The experimental set-up during nocturnal boundary-layer experiments.

Parameter	Instrument	Measuring height (m)
Turbulence	Trivane, fast thermocouple	20, 40, 80, 120, 160, 200
Wind speed	Cup anemometer	10, 20, 40, 80, 120, 160, 200
Wind direction	Wind vane	20, 40, 80, 120, 160, 200
Temperature	Ventilated thermocouples	10, 20, 40, 80, 120, 160, 200
Boundary-layer height	Acoustic sounder	

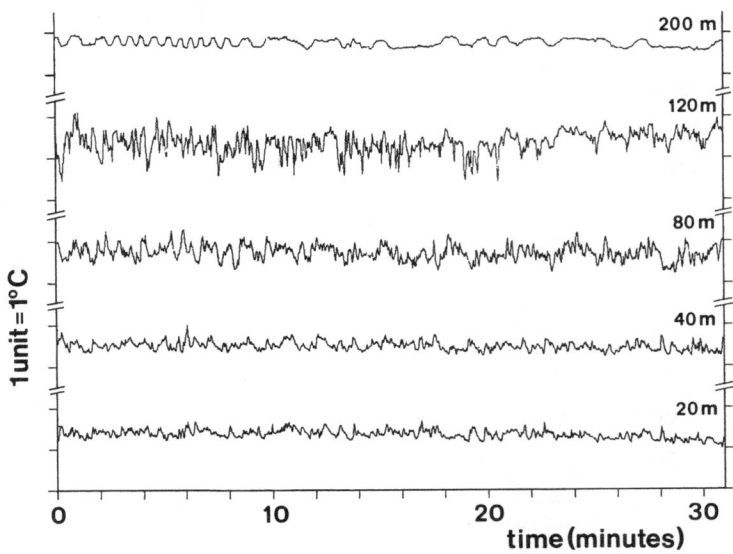

Fig. 1.   Temperature traces at several heights as a function of time, 20 February, 1978, 22:00–
22:30 GMT. Temp in °C. (Geostrophic wind ≈ 11 m/s).

on both sides of the trivane (1 m apart). In August 1983, only one fast-response thermocouple was used. In Figure 1, examples of traces of temperature as a function of time are shown for different heights. Observed vertical profiles of half-hourly mean potential temperature, wind speed and wind direction for the same period as in Figure 1 are shown in Figure 2.

All turbulent data were sampled with a frequency of 5 Hz and stored on magnetic tape. Variances and covariances of the velocity and temperature fluctuations were calculated over a time period of 30 min after removing a linear trend from the time series. By this procedure, only fluctuations with time scales ≤15 min contribute to the (co)variances. The average wind speed, wind direction and temperature were determined over concurrent 30 min periods.

Because only situations with continuous turbulence will be considered, half-hour runs with a geostrophic wind speed ≥ 5 m s^{-1} were selected. In such cases, the wind shear is large enough to maintain a continuous turbulent state. By using only observations for which the vertical velocity variance decreases continuously with height, we excluded boundary layers which are dominated by gravity waves (Driedonks and de Baas, 1983). Finally, we demanded that $h/L \geqslant 1$ (stability criterion), where h is the boundary-layer height determined with an acoustic sounder and L the Monin–Obukhov length. After this selection procedure, 62 half-hour runs remained for analysis of the structure parameter of temperature (Table II).

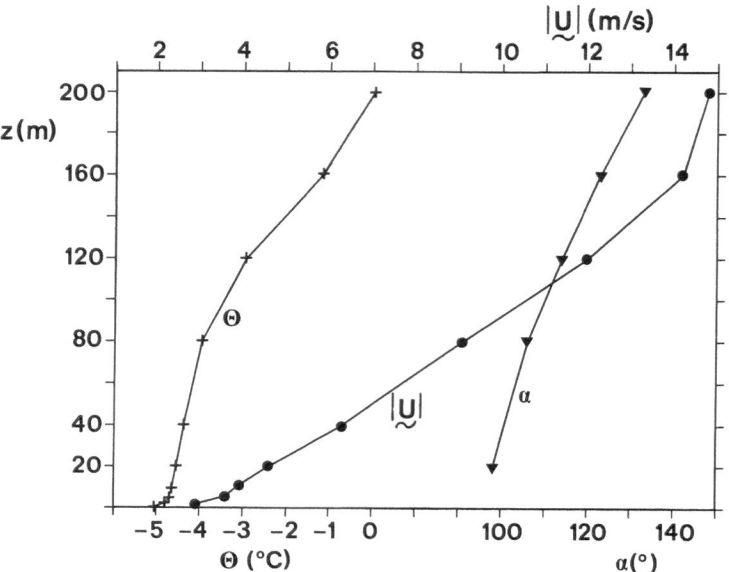

Fig. 2.   Measured vertical profiles of half-hourly mean potential temperature Θ, wind speed |U| and wind direction α at Cabauw, 20 February, 1978, 22:00–22:30 GMT. The height of the boundary layer was 150 m.

TABLE II

Half-hour runs used to determine structure parameter of temperature.

Date	Observation period begin (GMT)	end (GMT)	Number of half-hour runs
20/21 Feb 1978	19:00	06:00	17
19 May 1978	00:00	03:30	8
29/30 May 1978	23:00	01:00	4
31 May/1 Jun 1978	22:30	02:00	8
26 Sep 1978	01:00	04:00	7
9 Feb 1979	04:30	08:00	7
30/31 Aug 1983	20:00	01:30	11

3. The Structure Parameter of Temperature

The temperature structure function D_T is defined as the mean-square temperature difference between two microthermal probes:

$$D_T \equiv \overline{[T(x) - T(x + R)]^2}, \tag{1}$$

where $T(x)$ is the temperature at position x, and $T(x + R)$ that at $x + R$. The overbar indicates an ensemble average, which in practice is replaced by time averaging. On the assumptions of an inertial subrange and local isotropy and homogeneity, the relation between D_T and the structure parameter of temperature is given by (Tatarski, 1961):

$$C_T^2 = \frac{\overline{[T(x) - T(x + R)]^2}}{|R|^{2/3}} = D_T(R)R^{-2/3}. \tag{2}$$

C_T^2 is independent of R if the separation $|R|$ is within the inertial subrange.

C_T^2 can be measured with two temperature sensors separated by a distance R, or with one sensor using measurements at different instants of time. In the latter case, $T(x, t + \tau)$ is put equal to $T(x + R, t)$ with $R = U\tau$; U is the mean wind speed and τ is the time delay.

The spectrum for temperature fluctuations is given by (Hinze, 1975):

$$F_T(k) = 0.25 C_T^2 k^{-5/3}. \tag{3}$$

In this study, we used (3) to infer the temperature structure parameter C_T^2 from observed spectra. The error in the values of C_T^2 is estimated at 20%. Comparing (3) with Corrsin's inertial subrange form for temperature (Corrsin, 1951):

$$F_T(k) = 0.8 N \epsilon^{-1/3} k^{-5/3}, \tag{4}$$

leads to:

$$C_T^2 = 3.2 N \epsilon^{-1/3}, \tag{5}$$

where ϵ is the dissipation of turbulent kinetic energy and N is the dissipation of temperature variance. So, a fourth way to determine the structure parameter C_T^2 is by these molecular destruction terms. This method will be used in the models, to be described in the next section.

4. The Vertical Profile of the Temperature Structure Parameter: Models

Several models have been developed to describe the nocturnal boundary layer. The model of Nieuwstadt (1984, 1985) is a steady-state second-order closure model to study the vertical structure of the stable boundary layer. In Section 4a we shall discuss this model and focus on the vertical profile of the structure parameter C_T^2. Another approach is offered in Section 4b: once the vertical profiles of the Reynolds stress τ and the turbulent heat flux $\overline{w'\theta'}$ have been determined, one can determine the vertical profile of C_T^2 using similarity theory. Thirdly, Duynkerke and Driedonks (1987) developed a multilevel ensemble-averaged E–1 model to study the cloud-topped atmospheric boundary layer. With this model, the structure parameter C_T^2 can be calculated as a function of height. A discussion of this model will be given in Section 4c.

4a. THE MODEL OF NIEUWSTADT

To solve the equations that describe the evolution of the mean temperature Θ and the mean wind speed vector $\mathbf{U} = (U, V)$ in a horizontally homogeneous boundary layer, Nieuwstadt (1985) added as a closure hypothesis that the gradient Richardson number, Ri, and the flux Richardson number, Ri_f, are constant:

$$\text{Ri} \equiv \frac{g}{T} \frac{\partial \Theta}{\partial z} \bigg/ \left| \frac{\partial \mathbf{U}}{\partial z} \right|^2 = \text{constant}, \tag{6a}$$

$$\text{Ri}_f \equiv -\frac{g}{T} \overline{w'\theta'} \bigg/ \left(\boldsymbol{\tau} \cdot \frac{\partial \mathbf{U}}{\partial z} \right) = \text{constant}, \tag{6b}$$

where $\boldsymbol{\tau}$ is given by: $\boldsymbol{\tau} = [\overline{(-u'w')}, \overline{(-v'w')}]$. It is further assumed that there is a stationary, stable boundary layer.

With this model, the production terms of turbulent kinetic energy and temperature variance become:

$$-\overline{u'w'} \frac{\partial U}{\partial z} - \overline{v'w'} \frac{\partial V}{\partial z} = \frac{1}{k\text{Ri}_f} \frac{u_*^3}{L} (1 - z/h), \tag{7a}$$

$$-\overline{w'\theta'} \frac{\partial \Theta}{\partial z} = -\frac{\text{Ri}}{k\text{Ri}_f} \frac{T_*}{L} \overline{w'\theta_0'}, \tag{7b}$$

where u_* is the friction velocity and $\overline{w'\theta_0'}$ is the surface heat flux. The temperature scale T_* and the Obukhov-length L are defined by:

$$T_* = -\overline{w'\theta_0'}/u_*, \tag{8a}$$

$$L = -u_*^3/(kg/T\overline{w'\theta_0'}), \tag{8b}$$

with k the von Karman constant. With (7), the molecular destruction terms ϵ and N in Equation (5) can be found from the complete turbulent kinetic energy and temperature variance budgets (Businger, 1982) and neglecting the time variation, advection and flux divergence terms (Brost and Wyngaard, 1978):

$$\frac{kh\epsilon}{u_*^3} = \frac{1 - \text{Ri}_f}{\text{Ri}_f} \frac{h}{L} (1 - z/h), \tag{9a}$$

$$\frac{khN}{T_*^2 u_*} = \frac{\text{Ri}}{\text{Ri}_f^2} \frac{h}{L}. \tag{9b}$$

Note that the temperature dissipation N is independent of height.

Now we can use Equation (5) to find the vertical profile of the structure parameter of temperature:

$$\frac{C_T^2 (kh)^{2/3}}{T_*^2} = 3.2 \frac{\text{Ri}}{\text{Ri}_f^{5/3} (1 - \text{Ri}_f)^{1/3}} \left(\frac{h}{L} \right)^{2/3} (1 - z/h)^{-1/3}. \tag{10}$$

It follows that towards the surface, C_T^2 approaches a constant value. This value and consequently the whole profile is a function of the stability parameter h/L.

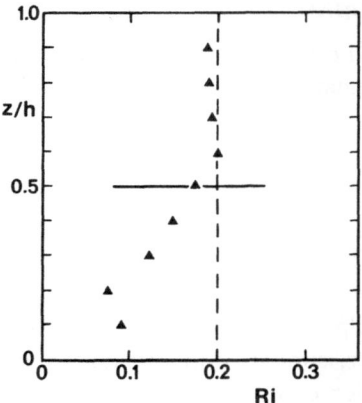

Fig. 3. The Richardson number as a function of non-dimensional height. Each point indicates the average of all observations within a given height interval. The horizontal bar indicates the standard deviation of the data (from Nieuwstadt, 1985).

The values for Ri and Ri_f are taken equal to 0.2, which seems to be a reasonable assumption for a major part of the boundary layer (Figure 3). Equation (10) also predicts that C_T^2 increases with height. The profile of C_T^2 is determined by the profile of ϵ solely, because, as noted above, N is constant with height. Figure 4 (thin lines) shows the calculated profiles of the dimensionless structure parameter $\text{CTN} = (C_T^2 (kh)^{2/3} / T_*^2)$ for different values of the stability parameter h/L. Note that for $z/h \to 1$, CTN becomes infinite.

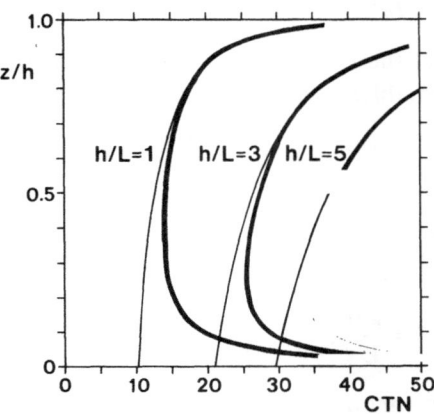

Fig. 4. Calculated profiles of the dimensionless structure parameter for different values of the stability parameter h/L. The thin lines are according to Nieuwstadt's model (10). The thick lines accord to (18) with the constants α_1 and α_2 of Nieuwstadt, i.e., $\alpha_1 = 1.5$ and $\alpha_2 = 1$.

4b. THE MODEL OF SORBJAN

In the Monin and Obukhov (1954) similarity theory of the surface layer, the non-dimensional wind shear and temperature gradients must be functions of $\zeta = z/L$ only:

$$\frac{kz}{u_*} \frac{\partial U}{\partial z} = \phi_m(\zeta) , \tag{11a}$$

$$\frac{kz}{T_*} \frac{\partial \Theta}{\partial z} = \phi_h(\zeta) . \tag{11b}$$

The theory is a great help in the analysis of the mean flow in the atmospheric boundary layer. However, similarity theory does not predict the shapes of the functions ϕ_m and ϕ_h, which can only be determined by experiments. See Yaglom (1977) for a review of the many formulas for ϕ_m and ϕ_h that have been proposed. A generally used version of the functions ϕ_m and ϕ_h is (Dyer, 1974):

$$\phi_m = \phi_h = 1 + 5\zeta . \tag{12}$$

Nieuwstadt (1984) introduced the idea of local scaling for the stable boundary layer. Dimensionless combinations of variables which are measured at the same height (therefore the term local is used) can be expressed as a function of a single parameter z/Λ solely, where Λ is the local Obukhov-length:

$$\Lambda = - \tau^{3/2}/(k \, g/T \, \overline{w'\theta'}) . \tag{13a}$$

The local values of the Reynolds stress $\tau(z)$ and the turbulent heat flux $\overline{w'\theta'}(z)$ define the local friction velocity $U_*(z)$ and temperature scale $t_*(z)$:

$$U_*(z) = \tau^{1/2} , \tag{13b}$$

$$t_*(z) = -\overline{w'\theta'}/\tau^{1/2} . \tag{13c}$$

Nieuwstadt showed that the Cabauw observations support local scaling.

Using these local variables, one can define the following similarity functions:

$$\Phi_m = \frac{kz}{U_*} \frac{\partial U}{\partial z} , \tag{14a}$$

$$\Phi_h = \frac{kz}{t_*} \frac{\partial \Theta}{\partial z} , \tag{14b}$$

$$\Phi_\epsilon = \Phi_m - Z = \frac{kz\epsilon}{U_*^3} , \tag{14c}$$

where $Z = z/\Lambda$. These functions must approach the Monin–Obukhov functions for $z \rightarrow 0$. Therefore, Sorbjan (1986) introduced the next hypothesis: "The form of the similarity functions Φ of Z in the outer layer (the part of the boundary layer above the surface layer) is identical to the form of Monin–Obukhov similarity functions ϕ of ζ in the surface layer". This leads to:

$$\Phi_m = \Phi_h = 1 + 5Z, \tag{15a}$$

$$\Phi_\epsilon = 1 + 4Z. \tag{15b}$$

Sorbjan assumed that the vertical profiles of stress and heat flux are given by:

$$\tau = u_*^2 (1 - z/h)^{\alpha_1}, \tag{16a}$$

$$\overline{w'\theta'} = \overline{w'\theta'_0}(1 - z/h)^{\alpha_2}. \tag{16b}$$

The constants α_1 and α_2 depend *inter alia* on the state of the stable boundary layer and terrain slope and must be determined empirically. Observations give values between 1 and 2 for α_1 and values between 1 and 3 for α_2 (Lacser and Arya, 1986). One further expects that the gradients of temperature and wind speed will remain finite throughout the boundary layer. With (14), (15) and (16), this leads to the following restriction for the vertical profiles: $\alpha_2 \geqslant \alpha_1$. Because of large scatter in the data, the constants α_1 and α_2 are rather difficult to obtain for one single night. Using all his observations, Nieuwstadt (1984) found $\alpha_1 = 1.5$ and $\alpha_2 = 1$, so the restriction $\alpha_2 \geqslant \alpha_1$ was not confirmed by his data set. The fact that these values are overall means might have caused this result.

Using (13) and (16), we find the following relations between local quantities and surface quantities:

$$U_* = u_*(1 - z/h)^{\alpha_1/2}, \tag{17a}$$

$$t_* = T_*(1 - z/h)^{\alpha_2 - \alpha_1/2}, \tag{17b}$$

$$\Lambda = L(1 - z/h)^{3/2\alpha_1 - \alpha_2}. \tag{17c}$$

With these relations and (14), (15) and (5), we obtain the vertical profile for the structure parameter of temperature in dimensionless form:

$$\frac{C_T^2(kh)^{2/3}}{T_*^2} = 3.2 \frac{(5\mu \, z/h + (1 - z/h)^{3/2\alpha_1 - \alpha_2})}{(4\mu \, z/h + (1 - z/h)^{3/2\alpha_1 - \alpha_2})^{1/3}}$$

$$\times \frac{(1 - z/h)^{8/3\alpha_2 - 2\alpha_1}}{(z/h)^{2/3}}, \tag{18}$$

where $\mu = h/L$ is the stability parameter.

In Figure 4, we show by the thick lines the profiles of the dimensionless structure parameter CTN for different values of the stability parameter h/L and the constants α_1 and α_2 of Nieuwstadt. For $z/h \geqslant 0.35$, (18) gives the same results as Nieuwstadt's model, which is only valid for $z/L \gg 1$. Using this restraint and the constants of Nieuwstadt in (18), we obtain:

$$\frac{C_T^2(kh)^{2/3}}{T_*^2} \approx 3.2 \, C(h/L)^{2/3}(1 - z/h)^{-1/3}, \tag{19}$$

where the constant $C = 5/4^{1/3} = 3.15$ equals the factor $\mathrm{Ri}/(\mathrm{Ri}_f^{5/3}(1 - \mathrm{Ri}_f)^{1/3})$ in

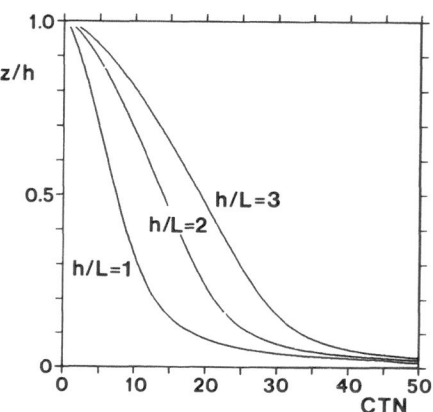

Fig. 5.   The profiles of the dimensionless structure parameter using Equation (18) with $\alpha_1 = \alpha_2 = 1$
for three values of h/L.

(10) with $\mathrm{Ri} = \mathrm{Ri}_f = 0.2$. It now becomes clear that it is the closure hypothesis $\mathrm{Ri} = \mathrm{Ri}_f = $ constant that caused the structure parameter following from (10) to become constant near the surface. However, observations show that near the surface the values of Ri and Ri_f deviate from 0.2 (Figure 3). Here, by using similarity functions, the influence of the surface layer is better represented. So, we expect that the shape of the structure parameter profile in that part of the boundary layer will be better described by (18) than by Nieuwstadt's model.

We note further that CTN always increases towards the surface irrespective of $\alpha_2 > \alpha_1$ or $\alpha_2 < \alpha_1$. However, for larger z/h, the shape of the profile does depend on the value of α_2/α_1. For $\alpha_2 > \frac{3}{4}\alpha_1$, CTN $\rightarrow 0$ at the top of the boundary layer, whereas for $\alpha_2 < \frac{3}{4}\alpha_1$, the structure parameter of temperature goes to infinity. With the earlier restriction of $\alpha_2 \geqslant \alpha_1$, it follows that (18) describes a structure parameter decreasing with height (Figure 5), whereas, adopting Nieuwstadt's values $\alpha_1 = 1.5$ and $\alpha_2 = 1$, CTN $\rightarrow \infty$ at the top of the boundary layer (Figure 4).

4c. THE MODEL OF DUYNKERKE AND DRIEDONKS

To study the cloud-topped atmospheric boundary layer, Duynkerke and Drie-donks (1987) developed a multilevel ensemble-averaged model. Tjemkes and Duynkerke (1989) show that this model can simulate the structure and evolution of the nocturnal boundary layer as well. In this type of model, the combined effect of all eddy sizes has to be parameterized. For this purpose, several turbulent closure hypotheses have been developed, which are mainly based on observational data from the clear-sky atmospheric boundary layer. In the model of Duynkerke and Driedonks, turbulence closure is formulated by using an equation for the turbulent kinetic energy and a diagnostic formulation for the length scale.

The ensemble-averaged equations describing the dynamics of the atmospheric boundary layer in horizontally homogeneous conditions are a more complete version of the set used by Nieuwstadt. Duynkerke and Driedonks added to these equations an equation that describes the evolution of total water, i.e., water vapor and liquid water. We adopted a low humidity in order to prevent cloud formation. Moreover, the humidity is taken independent of time and no evaporation has been considered.

The fluxes in the equations are expressed as:

$$- \overline{\phi' w'} = K_{m,h} \frac{\partial \phi}{\partial z}, \tag{20}$$

where ϕ is either a horizontal velocity component (K_m is used), or temperature or specific humidity (K_h). The exchange coefficients are calculated with:

$$K_{m,h} = c l_{m,h} E^{1/2}, \tag{21}$$

where c is a constant, $l_{m,h}$ a length scale and E the turbulent kinetic energy, determined with the complete turbulent kinetic energy budget (Businger, 1982). In the stable boundary layer, the length scale $l_{m,h}$ is determined by a suitable interpolation between two length scales; viz., a length scale for the surface layer:

$$l = kz/\phi_{m,h}, \tag{22a}$$

with $\phi_{m,h} = 1 + 5 \ z/L$, and a length scale for the stable layer:

$$l_s = c_s E^{1/2}/N, \tag{22b}$$

where $c_s = 0.36$ and N is the Brunt–Väisälä frequency given by:

$$N^2 = \frac{g}{\Theta_0} \frac{\partial \Theta}{\partial z}. \tag{22c}$$

Longwave radiative cooling of the atmosphere is calculated with the band-model of Tjemkes and Duynkerke (1989), the surface temperature with a model of Deardorff (1978). For the roughness length, a value of 0.15 m is adopted, which is typical for the Cabauw surroundings. The calculations were started with a neutral temperature profile.

Two situations, with geostrophic winds of 6 and 10 m/s have been simulated. In Figure 6, the curves are shown after 4, 6, 8 and 10 hr.

With a geostrophic wind of 6 m/s, there is a continuous decrease of the structure parameter with height. The stability parameter h/L for this case is about 3. With a geostrophic wind of 10 m/s, only in the beginning of the night there is a continuous decrease of the structure parameter. After about 6 hr, the structure parameter decreases with height in the lower half of the boundary layer, while in the upper half it reaches a maximum at $z/h \approx 0.7$, decreasing again down to a small value at the top of the boundary layer. During the night, the whole curve above $z/h \approx 0.1$ moves to the right. This implies an increase of the structure parameter in the entire boundary layer, which is most pronounced

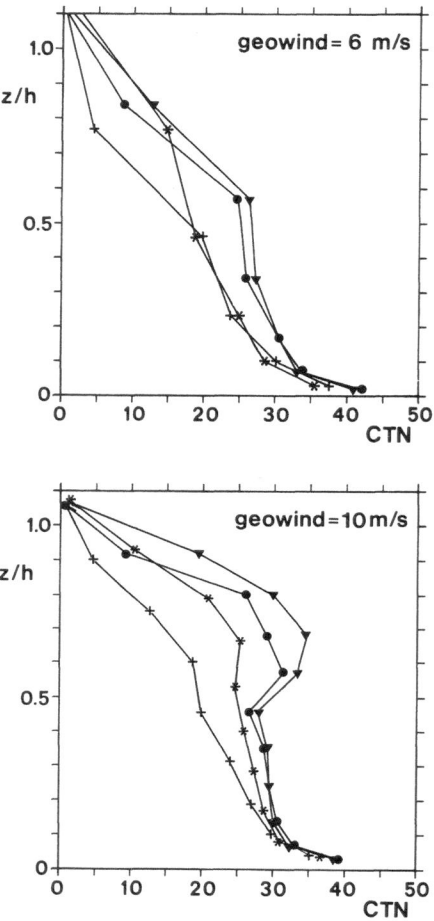

Fig. 6. Profiles of the dimensionless structure parameter calculated with the model of Duynkerke and Driedonks. Curves are shown after 4 hr (+), 6 hr (*), 8 hr (●) and 10 hr (▼), for geostrophic wind speeds of 6 and 10 m/s.

around $z/h \approx 0.6 - 0.7$. However, there is also a slight decrease of the stability parameter h/L from 2.5 to 2.0. We thus observe that CTN increases with decreasing h/L. In contrast, Nieuwstadt's model predicts an increase of the structure parameter with increasing h/L (Equation (10)). Moreover, the profiles of Nieuwstadt's model disagree with the model of Duynkerke and Driedonks. On the other hand, there is reasonable agreement between the profiles of the model of Duynkerke and Driedonks at a geostrophic wind speed of 6 m/s and the profile based on Sorbjan's hypothesis, Equation (18), with $\alpha_1 = \alpha_2 = 1$. However, the profiles with geostrophic wind $\mathbf{U}_g = 10$ m/s do not agree with (18) for any value of α_1 and α_2.

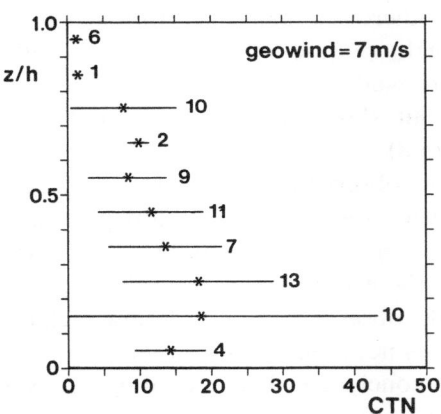

Fig. 7. Observations of the dimensionless structure parameter with data taken during the whole night for moderate geostrophic wind periods and only during the beginning of nights with a high geostrophic wind. The wind speed is the mean of the low geostrophic wind periods only. Each point indicates the average of all observations within a given height interval. The bar indicates the standard deviation of the data and the number represents the number of data points within each height interval.

5. Results and Discussion

In Figures 7 and 8, we show observations of the dimensionless structure parameter CTN for the periods mentioned in Table II. The mean geostrophic wind speed during the observation periods is indicated. From these figures it appears that the structure parameter is large near the surface and small at the top of the boundary layer. The profiles between surface layer and inversion layer can roughly be divided into two classes. First, there are observation periods when CTN decreases continuously with height. In these periods, the geostrophic wind

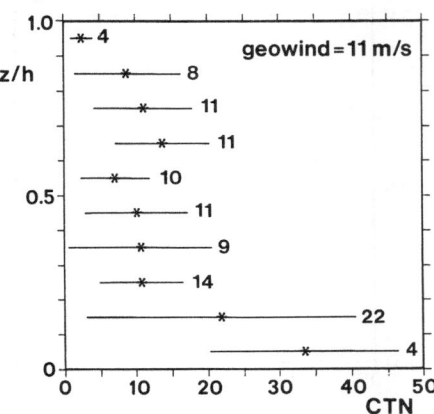

Fig. 8. As Figure 7, but for high geostrophic wind periods later at night. The mean geostrophic wind speed in these periods is indicated.

speed is about 7 m/s or there is a higher geostrophic wind speed but observations were made early at night (Figure 7). Second, in the other periods (with $U_g \approx$ 11 m/s and observations made later at night), CTN decreases in the lower half of the boundary layer, but shows an increase about $z/h \approx 0.7$, after which it decreases again (Figure 8).

In Figure 9, we show observations of CTN as a function of z/h where only those values of CTN were used with the stability parameter either in the interval $1.0 \le h/L < 1.5$ or $5.5 \le h/L < 7.0$. According to Nieuwstadt's expression (10), CTN should be about three times as large in the latter case as in the first case (Figure 4). However, the observations do not show such behaviour. The observed structure parameter does not seem to depend on the stability parameter.

In Section 4a, we concluded that according to Nieuwstadt's theory, the temperature structure parameter approaches a constant value at the surface and increases with height. The observations on the contrary show that the structure parameter increases near the surface and is small at the top of the boundary layer. The reason for the discrepancy in the surface layer is that Ri and Ri_f were assumed to be constant (0.2). Near the surface, Ri and Ri_f are smaller than 0.2 due to the high shear. At the top of the boundary layer, the assumption seems correct. The second assumption in Nieuwstadt's theory, that of stationarity, is likely to cause the theoretical profile to become infinite, which is not supported by the observations. As a consequence of this assumption, $\partial \Theta / \partial z$ is proportional to $(1 - z/h)^{-1}$. Because $\overline{w'\theta'}$ is found to be proportional to $(1 - z/h)$, the dissipation of temperature variance $N = -\overline{w'\theta'}(\partial \Theta / \partial z)$ is independent of height and consequently, the profile of CTN is determined by the profile of the dissipation of turbulent kinetic energy ϵ only (Section 4a). However, if $\partial \Theta / \partial z$ is proportional to $(1 - z/h)^{-1}$, then Θ should increase continuously with height to become infinite at the top of the boundary layer. Figure 10 shows that this does not occur. So, $\partial \Theta / \partial z$

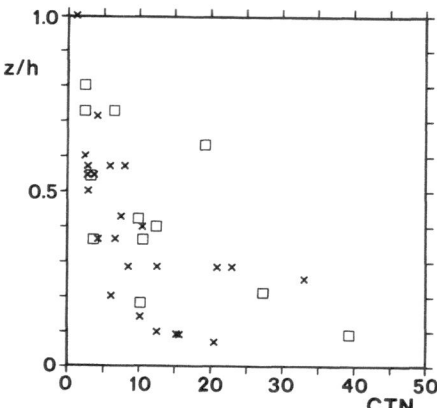

Fig. 9. Observations of the dimensionless structure parameter with the stability parameter either in the interval $1.0 \le h/L < 1.5$ (×) or $5.5 \le h/L < 7.0$ (□),

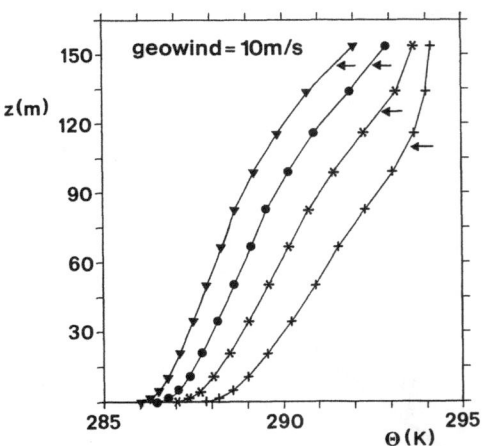

Fig. 10. Potential temperature profiles computed for two values of the geostrophic wind after 4 hr (+), 6 hr (*), 8 hr (●) and 10 hr (▼). The boundary-layer height, defined as the altitude where the heat flux is 5% of its surface value, is indicated by an arrow.

is not proportional to $(1 - z/h)^{-1}$. Therefore, N must be a function of height and the profile of CTN is determined by the profiles of ϵ and N.

Next, we turn to the model of Sorbjan. We first examine the dimensionless profiles of the temperature and wind gradient (14). Observations of these so-called similarity functions Φ_m and Φ_h are shown in Figures 11a and 11b. The line $1 + \beta z/\Lambda$ with $\beta = 5$ seems to fit reasonably well although close to the surface, β might be larger (Yaglom, 1977; Wieringa, 1980; Zhang et al., 1988).

The profile of the structure parameter in the surface layer is described acceptably by using similarity functions of Sorbjan. However, above this layer the profile is sensitive to the specific values of α_1 and α_2 of the stress and heat flux profiles, Equation (16). For a single night, these values could not be determined because of large scatter in the data. But whatever choices of α_1 and α_2 are made, the profile of CTN above the surface layer either continuously increases or decreases. So, even if one were able to determine these constants, one could only expect good agreement in moderate geostrophic wind speed cases. The model can not generate the maximum of CTN observed in the upper mixed layer for the high wind speed cases. Moreover, like the model of Nieuwstadt (Section 4a), the profile of CTN is also a function of h/L. The more stable the atmosphere, the larger is CTN. As mentioned earlier, the observations do not show such behaviour (Figure 9).

The best agreement with observations is found with the profiles calculated from the model of Duynkerke and Driedonks (1987). The calculated profiles show an increase near the surface. For medium geostrophic winds (~6 m/s), there is a continuous decrease of CTN towards the top of the boundary layer.

In the situation of a strong geostrophic wind, the model shows that, after a few hours, a local maximum of CTN develops about $z/h \sim 0.7$. This behaviour can

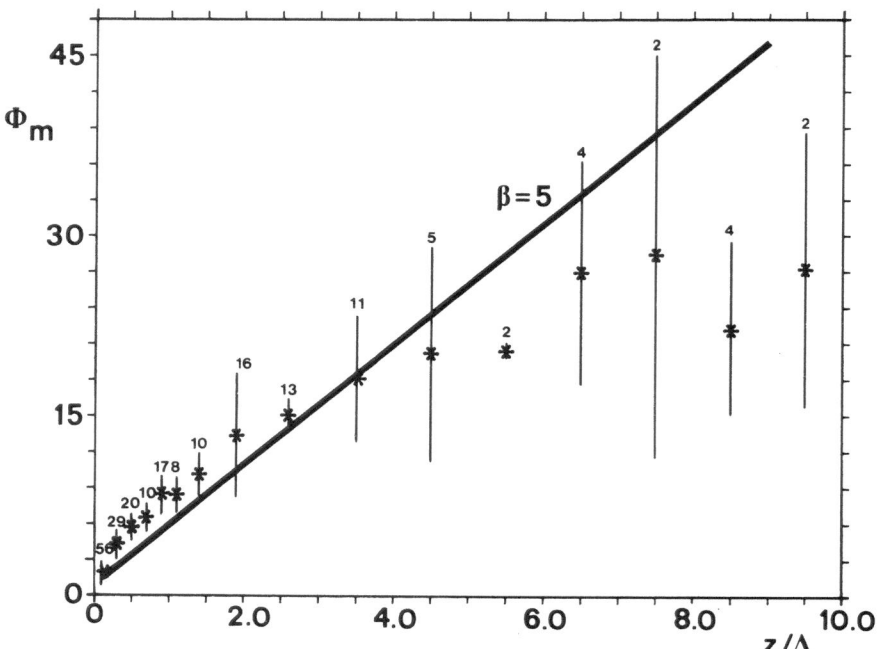

Fig. 11(a).   Dimensionless profile of wind gradient Φ_m as a function of the dimensionless height z/Λ. Each point indicates the average of all observations within a given height interval. The bar indicates the standard deviation of the data and the number represents the number of data points within each height interval. Also shown is the function $\Phi = 1 + \beta\, z/\Lambda$ with $\beta = 5$.

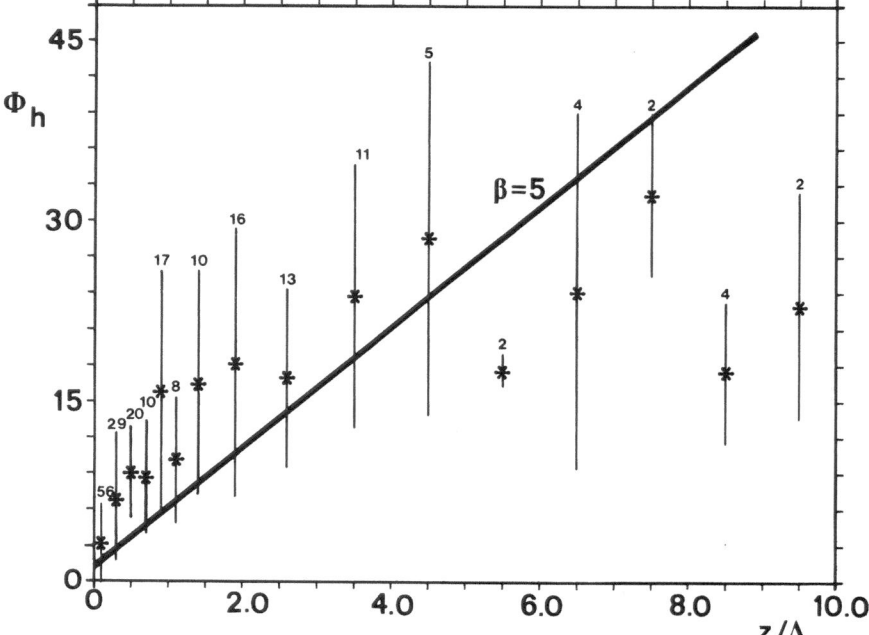

Fig. 11(b).   As Figure 11(a), but for the dimensionless temperature gradient Φ_h.

roughly be understood by using the profiles of potential temperature (Figure 10). There is a large temperature gradient close to the surface. In that region, there is considerable mechanical turbulence and CTN is large. At the beginning of the night, the vertical temperature gradient decreases continuously with height and so does CTN, which depends on it. However, after a few hours, we see an increase of the temperature gradient at larger height. This causes the increase of CTN at $z/h \approx 0.7$.

So, the shapes of the calculated profiles agree qualitatively with observed profiles. The reason why the values of CTN obtained from the model are about 2–3 times as high as those observed is not clear to us. We found that adopting a value larger than 5 for β makes the discrepancy only greater.

6. Conclusions

We have examined the structure parameter of temperature in the stable, nocturnal boundary layer with continuous turbulence. The data were gathered during clear nights from the meteorological mast at Cabauw, The Netherlands.

The vertical profile of the temperature structure parameter depends on the geostrophic wind speed. On nights with a moderate geostrophic wind speed and in the first few hours of nights with a high geostrophic wind speed, the structure parameter is large near the surface and decreases continuously to become small at the top of the boundary layer. On nights with a high geostrophic wind speed, the observations show, a few hours after transition, the development of a maximum of the structure parameter at about three quarters of the boundary-layer height.

Comparing these observations with three theoretical profiles leads to the following conclusions:

Because of the assumptions of stationarity and constant Ri and Ri_f, the model of Nieuwstadt (1985) is not suited to describe the profile of the structure parameter.

Using similarity functions following Sorbjan (1986), we found qualitative agreement between observations and model in the lower part of the boundary layer. Because the constants α_1 and α_2 in the stress and heat flux profiles could not be determined experimentally separately for each night, we can not simulate the profiles in the rest of the boundary layer. However, calculations with assumed values of α_1 and α_2 show that the maximum at $z/h \approx 0.7$ observed on nights with a high geostrophic wind can not be explained. Like the model of Nieuwstadt, the model using similarity functions shows that the profile of CTN is a function of the stability parameter. The observations do not show such a dependency.

The model of Duynkerke and Driedonks shows the right shape of the profile in moderate and high geostrophic wind speed situations. However, the computed value of the structure parameter is 2–3 times higher than observed.

Acknowledgements

The authors wish to thank Dr P. G. Duynkerke for calculating the profiles of the temperature structure parameter. We thank him and Dr A. P. van Ulden for useful comments on earlier versions of the manuscript.

References

Brost, R. A. and Wyngaard, J. C.: 1978, 'A Model Study of the Stably Stratified Planetary Boundary Layer', *J. Atmos. Sci.* **35**, 1427–1440.

Businger, J. A.: 1982, 'Equations and Concepts', in F. T. M. Nieuwstadt and H. van Dop (eds.), *Atmospheric Turbulence and Air Pollution Modelling*, D. Reidel Publ. Co., pp. 1–36.

Caughey, S. J., Wyngaard, J. C., and Kaimal, J. C.: 1979, 'Turbulence in the Evolving Stable Boundary Layer', *J. Atmos. Sci.* **36**, 1041–1052.

Champagne, F. H., Friehe, C. A., LaRue, J. C., and Wyngaard, J. C.: 1977, 'Flux Measurements, Flux Estimation Techniques, and Fine-Scale Turbulence Measurements in the Unstable Surface Layer Over Land", *J. Atmos. Sci.* **34**, 515–530.

Corrsin, S.: 1951, 'On the Spectrum of Isotropic Temperature Fluctuations in an Isotropic Turbulence', *J. Appl. Phys.* **22**, 469–473.

Coulman, C. E.: 1985, 'Fundamental and Applied Aspects of Astronomical Seeing', *Ann. Rev. Astron. Astrophys.* **23**, 19–57.

Coulter, R. L. and Wesely, M. L.: 1980, 'Estimates of Surface Heat Flux From Sodar and Laser Scintillation Measurements in the Unstable Boundary Layer', *J. Appl. Meteorol.* **19**, 1209–1222.

Deardorff, J. W.: 1978, 'Efficient Prediction of Ground Surface Temperature and Moisture, with Inclusion of a Layer of Vegetation', *J. Geophys. Res.* **83**, 1889–1903.

Driedonks, A. G. M. and de Baas, A. F.: 1983, 'Internal Waves in the Stably Stratified Atmospheric Boundary Layer', Preprints Sixth Symp. on Turbulence and Diffusion, Boston, Amer. Meteorol. Soc., 276–279.

Duynkerke, P. G. and Driedonks, A. G. M.: 1987, 'A Model for the Turbulent Structure of the Stratocumulus-Topped Atmospheric Boundary Layer', *J. Atmos. Sci.* **44**, 43–64.

Dyer, A. J.: 1974, 'A Review of Flux-Profile Relationships', *Boundary-Layer Meteorol.* **7**, 363–372.

Fairall, C. W.: 1987, 'A Top-Down and Bottom-up Diffusion Model of C_T^2 and C_Q^2 in the Entraining Convective Boundary Layer', *J. Atmos. Sci.* **44**, 1009–1017.

Finnigan, J. J. and Einaudi, F.: 1981, 'The Interaction Between an Internal Gravity Wave and the Planetary Boundary Layer, Part II: Effect of the Wave on the Turbulence Structure', *Quart. J. Roy. Meteorol. Soc.* **107**, 807–832.

Garratt, J. R. and Brost, R. A.: 1981, 'Radiative Cooling Effects within and Above the Nocturnal Boundary Layer', *J. Atmos. Sci.* **38**, 2730–2746.

Herben, M. H. A. J.: 1983, 'Amplitude Scintillations on the OTS-TM/$\overline{TM}$ Beacon', *Archiv für Elektronik und Übertragungstechnik* **37**, 130–132.

Hill, R. J. and Ochs, G. R.: 1983, 'Surface-Layer Micrometeorology by Optical Scintillation Techniques', *Remote Probing of the Atmosphere*, Technical Digest of the Topical Meeting on Optical Techniques for Remote Probing of the Atmosphere, Washington, D.C., Optical Society of America, TuC16.1–TuC16.4.

Hinze, J. O.: 1975, *Turbulence*, McGraw-Hill, New York.

Kondo, J., Kanechika, O., and Yasuda, N.: 1978, 'Heat and Momentum Transfer and Strong Stability in the Atmospheric Surface Layer, *J. Atmos. Sci.* **35**, 1012–1021.

Lacser, A. and Arya, S. P. S.: 1986, 'A Numerical Model Study of the Structure and Similarity Scaling of the Nocturnal Boundary Layer', *Boundary-Layer Meteorol.* **35**, 369–385.

Monin, A. S. and Obukhov, A. M.: 1954, 'Basic Laws of Turbulent Mixing in the Atmosphere Near the Ground', *Tr. Geophys. Inst. Ak. Nauk SSSR* **24**, 163–187.

Monna, W. A. A. and van der Vliet, J. G.: 1987, 'Facilities for Research and Weather Observations on the 213 m Tower at Cabauw and at Remote Locations', Scientific Reports WR-nr 87-5, KNMI, De Bilt, The Netherlands.

Nieuwstadt, F. T. M.: 1984, 'The Turbulent Structure of the Stable, Nocturnal Boundary Layer', *J. Atmos. Sci.* **41**, 2202–2216.

Nieuwstadt, F. T. M.: 1985, 'A Model for the Stationary, Stable Boundary Layer', in J. C. R. Hunt (ed.), *Turbulence and Diffusion in Stable Environments*, Clarendon Press, Oxford, England, 149–179.

Ochs, G. R. and Hill, R. J.: 1985, 'Optical-Scintillation Method of Measuring Inner Scale', *Appl. Opt.* **24**, 2430–2432.

Sorbjan, Z.: 1986, 'On Similarity in the Atmospheric Boundary Layer', *Boundary-Layer Meteorol.* **34**, 377–397.

Tatarski, V. I.: 1961, *Wave Propagation in a Turbulent Medium* (translated by R. A. Silverman), McGraw-Hill, New York.

Tjemkes, S. A. and Duynkerke, P. G.: 1989, 'The Nocturnal Boundary Layer. Model Calculations Compared with Observations', *J. Appl. Meteorol.* (to be published).

Wieringa, J.: 1967, 'Evaluation and Design of Windvanes', *J. Appl. Meteorol.* **6**, 1114–1122.

Wieringa, J.: 1972, 'Tilt Errors and Precipitation Effects in Trivane Measurements of Turbulent Fluxes over Water', *Boundary-Layer Meteorol.* **2**, 406–426.

Wieringa, J.: 1980, 'A Revaluation of the Kansas Mast Influence on Measurements of Stress and Cup Anemometer Overspeeding', *Boundary-Layer Meteorol.* **18**, 411–430.

Wyngaard, J. C. and LeMone, M. A.: 1980, 'Behaviour of the Refractive Index Structure Parameter in the Entraining Convective Boundary Layer', *J. Atmos. Sci.* **37**, 1573–1585.

Yaglom, A. M.: 1977, 'Comments on Wind and Temperature Flux-Profile Relationships', *Boundary-Layer Meteorol.* **11**, 89–102.

Zhang, S. F., Oncley, S. P., and Businger, J. A.: 1988, 'A Critical Evaluation of the von Karman Constant from a New Atmospheric Surface Layer Experiment', Preprints Eighth Symp. on Turbulence and Diffusion, San Diego, Amer. Meteorol. Soc., 148–150.

TIME SERIES ANALYSES OF CONCENTRATION AND WIND FLUCTUATIONS

STEVEN R. HANNA and ELIZABETH M. INSLEY

Sigma Research Corporation, 234 Littleton Road, Suite 2E, Westford, MA 01886, U.S.A.

(Received in final form 16 September, 1988)

Abstract. Analyses of concentration fluctuation (C') spectra from boundary-layer smoke plume experiments at six separate locations show that the spectra from these experiments generally exhibit an inertial subrange at high frequencies with a slope of $-5/3$ and indicate peak energy at a time period of about 50 to 100 s. These periods of peak energy are a factor of two to five less than those for the peak of the wind speed fluctuation (u' or v') spectra. A general spectral formula fits normalized spectra from the U.S. and Australia, where the frequency, n, is made dimensionless by multiplying by the plume dispersion parameter, σ_y, and dividing by the wind speed, u. Peak energy occurs at a dimensionless frequency of $n\sigma_y/u$ equal to about 0.15. The Kolmogorov constant in the inertial subrange is estimated from a set of averaged spectra. Cross-spectra indicate little relation between concentration and wind fluctuations. However, most of the correlation that exists is due to periods larger than about 10 or 20 s.

1. Purpose of Study

The boundary layer of the atmosphere is characterized by a relatively high degree of turbulence, with fluctuations in one-second averages of variables such as wind speed typically having magnitudes roughly equal to 10 to 100% of the mean value. When pollutants are emitted into this boundary-layer velocity field, they are carried about by turbulent eddies and also are observed to exhibit fluctuations of the same order of magnitude as their mean value (Csanady, 1973). These fluctuations are of great practical importance when the pollutant is a highly toxic or flammable material. Most dispersion models are capable of predicting the mean pollutant concentration, but provide no guidance on the probability of fluctuations from this mean.

Recently there have been several studies completed of the probability distribution of concentration fluctuations. Sykes (1984), Hanna (1984) and Sawford *et al.* (1986) agree that the distribution of concentration fluctuations is skewed towards higher values, but disagree on the exact form of the optimum distribution function (exponential, log-normal, or clipped normal). These authors, as well as several other researchers, have suggested analytical formulas for the mean, $\bar{C}$, and the standard deviation, σ_c, of the concentration fluctuations. They find that the fluctuation intensity, $\sigma_c/\bar{C}$, varies with distance from the source, position within the plume, averaging time, and the time scale of atmospheric turbulence. However, observations or models of the mean, the variance, or the probability distribution can provide no information on the eddy sizes that cause the observed fluctuations.

With the advent of practical instruments for measuring high-frequency (1 Hz or better) variations in concentration (e.g., Hadjitafi and Wilson, 1979; and Jones, 1983), a few field and laboratory experiments have been completed in which time series of pollutant concentrations have been observed. In most cases, concurrent observations of wind and temperature fluctuations have also been obtained. Consequently, it is now possible to investigate the spectral characteristics of concentration time series, such as the variation of fluctuation intensities with the period of turbulent eddies, and the relation between the wind and concentration time series. We have acquired six of these independent experimental data sets, and have applied time series analysis procedures to the available concentration and wind fluctuation data.

The purpose of the study is to answer the following questions:

(1) Can the concentration energy spectrum be represented by a generalized analytical formula?
(2) Does the concentration energy spectrum have an inertial subrange and, if so, can the constant in the Kolmogorov formula be estimated?
(3) What is the relation between the periods with peak energy in the concentration and wind spectra?
(4) What is the shape of the cross-spectrum between concentration and wind fluctuations?

The following sections describe the theoretical basis for concentration fluctuation spectra, the data bases that were analyzed, the time series analysis procedures, and the spectra and cross-spectra results.

2. Theoretical Basis for Concentration Fluctuation Spectra

It is assumed that a time series of concentration fluctuations is available with a total sampling length of about one hour and a sampling interval or averaging time on the order of one second (i.e., the record consists of a few thousand data points). The moments of the time series can be calculated (mean, variance, skewness, kurtosis, etc.), but they reveal nothing about the time periods of the variations in concentration. To estimate toxic response or flammability of hazardous pollutants, it is better to know whether the dominant period of the fluctuations is 5, 50, or 500 s. Furthermore, if the concentration fluctuations are somehow tied to wind speed fluctuations, it may be possible to predict the spectral characteristics of concentration fluctuations knowing only the characteristics of wind speed fluctuations.

The idea for this research was formed while viewing videotapes of smoke plumes from ground-level continuous sources at Dugway Proving Ground. During daytime experiments with light winds and strong surface heating, portions of the continuous plume were carried upward by convective eddies so that the plume looked more like a series of misshapen teardrops (see Figure 1). The

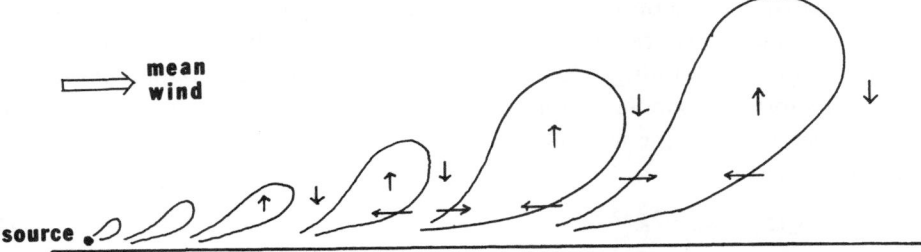

Fig. 1. Schematic drawing of the sideview of a smoke plume in a convective atmosphere with $\sigma_w/u > 1$, where σ_w is the standard deviation of vertical wind fluctuations and u is the mean wind speed. The small arrows represent turbulent velocity vectors, as perturbations from the mean wind vector.

postulated turbulent wind fluctuation field that would lead to these shapes is drawn on the figure. If this relationship were true, then high values of concentration would be correlated with low values of wind speed and positive values of vertical velocity (i.e., updrafts). As Sawford *et al.* (1985) point out, the scale of atmospheric wind fluctuations that influence the plume fluctuations will change as the plume size increases. Figure 2 illustrates a plume imbedded in a field of eddies of uniform size, showing that when the plume is smaller than the eddies, the eddies advect (meander) the plume back and forth, but when the plume is larger than the eddies, they no longer advect the whole plume but merely cause minor fluctuations deep within the plume.

Often it is possible to estimate the dominant period in a time series by visual inspection of the time series (i.e., the graph of concentration plotted versus time). But usually there are so many time periods acting that the time series looks like a random set of fluctuations. Spectral analysis can be used to estimate the fraction of the total variance or energy that is due to eddy periods in a certain narrow range. Lumley and Panofsky (1964) introduced the concept of atmospheric spectral analysis, and Panofsky and Dutton (1984, pp. 169–209) provided a

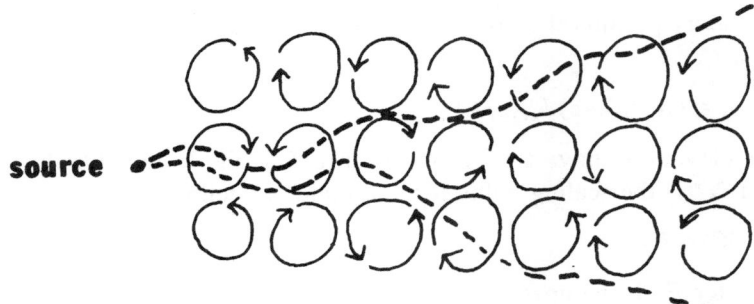

Fig. 2. Illustration of the influence of eddies of a given size on a smoke plume. Initially, the eddies transport the plume laterally; later, they diffuse the plume.

detailed discussion of theoretical aspects and gave many examples. There is much interest in the inertial subrange of eddies, where energy is being transferred from the larger energy-generating eddies to the smaller energy-dissipating eddies. In the inertial subrange, simple scaling relations may be valid. For example, much work has been done on the inertial subrange of wind speed spectra, which can be expressed in the form:

$$kS_u(k) = a_u \epsilon^{2/3} k^{-2/3} , \tag{1}$$

where k is wavenumber (in cycles/m), S is wind speed spectral energy per unit wavenumber (in m^3/sec^2), a_u is the Kolmogorov constant (about 0.5), and ϵ is the eddy dissipation rate (in m^2/sec^3). Taylor's "frozen-turbulence" hypothesis is invoked to convert from wavenumber, k, to frequency, n:

$$k = n/u , \tag{2}$$

Taylor's hypothesis has been shown to be valid for wind spectra, but has never been demonstrated for concentration spectra.

An inertial subrange can be postulated for fluctuations of any conservative variable, x, in the atmospheric boundary layer:

$$nS_x(n) = a_x \epsilon_x \epsilon^{-1/3}(n/u)^{-2/3} , \tag{3}$$

where ϵ_x is the dissipation rate of fluctuations of x by molecular forces, ϵ is the eddy dissipation rate of kinetic energy, and a_x is the Kolmogorov constant appropriate for x (Panofsky and Dutton, 1984). It is desirable to use scaling parameters to normalize Equation (3) so that it is universally valid. Monin–Obukhov similarity theory is used for variables, x, whose statistics are spatially homogeneous. For example, the wind speed fluctuation variance, σ_u^2, the mean wind speed, u, and the momentum and heat fluxes would be spatially homogeneous (constant) at a height of 2 m over a broad grassy field. Other variables that satisfy this criterion are the pressure, the temperature, and the relative humidity. Equation (3) can be normalized by the variance, σ_x^2:

$$nS_x(n)/\sigma_x^2 = a_x \epsilon_x \epsilon^{-1/3}(n/u)^{-2/3}/\sigma_x^2 . \tag{4}$$

Similarity theory permits the dissipation rate of fluctuations in substance x to be expressed as:

$$\epsilon_x = (X^{*3}/0.4z)\phi_{ex}(z/L) , \tag{5}$$

where z is elevation above the ground (in m), L is the Monin–Obukhov length (in m), and X^* is the scaling value for parameter x, defined by

$$X^* = F_*/u_* . \tag{6}$$

The parameter F_* is the upward turbulent vertical flux of x (also defined as $\overline{w'x'}$, where w' and x' are fluctuations in vertical speed and x and the overbar indicates a time average) and u_* is the friction velocity ($u_* = (-\overline{u'w'})^{1/2}$). If x is the turbulent velocity component, u, then $X^* = u_*$ and $\phi_{\epsilon u}(z/L) = 1 - z/L$ (Panofsky and Dutton, 1984). Consequently $\epsilon = (u_*^3/0.4z)\ \phi_{\epsilon u}(z/L)$ in Equation (4). The generalized stability function $\phi_{\epsilon x}(z/L)$ is defined so that it equals 1.0 in neutral conditions $(z/L = 0)$.

Panofsky and Dutton suggest formulas for the scaling relations given above for scalars such as temperature, relative humidity, and pressure. However, the turbulent flux, F_c, of pollutant concentration, C (units $\mu g/m^3$ or ppm), in a smoke plume is spatially homogeneous only if the pollutant is uniformly emitted over a broad area source with area 1 km^2 or greater. For example, radon may satisfy this criterion, since it is emitted nearly uniformly from the ground. If the pollutant is emitted from a point source, the turbulent flux, F_c, will be directed radially outward from the plume axis and thus can change sign and magnitude depending on the location of the monitor within the plume. Furthermore, the mean and variance in concentration fluctuations, $\bar{C}$ and σ_c^2, vary with downwind distance and with crosswind position. Consequently it is unlikely that Monin–Obukhov similarity relations will apply to concentration fluctuation spectra. It is possible that some sort of scaling relations will be valid locally. Consider the total variance, σ_c^2, as one scale. Assume that the dissipation rate of scalar variance, ϵ_c, is given by the relation

$$\epsilon_c \propto \sigma_c^2/\tau = \sigma_c^2 \sigma_w/l_c , \qquad (7)$$

where τ is a time scale for the fluctuations at the low-frequency end of the inertial subrange for the scalar energy, l_c is the length scale of these fluctuations, and σ_w is a velocity characteristic of turbulence fluctuations. In addition to Equation (7), the following well-known scaling law is used:

$$\epsilon \propto \sigma_w^3/l_v , \qquad (8)$$

where l_v is the length scale of velocity fluctuations at the low-frequency end of the inertial subrange for the kinetic energy. With these definitions, Equation (4) becomes:

$$nS_c(n)/\sigma_c^2 = a_c(l_v/l_c)^{1/3}(nl_c/u)^{-2/3} . \qquad (9)$$

It can be assumed that $l_c \approx \sigma_y$, the lateral dispersion parameter. The following generalized relation can be postulated:

$$nS_c(n)/\sigma_c^2 = f(l_v/\sigma_y, n\sigma_y/u) , \qquad (10)$$

where f represents a universal function. Thus, at constant l_v/σ_y, if $nS_c(n)/\sigma_c^2$ is plotted versus $n\sigma_y/u$, it is hoped that all observations will fall along some general curve.

The length scale l_v may be proportional to mixing depth, h, during well-mixed convective conditions, and may be proportional to a representative height of the plume during neutral or stable conditions. The mixing depth was not observed during any of the field experiments that were studied. For ground-level sources, the vertical dispersion parameter, σ_z, would be a measure of the plume height, in which case the ratio, l_v/σ_y, would equal σ_z/σ_y. Gifford (1968) shows that this ratio is a constant on the order of unity during unstable conditions. Consequently, we assume that $(l_v/\sigma_y)^{1/3} = (\sigma_z/\sigma_y)^{1/3} = 1.0$, which would suggest that the normalized spectral density, $nS_c(n)/\sigma_c^2$, is a function only of $n\sigma_y/u$.

Wilson and Simms (1985) and Hanna (1986) have postulated that the simple Markov spectrum may be valid for concentration fluctuations:

$$nS_c(n)/\sigma_c^2 = (2/\pi)nT_{pc}/(1+(nT_{pc})^2),\qquad(11)$$

where T_{pc} is the period with peak spectral energy. This formula is derived from the exponential autocorrelogram:

$$R(t) = \exp(-t/T_c),\qquad(12)$$

where the integral time scale, T_c, equals $T_{pc}/2\pi$ for sinusoidal fluctuations. The large $-n$ asymptote of Equation (11) is $nS_c(n) \propto n^{-1}$ rather than the $nS_c(n) \propto n^{-2/3}$ relation of Equation (9). Both Equations (9) and (11) will be tested with the field data.

3. Description of Data

During the past five years, a number of fast-response concentration fluctuation data sets have appeared in the literature. The data tapes and descriptions of the experiments were requested from each principal investigator. In a few cases, the data could not be used because of problems such as poorly formatted data tapes, but in most cases it was possible to analyze the concentration fluctuation time series. In most data sets, concurrent time series of wind speed fluctuations were also available. A summary of the characteristics of the data sets is given in Table I, and brief descriptions of the data are given below. Except for the Deardorff–Willis data set, all source releases were non-buoyant.

Smoke Week III. The U.S. Army sponsors an annual field experiment in which smoke/obscurant devices are tested and concurrent fast-response aerometric and meteorological data are taken. Nearly all of the sources are complicated multiple releases of time-varying emissions. Only a few experiments employ simplified continuous releases of smoke suitable for our analysis. Trial 2 in Smoke Week III took place at Eglin AFB, Florida, and has been analyzed previously by Hanna (1984). The fog oil source was located at a height of 1 m at a distance of about 70 m upwind of a line of aerosol photometers oriented across the plume. Point measurements of particle concentrations were made by each instrument at a height of 1 m. In our analysis, spectra were averaged over the five monitors that were nearest to the center of the plume. A meteorological tower was located at the mid-point of the line. The experiment took place during the daytime with

TABLE I

Characteristics of Experiments

Test name	Downwind distance (m)	Number of runs	Duration of runs	Averaging time (sec)	Average wind Speed (m/s)	Wind Speed range (m/s)	Stability
Smoke Week III (fog oil):							
Trial 2	70	1	300 s	1	3.0	0.8–4.6	Moderately unstable
Dugway Trials (fog oil):							
Trial T091	50	1	3560 s	1	6.6	2.9–11.4	Nearly neutral
Trial T011	50	1	3400 s	1	3.0	0.5–6.5	Moderately unstable
WSU – Palouse Wheat Field Study 1985 (SF6):							
P3	100	1	10 min	0.5	11.6	5.2–18.7	Nearly neutral
P18	100	1	20 min	0.5	2.1	1.1–3.0	Stable
WSU – Hanford Diffusion Grid							
Unstable	192	1	34 min	0.5	2.2	0.5–4.5	Unstable
Sawford – 2 Source Project (SF6, P):							
	25	23	1 hr	6	4.8	1.5–9.0	Slightly unstable
	50	6	1 hr	6	5.6	4.5–7.5	Slightly unstable
	100	15	1 hr	6	4.1	2.0–6.5	Slightly unstable
Deardorff-Willis Tank							
	500	1			5		Unstable
	(Instantaneous cross-section of plume)						

light-to-moderate winds. Only 300 s of data were available due to the short duration of the fog oil release.

Dugway Trials. Fog oil was also used as source in field tests at Dugway Proving Ground (Bowers and Black, 1985). The source was again located at a height of 1 m, but in this case the continuous release extended over a 1-hr period. Cross-wind integrated concentrations at an elevation of 1 m and a distance of 50 m downwind (at the point the observed line crossed the plume centerline) were recorded by a transmissometer looking perpendicular to the plume center-line. A meteorological tower was located close to the plume centerline near the line-of-sight of the transmissometer. All data were available as 1-s averages. Two trials were analyzed: one during nearly-neutral conditions and another during unstable conditions (this trial inspired the schematic drawing in Figure 1).

WSU – Palouse Wheat Field Study. Lamb *et al.* (1985) discuss an SF_6 tracer study conducted by Washington State University (WSU) at their Palouse wheat field experimental site. Fast response (20 Hz) point measurements of SF_6 concentration and wind fluctuations were made at a single site at a distance of about 100 m from the source during several runs. Source height and monitor height were each about 1.5 m. The duration of the runs was on the order of 10 to 20 min. The 20 Hz data were averaged to produce a data record consisting of 2 Hz averages. Otherwise the storage limits in the spectral analysis program would have been exceeded. Because there was only one monitor, it was difficult to assure that the plume would blow right over the monitor, and during several runs the sporadic nature of the concentration time series suggested that the monitor was off the edge of the time-mean plume. We selected two runs where the data indicated that the time-mean plume was over the monitor.

WSU – Hanford Diffusion Grid. Peterson *et al.* (1988) discuss the results of another test of the WSU system at the Hanford, WA, diffusion grid. The instrument set-up was the same as above, except that the source and the concentration monitor were at a height of 3.9 m and the monitor was at a distance of 192 m from the source. The 10 Hz data were averaged to produce a data record consisting of 1 Hz averages.

Sawford – 2 Source Projects. Sawford *et al.* (1985) released SF_6 and P from two point sources located about 10 m apart at their field site in Australia and studied the relations between the time series of the two tracers observed at monitors located at a distance downwind of 25, 50, or 100 m. Source height and monitor height were both about 1.5 m. Our analysis was limited to the SF_6 data, which were characterized by an averaging time of 6 s and a sampling time of 3000 s. Forty-four separate runs were analyzed. Unfortunately, concurrent wind fluctuation data were not available.

Deardorff–Willis Tank. Deardorff and Willis (1987) have used a 1 m^3 tank of water to conduct several innovative studies of turbulence and dispersion in a convective boundary layer. In this experiment, they towed a stack through the tank and scanned the buoyant plume with a laser at a fixed position relative to the tank to produce nearly instantaneous cross-sections of concentrations at six different downwind distances. The effective averaging distance of the laser was about 10 m (scaled to full size). The ratio, σ_w/u, was equal to about 0.75, where σ_w is the standard deviation of vertical velocity fluctuations and u is the speed that the stack is towed through the tank. The dimensionless plume buoyancy parameter (ratio of heat flux in the plume to the convective heat flux in the boundary layer), F_*, equaled 0.116, and the downwind distance was 3.57 times the length scale of the dominant large convective eddies. These concentration data were used to calculate space spectra rather than time spectra. Concurrent velocity fluctuation data were not available.

4. Time Series Analysis Procedure

A generalized software package for the analysis of time series using the Fast Fourier Transform procedure was obtained from the National Center for Atmospheric Research (NCAR) (B. Stankov, personal communication, 1986) and was modified by R. Yamartino for use on an IBM PC/AT microcomputer. The text by Bendat and Piersol (1980) was used to develop these modifications. Time series for one to eight concurrent variables can be input to this package. In order to fit existing storage space in a PC/AT, the total number of data points is limited to 2048 or less, if eight variables are studied. The limit is 4096 if five variables or less are studied. Standard procedures such as detrending, smoothing, and tapering can be applied to the data. In this analysis, the time series were detrended (a linear trend removed) and tapered (beginning and ending 10% of the time series tapered to zero). Smoothing was applied only to final graphs. Output consists of tables and graphs for single variables (original time series, autocorrelogram, and spectrum) and for pairs of variables (co-spectrum, quadrature spectrum, phase angle, and coherence). A separate program was used to calculate the cross-correlogram.

Panofsky and Brier (1968) provide derivations and discussions of the interpretation of each of these plots, which are briefly summarized below.

An *Autocorrelogram* shows the correlation between a variable and itself as a function of time lag. It is dimensionless and ranges between -1.0 and $+1.0$.

A *Spectrum* shows the relative contribution of eddies or oscillations of various frequencies to the total energy or mean square variability. It has the units of X^2 per unit frequency, where X is in units of the primary variable (concentration in, say, ppm, or wind speed in m/s). Frequency is given in units of cycles/s.

A *Cross-Correlogram* shows the correlation between one variable and another variable as a function of time lag. It is dimensionless and ranges from -1.0 to $+1.0$.

A *Co-Spectrum* measures the contribution of oscillations of different frequencies to the total cross-covariance at lag zero between two time series. It has the units of XY per unit frequency, where X and Y are the units of the two variables being analyzed.

A *Quadrature Spectrum* measures the contribution of different oscillations to the total covariance obtained when the harmonics of one of the time series are delayed by a quarter period. It has the units of XY per unit frequency, where X and Y are the units of the two variables being analyzed.

The *Phase Angle* shows the angle by which two time series are out of phase at each harmonic. It ranges from -180 to $+180°$ and is plotted as a function of frequency.

The *Coherence* shows how good the relationship is between two variables for various frequencies. It varies from 0 to 1.0 and is dimensionless.

5. Results of Time Series Analysis

The time series analysis software package is capable of producing large quantities of plots and tables. For example, if five concurrent time series are being analyzed, there are 55 plots produced for all possible types of graphs and combinations of variables. Because little information was contained in many of the plots produced in this study, the results discussed below are limited to the more interesting plots. Emphasis is on the shape of the concentration spectra, the dominant time scale and the slope of the inertial subrange in the concentration spectra, the magnitude of the cross-correlogram, and the dominant time scale in the cospectra.

5.1. DOMINANT TIME SCALE OF CONCENTRATION SPECTRA

Concentration spectra were calculated for each of the data sets in Table I. Except for the convective tank data, which are expressed as functions of position, all of the data were taken for similar conditions:

- Continuous point source located near the ground.
- Concentration measurements near the ground at downwind distances on the order of 100 m.
- Moderate wind speeds.
- Mostly daytime conditions (a few were during the night, though).

Consequently it is expected that the normalized spectra, $nS_c(n)/\sigma_c^2$ would exhibit similar shapes when plotted versus frequency, n. When $nS_c(n)/\sigma_c^2$ is plotted versus ln n, the area under the curve should equal 1.0, and the peak of the curve should occur at a frequency associated with maximum energy. In fact it is found that $nS_c(n)/\sigma_c^2$ reaches a maximum of about 0.3 at a frequency of about 0.01 to 0.02 cycles/sec or a time period of about 50 to 100 s. Concurrent wind speed spectra have peaks at time periods of about 200 to 600 s, and vertical velocity spectra have peaks at periods of about 1 to 5 s. Thus the concentration spectra have shapes similar to the wind speed spectra, but are shifted to higher frequencies by a factor of about five. If the plumes were sampled farther downwind, where they would encompass a larger range of eddy sizes, it would be expected that the time scales of the concentration spectra would increase, as found by Sawford *et al.* (1985).

Although individual spectral shapes can be quite variable, a smoother curve is obtained when several similar spectra are combined. Figure 3 shows the distribution of spectra (after normalization by the area or σ_c^2) for the six experimental data sets listed in Table I, where the normalized frequency, $n\sigma_y/u$, is plotted on the abscissa. Because the figure would be a mess if all 51 spectra were plotted

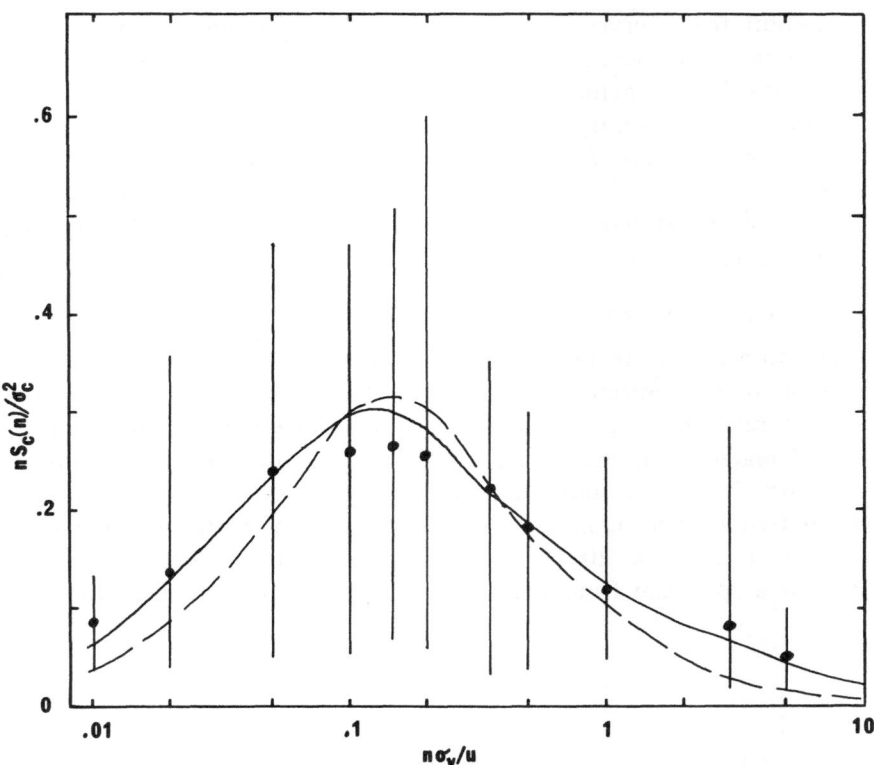

Fig. 3. Composite power spectrum for concentration fluctuations from all data in Table I, plotted in dimensionless coordinates as $nS_c(n)/\sigma_c^2$ versus $n\sigma_y/u$. The median and the range from 51 spectra are given at representative frequencies. Analytical curves from Equation 14 (solid line) and Equation 13 (dashed line) are drawn.

at once, the median and the range over all the spectra are given at 11 representative frequencies. It is seen that the median points follow a fairly smooth curve with peak energy at a normalized frequency of $n\sigma_y/u$ equal to about 0.15. Two empirical curves are also drawn on the figure. One represents a Markov spectrum (Equation 11) with an integral time scale, T_c, of σ_y/u. The time scale of peak spectral energy, T_{pc} is then $2\pi\sigma_y/u$.

$$\text{Markov: } nS_c(n)/\sigma_c^2 = 4(n\sigma_y/u)/(1 + (2\pi\sigma_y/u)^2) \,. \tag{13}$$

This curve tends to underestimate the spectral energy on the figure at high and low frequencies.

An alternative empirical equation that has the proper slope in the inertial subrange is the following:

$$nS_c(n)/\sigma_c^2 = 6.5(n\sigma_y/u)/(1 + 54(n\sigma_y/u)^{5/3}) \,. \tag{14}$$

This curve provides a better fit than Equation (13) to the points in Figure 3. It is implicit in both these curves that the plume scale, σ_y, is much less than the scale of atmospheric turbulence, Λ_a, and therefore Λ_a does not influence the concentration spectra. In the daytime boundary layer, Λ_a is approximately equal to the mixing depth, which is on the order of 1000 m. Consequently σ_y is two orders of magnitude smaller than Λ_a for these experiments. In the case when σ_y is the same order as Λ_a, Hanna (1986) suggests an empirical formula that is a linear combination of two Markov spectra – one with a time scale of σ_y/u and another with a time scale of Λ_a/u.

5.2. CHARACTERISTICS OF INERTIAL SUBRANGE

According to similarity theory, the relation $S_c(n) \propto n^{-5/3}$ is valid in the inertial subrange of the concentration spectrum (see Equation 9). Garvey *et al.* (1982) verified this relation using earlier U.S. Army Smoke Week data. Each of the spectra calculated from the data in Table I were analyzed in an attempt to determine whether this relation was valid. In most cases, the $-5/3$ power law was evident at frequencies, n, greater than about 0.05 or 0.10 cycles/s (i.e., time periods less than 10 or 20 s). Figure 4 contains an example of the inertial subrange of a spectrum from the Dugway Proving Ground field trials, which

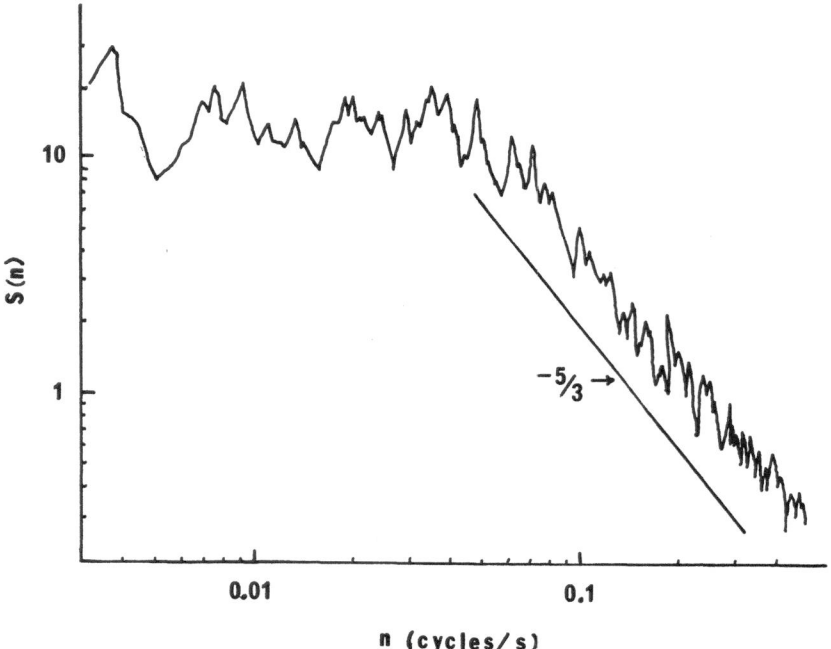

Fig. 4.   Power spectrum plotted versus frequency for Dugway proving ground cross-wind integrated concentration fluctuations in trial T091. The curve follows a $-5/3$ slope for periods less than about 20 s.

exhibits a particularly well-behaved curve. Generally the spectra contained more scatter than is seen in this figure.

The constant, a_c, in the following equation can be estimated from the data:

$$nS_c(n)/\sigma_c^2 = a_c(n\sigma_y/u)^{-2/3} . \tag{15}$$

Figure 3 illustrates that $nS_c(n)/\sigma_c^2$ equals about 0.12 at a normalized frequency, n, of unity for the median of 51 spectra. With these values for the parameters in Equation (13), the constant, a_c, would equal about 0.12. Because we have assumed that the term, l_v/σ_y, in Equation (9) equals unity, the constant a_c derived from Equation (15) is valid only when this condition is satisfied.

The Deardorff and Willis (1987) convective tank data were included in Figure 3. They can be used independently to derive the constant, a_c, although it is necessary to rewrite Equation (15) as a function of wavenumber, k, which equals n/u if Taylor's frozen turbulence hypothesis is valid:

$$kS_c(k)/\sigma_c^2 = a_c(k\sigma_y)^{-2/3} . \tag{16}$$

There are not enough runs to generate time series of concentration fluctuations from these data. We calculate a spatial spectrum $S_c(k)$ using an instantaneous cross-section of the plume taken at a position such that a few hundred observation points are well within the plume. Consequently the concentration distribution can be assumed to be spatially homogeneous over these points. Spectra are calculated over horizontal rows of points well within the plume and then the spectra over 11 rows are combined to give a composite spectrum. It is found that this composite concentration spectrum exhibits an inertial subrange at wave numbers, k, greater than about $5/h$ (or wavelengths less than $0.2h$), where h is the mixing depth. Thus the inertial subrange begins at a wavelength equal to about 10% of the wavelength of peak turbulent energy, $1.5h$ (Kaimal et al., 1982). This composite spectrum from a single cross-section of the plume in the convective tank yields a constant, a_c, of 0.36, which is about a factor of three larger than the value calculated from the median spectrum in Figure 3. It is expected that the calculated a_c would be larger during convective conditions, since the term $\phi_{\epsilon c}(z/L)$ in Equation (5) and the term l_v/σ_y in Equation (9) have been neglected. However, the difference could also be due to the fact that the convective tank data represent only a single realization, and there is large uncertainty in the positioning of the inertial subrange in a single experiment.

5.3. CROSS-CORRELATIONS AND CROSS-SPECTRA

Visual impressions from video tapes of smoke plumes would lead to the hypothesis that positive excursions in concentration fluctuations are correlated with positive excursions in turbulent vertical velocity. However, these visual effects are relatively short-lived, and when averaged over an hour, they contribute little to the total correlation. The cross-correlations between concentration fluctuations and turbulent wind speed fluctuations (e.g., $r_{cu} = \overline{c'u'}/\sigma_c\sigma_u$ or $r_{cw} =$

$\overline{c'w'}/\sigma_c\sigma_w$) at zero time lag are typically about 0.1 or 0.2. At non-zero time lags, the cross-correlations vary back and forth about zero, with a magnitude seldom exceeding 0.1. For comparison, the cross-correlation between u' and w' ($r_{uw} = \overline{u'w'}/\sigma_u\sigma_w = -u_*^2/\sigma_u\sigma_w$) at zero time lag is equal to about -0.4, where u_* is the friction velocity and it is assumed that σ_w/u_* equals 1.3 and σ_u/u_* equals 2.0. The cross-correlations r_{wc} and r_{vc} are proportional to the fluxes of plume material away from the center of the plume, and change signs depending on which side of the plume the monitor is located. None of the cross-correlation plots as a function of time lag is reproduced here because they are so random and have such small correlations.

The cross-spectra, on the other hand, do yield some useful, although not unexpected results. It is known that small eddies in the atmosphere are isotropic (i.e., with no preferred direction). Consequently the cospectra and quadrature spectra for pairs of the variables u', v', w', and c' are nearly zero for large frequencies (small time periods). The low frequency limit of this range corresponds roughly to the limit of the inertial subrange (about $n \sim 0.1$ s^{-1} or period ~ 10 s). This conclusion is well known for $u' - w'$ cross-spectra (Panofsky and Dutton, 1984), since there is observed to be little coherence between u' and w' for eddies in the inertial subrange, and the major contribution to the $\overline{u'w'}$ product

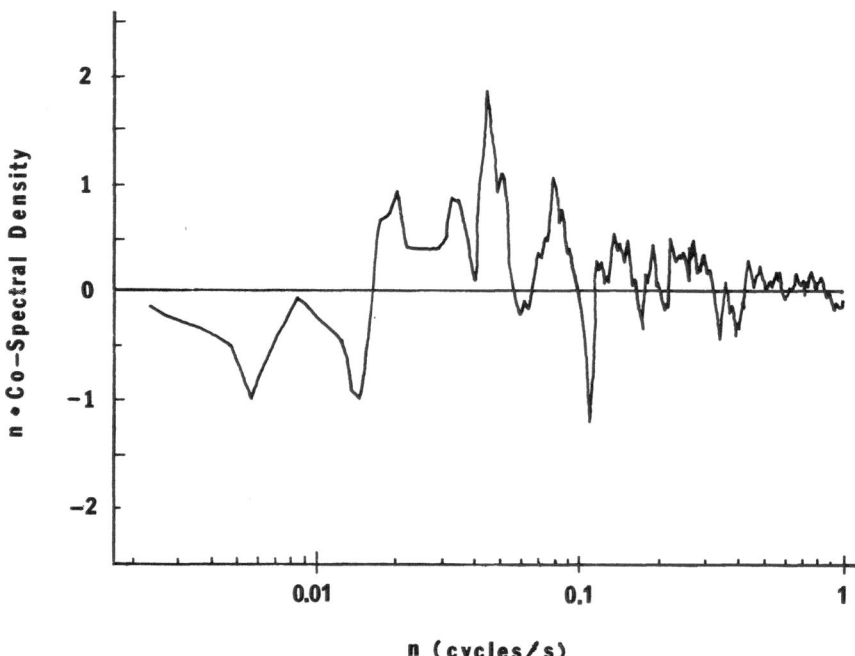

Fig. 5. Co-spectrum between concentration fluctuations and wind direction fluctuations for WSU run P3.

(i.e., the momentum flux) comes from eddies with relatively long time periods. Cross-spectra calculated between concentration and wind velocity fluctuations for the data in Table I agree with this hypothesis. Figure 5 is an example of a cospectrum from WSU Run P3, illustrating the lack of energy at high frequencies. The area under the curve is proportional to the cross-correlation at lag zero. Quadrature spectra (the out-of-phase component of the cross-spectra) yield similar results to those seen in Figure 5.

The ratio of the quadrature to the co-spectrum at a given frequency can be used to calculate the phase angle between the concentration and wind fluctuations at that frequency. These phase angles tend to flip back and forth from negative to positive with no systematic patterns discernible in the data. Coherence plots were also generated from the co-spectrum and the quadrature spectrum, and show the contribution of each frequency to the total cross-correlation between the variables (ranges from 0 to 1). The $C' - u'$ coherence plots illustrate a weak peak near the dominant meandering period, which tends to be about 100 to 200 s for these data, and yield little correlation at periods less than about 5 s.

6. Conclusions

Field and laboratory data sets containing fast-response measurements of concentration fluctuations were acquired. Of the six separate data sets, four also contained fast response measurements of wind velocity fluctuations. A time series/spectral analysis procedure was modified for use on the IBM PC/AT (copies of this computer program are available from the authors). This procedure calculates autocorrelograms, spectra, cospectra, quadrature spectra, coherence, and phase angle as a function of frequency for up to 2048 data points for each of up to 10 variables.

Nearly all of the field data sets were obtained under similar conditions – continuous point sources near the ground, observations near the ground at a downwind distance of about 100 m, and daytime conditions with moderate winds. Sampling times ranged from a few minutes to one hour, and averaging times were used that ranged from 0.5 to 6 s. For these conditions, an inertial subrange existed in the concentration spectra for time periods less than about 10 s. When written in the form $nS_c(n)/\sigma_c^2 = a_c(n\sigma_y/u)^{-2/3}$, the Kolmogorov inertial subrange formula is valid when the constant, a_c, is set equal to 0.12. The entire shape of the observed spectra can be fit by an analytical formula, where the normalized frequency associated with peak spectral energy is about $n\sigma_y/u = 0.15$. The time scale of the concentration spectra is typically about a factor of two to five less than the time scale of the concurrent wind speed spectra, and is a factor of 10 to 20 more than the time scale of the concurrent vertical velocity spectra.

A spatial concentration spectrum was calculated using data from an instantaneous cross-section of a plume in a laboratory convective tank. This spectrum exhibited an inertial subrange at wavelengths less than $0.1h$, where h is the mixing depth. The Kolmogorov constant, a_c, calculated from these data, was a factor of three higher than that calculated from the field data.

Cospectra, quadrature spectra, coherence, and phase angles were calculated between concurrent concentration and wind velocity time series. At most sites, the meteorological and concentration monitoring instruments were colocated. There was much variability in these plots, due to the tendency of a given time series to be dominated by one or two obvious sinusoidal patterns. The only clear conclusion is that small eddies with time scales less than about 10 s contribute little towards the relations between the time series.

Despite our attempts to acquire a comprehensive set of all publicly-available concentration fluctuation data sets, these data sets are quite limited in their characteristics. In the future, concentration fluctuation data should be collected for a wide range of source heights, monitor positions, downwind distances, and meteorological conditions. The validity of Taylor's frozen turbulence hypothesis should be tested. Furthermore, the cross-correlations are strengthened if the concentration and meteorological instruments are co-located. However, even though the available data sets are limited, the analysis represents a first step towards the development of generalized spectral relations for concentrations fluctuations.

Acknowledgements

This research was supported by the Army Research Office, with Dr Walter Bach as project monitor. Dr Robert Yamartino of Sigma Research Corporation developed the spectral software package for the IBM PC. We appreciate the cooperation of the following scientists who sent us copies of data from their experiments: Dr Brian Lamb, Mr James Bowers, Dr Brian Sawford, Ms Holly Peterson, and Dr James Deardorff.

References

Bendat, J. S. and Piersol, A. G.: 1980, *Engineering Applications of Correlation and Spectral Analysis*, John Wiley, New York.

Bowers, J. F. and Black, R. B.: 1985, 'Test Report – Product Improved M3A3 (M3A3E2) Smoke Generator (Mobile Applications)', U.S. Army Dugway Proving Ground, Dugway, UT 84022-5000.

Deardorff, J. W. and Willis, G. E.: 1984, 'Ground Level Concentration Fluctuations from a Buoyant and a Non-Buoyant Source within a Laboratory Connectively Mixed Layer', *Atmos. Environ.* **18**, 1297–1309.

Deardorff, J. W. and Willis, G. E.: 1989, 'Concentration Fluctuation within a Laboratory Convectively Mixed Layer', to appear in *J. Clim. and Appl. Meteorol.*

Fackrell, J. E. and Robins, A. G.: 1982, 'Concentration Fluctuations and Fluxes in Plumes from Point Sources in a Turbulent Boundary Layer', *J. Fluid Mech.* **117**, 1–26.

Garvey, D. M., Walters, D. L., Fernandez, G., and Pinnick, R. G.: 1982, 'Experimental Verification of the Kolmogorov Power Law from Aerosol Concentration Fluctuations', *Proceedings of the Smoke*

Obscurants Symposium VI (Editors R. E. Elkins and R. H. Kohl), PM Smoke/Obscurants, Attn; DRCPM-SMK-T, Aberdeen Proving Ground, MD, 21005, 397–405.

Gifford, F. A., 1968: 'An Outline of Theories of Diffusion in the Lower Layers of the Atmosphere', *Meteorology and Atomic Energy* – 1968, D. H. Slade (ed.), USAEC Report T10-24190, USAEC, NTIS, 66–116.

Hadjitofi, A. and Wilson, M. J. G.: 1979, 'Fast-Response Measurements of Air Pollution', *Atmos. Environ.* **13**, 755–760.

Hanna, S. R.: 1984, 'The Exponential Probability Density Function and Concentration Fluctuations in Smoke Plumes', *Boundary Layer Meteorol.* **29**, 361–375.

Hanna, S. R.: 1986, 'Spectra of Concentration Fluctuations: The Two Time Scales of a Meandering Plume', *Atmos. Environ.* **20**, 1131–1137.

Jones, C. D.: 1983, 'On the Structure of Instantaneous Plumes in the Atmosphere', *J. Haz. Mat.* **7**, 87–112.

Kaimal, J. C., Eversole, R. A., Lenschow, D. H., Stankov, B. B., Kahn, P. H., and Businger, J. A.: 1982, 'Spectral Characteristics of the Convective Boundary Layer over Uneven Terrain', *J. Atmos. Sci.* **39**, 1098–1114.

Lamb, B., Peterson, H., Campbell, M., and Stock, D.: 1985, 'Concentration Fluctuations in Dispersing Tracer Plumes', *Seventh Symp. on Turb. and Diff.*, AMS, 331–334.

Lewellen, W. S. and Sykes, R. I.: 1986, 'Analysis of Concentration Fluctuations from LIDAR Observations of Atmospheric Plumes', *J. of Clim. and Appl. Meteorol.* **25**, 1145–1154.

Lumley, J. L. and Panofsky, H. A.: 1968, *The Structure of Atmospheric Turbulence*, Interscience, New York, 239 pp.

Panofsky, H. A. and Brier, G. W.: 1968, *Some Applications of Statistics to Meteorology*, Penn. State Univ. Press., 224 pp.

Panofsky, H. A. and Dutton, J. A.: 1984, *Atmospheric Turbulence*, John Wiley, New York, pp. 169–209.

Peterson, H. G., Lamb, B. K., and Stock, D. E.: 1988, 'Plume Concentration and Velocity Fluctuations during Convective and Stable Conditions', *Eighth Symp. on Turb. and Diff.*, AMS, 341–344.

Sawford, B. L.: 1984, 'Concentraton Statistics for Surface Plumes in the Atmospheric Surface Layer', *Seventh Symp. on Turb. and Diff.*, AMS, 323–326.

Sawford, B. L.: 1987, 'Conditional Concentration Statistics for Surface Plumes in the Atmospheric Boundary Layer', *Boundary Layer Meteorol.* **38**, 209–223.

Sawford, B. L., Frost, C. C., and Allan, T. C.: 1985, 'Atmospheric Boundary-Layer Measurements of Concentration Statistics from Isolated and Multiple Sources', *Boundary Layer Meteorol.* **31**, 249–268.

Sawford, B. L. and Stapountzis, H.: 1987, 'Concentration Fluctuations According to Fluctuating Plume Models in One and Two Dimensions', *Boundary-Layer Meteorol.* **37**, 89–100.

Sykes, R. I.: 1984, 'The Variance in Time-Averaged Samples from an Intermittent Plume', *Atmos. Environ.* **18**, 121–123.

Wilson, D. J. and Simms, B. W.: 1985, 'Exposure Time Effects on Concentration Fluctuations in Plumes', Report No. 47, Dept. of Mech. Eng., University of Alberta, Edmonton, Alberta, Canada.

THE SPECTRAL VELOCITY TENSOR FOR
HOMOGENEOUS BOUNDARY-LAYER TURBULENCE

L. KRISTENSEN[1] and D. H. LENSCHOW

National Center for Atmospheric Research[2], Boulder, CO, U.S.A.

P. KIRKEGAARD and M. COURTNEY

Risø National Laboratory, Roskilde, Denmark

(Received 16 September, 1988)

Abstract. We have postulated a simple model for the spectral tensor $\Phi_{ij}(\mathbf{k})$ of an anisotropic, but homogeneous turbulent velocity field. It is a simple generalization of the spectral tensor $\Phi_{ij}^{iso}(\mathbf{k})$ for isotropic turbulence and we show how in the limit of isotropy, $\Phi_{ij}(\mathbf{k})$ becomes equal to $\Phi_{ij}^{iso}(\mathbf{k})$. Whereas $\Phi_{ij}^{iso}(\mathbf{k})$ is determined entirely by one scalar function of $k = |\mathbf{k}|$, namely the energy spectrum, we need three independent scalar functions of k to specify $\Phi_{ij}(\mathbf{k})$. We show how it is possible by means of the three stream-wise velocity component spectra to determine the three scalar functions in $\Phi_{ij}(\mathbf{k})$ by solving two uncoupled, ordinary linear differential equations of first and second order. The analytic form of the component spectra each has a set of three parameters: the variance and the integral length scale of the velocity component and a dimensionless parameter, which governs the curvature of the spectrum in the transition domain from the inertial subrange towards lower wave numbers. When the three sets of parameters are the same, the three spectra correspond to isotropic turbulence and they are all interrelated and related to the energy spectrum. We show how it is possible to obtain these spectral forms in the neutral surface layer and in the convective boundary layer from data reported in the literature. The spectral tensor is used to predict the lateral coherences for all three velocity components and these predictions are compared with coherences obtained in two experiments, one using three masts at a horizontally homogeneous site in Denmark and one employing two aircraft flying in formation over eastern Colorado. Comparison shows reasonable agreement although with considerable experimental scatter.

1. Introduction

Many observations of one-dimensional velocity component spectra in the atmospheric boundary layer have been reported in the literature (e.g., Panofsky and Dutton, 1984). As a result, we have a general understanding of the shapes of the component spectra, particularly for the surface layer and the convective boundary layer (CBL). On the other hand, except for isotropic turbulence, the one-dimensional spectra give us little insight into the spatial structure of turbulence. Furthermore, there are only a few observations of the spatial structure of turbulence (e.g., Panofsky and Dutton, 1984). Yet, there are important practical problems that require some understanding of spatial variability. An example is computing the effects of wind loads on structures, where the spectral

[1] Permanent affiliation: Risø National Laboratory, Roskilde.
[2] The National Center for Atmospheric Research is sponsored by the National Science Foundation.

Boundary-Layer Meteorology **47**: 149–193, 1989.
© 1989 *Kluwer Academic Publishers.*

coherences are used to estimate the spatial variability of the wind force on the structures.

Kristensen and Jensen (1979) found general expressions for the spectral coherences with the assumption that the spatial displacement was small compared to the scale of the turbulence so that they could assume local isotropy. In this case, there are no scale parameters to consider. When displacements are not negligible compared to the scale of the turbulence, Kristensen and Jensen (1979) showed qualitatively, using the von Kármán spectrum and assuming isotropy, that the coherences depend upon the ratio between the scale of the turbulence and the displacement in such a way that the coherence does not approach unity for $kD \to 0$, where k is the wave number and D is the displacement perpendicular to the mean wind direction. This result seems to contradict many measurements as well as previous suggestions in the literature, which have been reviewed by Panofsky and Dutton (1984).

Because of this limitation in understanding the spatial variability of turbulence, in this paper we construct a kinematic model of turbulence which is not limited to isotropic turbulence in order to develop a quantitative tool for calculating lateral spectral coherence in the case where the displacement is not small compared to the scale of the turbulence. This model requires that we develop more generally a model for the spectral tensor $\Phi_{ij}(\mathbf{k})$ of the turbulent velocity based on knowledge of the behavior of the one-dimensional spectra. From a practical point of view, it is advantageous not to have to measure spatial turbulence statistics at many points in order to obtain knowledge of the spatial structure of the turbulence. If the model we present here is sufficiently accurate, measurement of the three one-dimensional spectra at one point is sufficient.

We first develop a simplified spectral model of turbulence which is not isotropic, but contains the isotropic case as a limit. Next we show that, from knowledge of the one-dimensional spectra at one point, it is possible to specify $\Phi_{ij}(\mathbf{k})$ completely under the assumption that the turbulence is homogeneous in all directions. We suggest a simple analytical form of the three-component spectra which has the property of isotropy as a limiting case and we show that there is an analytical solution for $\Phi_{ij}(\mathbf{k})$. We then evaluate the parameters in the analytical component spectra using empirical relations obtained from observational data. We first consider the neutral surface layer, then the CBL. Finally, we apply the model to real data obtained from spatially separated cup anemometers in the surface layer (Courtney, 1987, 1988) and from two aircraft flying in formation in the CBL (Lenschow and Kristensen, 1988).

2. The Tensor

We consider an incompressible, turbulent velocity field which is homogeneous in all directions. For this field, the ensemble mean $\mathbf{U} = \langle \mathbf{u}(\mathbf{x}) \rangle$ is constant in space and the autocovariance tensor

$$R_{ij}(\mathbf{r}) = \langle (u_i(\mathbf{x}) - U_i)(u_j(\mathbf{x}+\mathbf{r}) - U_j) \rangle, \quad i, j = 1, 2, 3, \tag{1}$$

will be a function of the displacement vector $\mathbf{r}$ only. We assume that $R_{ij}(\mathbf{r})$ is symmetric, i.e.,

$$R_{ij}(-\mathbf{r}) = R_{ji}(\mathbf{r}) = R_{ij}(\mathbf{r}), \tag{2}$$

which is the same as stating that the tensor is invariant to reflections (and translations). We do not assume that the velocity field is fully isotropic in the sense that there is no preferred direction in space. We allow for different length scales in each of the two horizontal and vertical directions. However, since we assume homogeneity in all directions, we do not consider the possibility of wind shear. By convention, we let the 1-direction, characterized by the unit vector $\mathbf{i}_1$, be along the mean wind direction and the 3-direction, unit vector $\mathbf{i}_3$, vertical. The 2-direction is defined by the unit vector

$$\mathbf{i}_2 = \mathbf{i}_3 \times \mathbf{i}_1. \tag{3}$$

Under the conditions stated above, an appropriate form for the spectral tensor, defined by the identity

$$\Phi_{ij}(\mathbf{k}) \equiv \frac{1}{(2\pi)^3} \oint R_{ij}(\mathbf{r}) \, e^{-i\mathbf{k}\cdot\mathbf{r}} d^3 r, \tag{4}$$

is

$$\Phi_{ij}(\mathbf{k}) = \sum_{l=1}^{3} A_l(k) \left\{ \delta_{li} - \frac{k_l k_i}{k^2} \right\} \left\{ \delta_{lj} - \frac{k_l k_j}{k^2} \right\}. \tag{5}$$

Here $A_1(k)$, $A_2(k)$ and $A_3(k)$ are real functions of the magnitude k of the wave-number vector $\mathbf{k}$.

It is interesting to compare (5) with the model presented by Kristensen et al. (1983). In that case it was assumed that the spectral tensor was axisymmetric with respect to the vertical direction. In fact, the two models coincide if we let $A(\mathbf{k}) = A_1(k) = A_2(k)$ and $B(\mathbf{k}) = A_3(k) - A(\mathbf{k})$.

We see that the principal directions of the tensor (5) are along the coordinate axes in $\mathbf{k}$-space. This is consistent with the homogeneity property of the velocity field. Further, the tensor is symmetric, $\Phi_{ij}(\mathbf{k}) = \Phi_{ji}(\mathbf{k})$, and a fortiori Hermitian, i.e.,

$$\Phi_{ij}^*(\mathbf{k}) = \Phi_{ji}(\mathbf{k}), \tag{6}$$

in accordance with the symmetric autocovariance function. Finally, the incompressibility property of the atmosphere can be formulated as (e.g., see Lumley and Panofsky, 1964)

$$\sum_{j=1}^{3} \Phi_{ij}(\mathbf{k}) k_j = 0. \tag{7}$$

In other words, $\mathbf{k}$ is itself an eigenvector of $\Phi_{ij}(\mathbf{k})$ corresponding to the eigenvalue 0, and $\Phi_{ij}(\mathbf{k})$ cannot possibly be positive definite, since this would imply that the left-hand side of (7) is always positive. However, we must insist that $\Phi_{ij}(\mathbf{k})$ be positive semidefinite, which means that

$$P = \sum_{i=1}^{3} \sum_{j=1}^{3} \Phi_{ij}(\mathbf{k}) a_i a_j \geq 0, \tag{8}$$

for any vector a_i in $\mathbf{k}$-space. Kristensen *et al.* (1983, p. 70) explained why (8) must hold; they also pointed out that, as a consequence, $\Phi_{ii}(\mathbf{k}) \geq 0$, $i = 1, 2, 3$, and that Schwarz's inequality

$$|\Phi_{ij}(\mathbf{k})|^2 \leq \Phi_{ii}(\mathbf{k}) \Phi_{jj}(\mathbf{k}) \tag{9}$$

holds. This means that all the eigenvalues of $\Phi_{ij}(\mathbf{k})$ are non-negative. It is thus a genuine covariance tensor, so that if we, for instance, use it to compute coherences, which will be discussed later, they will be bounded by one.

Notice that the structure of the tensor $\Phi_{ij}(\mathbf{k})$ depends on $\mathbf{k}$. Ideally we would like (8) to hold for any vector $\mathbf{k}$. In that case, we should say that $\Phi_{ij}(\mathbf{k})$ is uniformly positive semidefinite throughout $\mathbf{k}$-space. We can show that $\Phi_{ij}(\mathbf{k})$ is positive semidefinite over the sphere $|\mathbf{k}| = k$ if and only if

$$A_i(k) \geq 0, \quad i = 1, 2, 3. \tag{10}$$

Substituting (5) into (8), we get

$$P = \sum_{l=1}^{3} A_l(k) \left\{ a_l - \frac{k_l \mathbf{a} \cdot \mathbf{k}}{k^2} \right\}^2. \tag{11}$$

Obviously, if (10) is fulfilled, then $P \geq 0$. Conversely, if for example $A_1(k) < 0$, and we choose the vectors $\mathbf{k}$ and $\mathbf{a}$ such that $k_i = (1 - \delta_{1i}) k / \sqrt{2}$ and $a_i = \delta_{1i}$, then P would be negative.

The energy spectrum $E(k)$ is defined as

$$E(k) \equiv \frac{k^2}{2} \int_0^{\pi} \sin \vartheta \, d\vartheta \int_0^{2\pi} d\phi \sum_{i=1}^{3} \Phi_{ii}(\mathbf{k}), \tag{12}$$

where we have integrated the trace of the spectral tensor in wave-number space over a sphere of radius k. The double integration can be performed quite easily and we get

$$E(k) = \frac{4\pi}{3} k^2 \{A_1(k) + A_2(k) + A_3(k)\}. \tag{13}$$

If the three A-functions were identical, we would have isotropy. Then

$$A_1(k) = A_2(k) = A_3(k) \equiv A(k) = \frac{E(k)}{4\pi k^2}. \tag{14}$$

Substituting into (5) we find that the summation over l can be carried out and we obtain the well-known expression for the isotropic spectral tensor $\Phi_{ij}^{iso}(\mathbf{k})$:

$$\Phi_{ij}(\mathbf{k}) = \frac{E(k)}{4\pi k^2}\left\{\delta_{ij} - \frac{k_i k_j}{k^2}\right\} = \Phi_{ij}^{iso}(\mathbf{k}) .\tag{15}$$

In other words, isotropy is included in the model as a special case.

We want to relate $A_1(k)$, $A_2(k)$ and $A_3(k)$ to the three one-dimensional spectra $F_L(k)$ (longitudinal velocity component), $F_T(k)$ (transversal velocity component) and $F_V(k)$ (vertical velocity component) determined at one point. The one-dimensional spectra $F_{ii}^1(k_1) \equiv F_{ii}(k_1 \mathbf{i}_1)$ with the wave-number vector in the longitudinal or 1-direction can be derived from $\Phi_{ii}(\mathbf{k})$ as (Lumley and Panofsky, 1964)

$$F_{ii}^1(k_1) = \int_{-\infty}^{\infty} dk_2 \int_{-\infty}^{\infty} dk_3 \Phi_{ii}(\mathbf{k}) .\tag{16}$$

Using the plane polar coordinates κ and Θ defined by

$$\begin{Bmatrix} k_2 \\ k_3 \end{Bmatrix} = \begin{Bmatrix} \kappa\cos\Theta \\ \kappa\sin\Theta \end{Bmatrix},\tag{17}$$

we have for the longitudinal component

$$\Phi_{11}(\mathbf{k}) = A_1(k)\left\{1 - \frac{k_1^2}{k^2}\right\}^2 + A_2(k)\frac{k_1^2 k_2^2}{k^4} + A_3(k)\frac{k_3^2 k_1^2}{k^4}$$

$$= A_1(k)\frac{\kappa^4}{(k_1^2 + \kappa^2)^2}$$

$$+ A_2(k)\frac{k_1^2 \kappa^2}{(k_1^2 + \kappa^2)^2}\cos^2\Theta + A_3(k)\frac{k_1^2 \kappa^2}{(k_1^2 + \kappa^2)^2}\sin^2\Theta ,\tag{18}$$

so that, after the integration over Θ has been carried out,

$$F_L(k) \equiv F_{11}^1(k) =$$

$$2\pi \int_k^{\infty} (K^2 - k^2)\{(K^2 - k^2)A_1(K) + \tfrac{1}{2}k^2(A_2(K) + A_3(K))\}\frac{dK}{K^3} .\tag{19}$$

Similarly, we obtain for the transverse and vertical one-dimensional spectra

$$F_T(k) \equiv F_{22}^1(k) = \pi \int_k^{\infty} \{k^2(K^2 - k^2)A_1(K) +$$

$$\tfrac{1}{4}(3K^4 + 2K^2 k^2 + 3k^4)A_2(K) + \tfrac{1}{4}(K^2 - k^2)^2 A_3(K)\}\frac{dK}{K^3}\tag{20}$$

and

$$F_V(k) \equiv F_{33}^1(k) = \pi \int_k^\infty \{k^2(K^2 - k^2)A_1(K) +$$

$$\tfrac{1}{4}(K^2 - k^2)^2 A_2(K) + \tfrac{1}{4}(3K^4 + 2K^2k^2 + 3k^4)A_3(K)\} \frac{\mathrm{d}K}{K^3}. \tag{21}$$

If the turbulence is isotropic, then there are two relations between the three one-dimensional spectra and the first derivative $F_L'(k)$ of the longitudinal spectrum: $F_T(k) = F_V(k)$ and $2F_T(k) = F_L(k) - kF_L'(k)$ (see, e.g., Panofsky and Dutton, 1984). In our case, this is of course not true, but it is natural to let the relations inspire us to define the residuals

$$H(k) = F_T(k) - F_V(k), \tag{22}$$

and

$$J(k) = F_L(k) - kF_L'(k) - (F_T(k) + F_V(k)), \tag{23}$$

which in principle can be evaluated from measured data. From (19), (20) and (21), we obtain the equations

$$H(k) = \frac{\pi}{2} \int_k^\infty (A_2(K) - A_3(K))(K^2 + k^2)^2 \frac{\mathrm{d}K}{K^3}, \tag{24}$$

and

$$J(k) = 2\pi \int_k^\infty \left(A_1(K) - \frac{A_2(K) + A_3(K)}{2} \right)(K^2 - k^2)(K^2 + 2k^2) \frac{\mathrm{d}K}{K^3}. \tag{25}$$

If we know the one-dimensional spectra, it should thus be possible to determine two relations between $A_1(k)$, $A_2(k)$ and $A_3(k)$ by solving the two Equations (24) and (25). We proceed by introducing new variables:

$$\begin{Bmatrix} s \\ t \end{Bmatrix} = \begin{Bmatrix} k^{-2} \\ K^{-2} \end{Bmatrix}, \tag{26}$$

$$\begin{Bmatrix} f(s) \\ g(s) \end{Bmatrix} = \begin{Bmatrix} H(s^{-1/2})s^2 \\ J(s^{-1/2})s^2 \end{Bmatrix} \tag{27}$$

and

$$\begin{Bmatrix} \alpha(t) \\ \beta(t) \end{Bmatrix} = \pi \begin{Bmatrix} A_2(t^{-1/2}) - A_3(t^{-1/2}) \\ A_1(t^{-1/2}) - \{A_2(t^{-1/2}) + A_3(t^{-1/2})\}/2 \end{Bmatrix}. \tag{28}$$

Equations (24) and (25) can now be reformulated into

$$\frac{1}{4} \int_0^s (s+t)^2 \alpha(t) \frac{dt}{t^2} = f(s) \tag{29}$$

and

$$\int_0^s (s-t)(s+2t)\beta(t) \frac{dt}{t^2} = g(s) . \tag{30}$$

By differentiation three times, we obtain the following ordinary differential equations

$$\alpha''(s) + \frac{1}{s} \alpha'(s) - \frac{1}{2s^2} \alpha(s) = f'''(s) , \tag{31}$$

$$\frac{3}{s} \beta'(s) - \frac{1}{s^2} \beta(s) = g'''(s) . \tag{32}$$

To solve (31), we use the standard technique, which is to first solve the homogeneous equation

$$\alpha''(s) + \frac{1}{s} \alpha'(s) - \frac{1}{2s^2} \alpha(s) = 0 . \tag{33}$$

This Euler equation has the independent solutions

$$\begin{Bmatrix} \alpha_+(s) \\ \alpha_-(s) \end{Bmatrix} = \begin{Bmatrix} s^{+1/\sqrt{2}} \\ s^{-1/\sqrt{2}} \end{Bmatrix} . \tag{34}$$

The general solution to (31) is then

$$\alpha(s) = - \alpha_+(s) \int \frac{\alpha_-(s) f'''(s)}{W(s)} ds$$

$$+ \alpha_-(s) \int \frac{\alpha_+(s) f'''(s)}{W(s)} ds , \tag{35}$$

where

$$W(s) = \begin{vmatrix} \alpha_+(s) & \alpha_-(s) \\ \alpha'_+(s) & \alpha'_-(s) \end{vmatrix} = -\frac{\sqrt{2}}{s} , \tag{36}$$

is the Wronskian. Thus we get

$$\alpha(s) = \frac{s^{+1/\sqrt{2}}}{\sqrt{2}} \int s^{1-1/\sqrt{2}} f'''(s) ds$$

$$- \frac{s^{-1/\sqrt{2}}}{\sqrt{2}} \int s^{1+1/\sqrt{2}} f'''(s) ds. \tag{37}$$

The general solution to (32) is

$$\beta(s) = \frac{s^{1/3}}{3} \int s^{2/3} g'''(s) \, ds .$$ (38)

In order to find the definite solutions, we specify that the turbulent energy

$$\mathscr{E} = \int_0^\infty E(k) \, dk$$ (39)

is finite. To demonstrate this, we assume that the functions $A_i(k)$ are power laws in the limit $k \to \infty$, i.e., $s \to 0$. Let

$$\max_{k \to \infty}(A_1(k), A_2(k), A_3(k)) \sim k^{-2\lambda} .$$ (40)

In view of (13), we conclude that if $\mathscr{E}$ is finite, then

$$\lambda > \tfrac{3}{2} .$$ (41)

Using (24) through (30), we see that

$$\left\{ \begin{matrix} H(k) \\ J(k) \end{matrix} \right\} \sim k^{-2\lambda+2} ,$$ (42)

$$\left\{ \begin{matrix} f(s) \\ g(s) \end{matrix} \right\} \sim s^{\lambda+1}$$ (43)

and

$$\left\{ \begin{matrix} \alpha(s) \\ \beta(s) \end{matrix} \right\} \sim s^{\lambda} .$$ (44)

Applying (41), (43) and (44) to the indefinite solutions (37) and (38), we obtain the definite solutions

$$\alpha(s) = \frac{s^{+1/\sqrt{2}}}{\sqrt{2}} \int_0^s t^{1-1/\sqrt{2}} f'''(t) \, dt$$

$$- \frac{s^{-1/\sqrt{2}}}{\sqrt{2}} \int_0^s t^{1+1/\sqrt{2}} f'''(t) \, dt$$ (45)

and

$$\beta(s) = \frac{s^{1/3}}{3} \int_0^s t^{2/3} g'''(t) \, dt .$$ (46)

With the aid of (26)–(28), we now have two equations for the determination of A_1, A_2 and A_3. With the help of (19), we obtain a third equation

$$A_2(k) + A_3(k) = \frac{k}{2\pi} \frac{d}{dk} \frac{1}{k} \frac{dF_L}{dk} - \frac{4k^2}{\pi} \int_0^{1/k^2} \beta(s)\, ds, \tag{47}$$

which, after substituting (46) and integrating by parts, may be written

$$3A_1(k) - \frac{A_2(k) + A_3(k)}{2} = \frac{k}{2\pi} \frac{d}{dk} \frac{1}{k} \frac{dF_L}{dk} + \frac{1}{\pi} s^{-1} \int_0^s t^2 g'''(t)\, dt. \tag{48}$$

We can now solve (48), (26) and (28), in combination with (45) and (46), for the functions $A_1(k)$, $A_2(k)$ and $A_3(k)$. The result is

$$A_1(k) = \frac{k}{4\pi} \frac{d}{dk} \frac{1}{k} \frac{dF_L}{dk}$$
$$+ \frac{1}{2\pi} \left\{ s^{-1} \int_0^s t^2 g'''(t)\, dt - \frac{s^{1/3}}{3} \int_0^s t^{2/3} g'''(t)\, dt \right\}, \tag{49}$$

$$A_2(k) = \frac{k}{4\pi} \frac{d}{dk} \frac{1}{k} \frac{dF_L}{dk}$$
$$+ \frac{1}{2\pi} \left\{ \frac{s^{1/\sqrt{2}}}{\sqrt{2}} \int_0^s t^{1-1/\sqrt{2}} f'''(t)\, dt - \frac{s^{-1/\sqrt{2}}}{\sqrt{2}} \int_0^s t^{1+1/\sqrt{2}} f'''(t)\, dt \right.$$
$$\left. + s^{-1} \int_0^s t^2 g'''(t)\, dt - s^{1/3} \int_0^s t^{2/3} g'''(t)\, dt \right\}, \tag{50}$$

$$A_3(k) = \frac{k}{4\pi} \frac{d}{dk} \frac{1}{k} \frac{dF_L}{dk}$$
$$- \frac{1}{2\pi} \left\{ \frac{s^{1/\sqrt{2}}}{\sqrt{2}} \int_0^s t^{1-1/\sqrt{2}} f'''(t)\, dt - \frac{s^{-1/\sqrt{2}}}{\sqrt{2}} \int_0^s t^{1+1/\sqrt{2}} f'''(t)\, dt \right.$$
$$\left. - s^{-1} \int_0^s t^2 g'''(t)\, dt + s^{1/3} \int_0^s t^{2/3} g'''(t)\, dt \right\}. \tag{51}$$

We note that in the limit for $k \to \infty$, the three functions $A_1(k)$, $A_2(k)$ and $A_3(k)$ become identical as they must when we approach the wave number domain of local isotropy.

Finally, we get by substituting in (13):

$$E(k) = k^3 \frac{d}{dk} \frac{1}{k} \frac{dF_L}{dk}$$

$$+ 2k^4 \int_0^{1/k^2} s^2 g'''(s) \, ds - \frac{14}{9} k^{4/3} \int_0^{1/k^2} s^{2/3} g'''(s) \, ds . \tag{52}$$

Thus, we have obtained a closed expression for the energy spectrum in terms of the second derivative of the longitudinal spectrum and a function that is related to the anisotropy of the turbulence.

3. The Spectra

The purpose of this investigation is to determine the spectral tensor from the measured one-dimensional spectra $F_L(k)$, $F_T(k)$ and $F_V(k)$. The problem is therefore to present the spectra in a form suitable to determine $H(k)$ and $J(k)$ and consistent with the scheme developed in Section 2. In particular, it is necessary for the spectra to be four times differentiable.

We have decided to fit the measured data to spectral forms that are consistent with isotropic turbulence as a special case. The form that we have selected for our model has the property that both the length scale and the rate at which the spectra change from the region of local isotropy to the energy-containing anisotropic region at low wave-number are specified. We start with $F_L(k)$:

$$F_L(k) = \frac{l_L \sigma_L^2}{\pi} \frac{1}{\left\{ 1 + \left(\dfrac{l_L k}{a(\mu_L)} \right)^{2\mu_L} \right\}^{5/6\mu_L}} . \tag{53}$$

Here σ_L^2 is the variance of the longitudinal velocity component,

$$l_L = \frac{1}{\sigma_L^2} \int_0^\infty R_L(r\mathbf{i}_1) \, dr \tag{54}$$

the longitudinal integral length scale, μ_L a dimensionless constant governing the transition from the local isotropy domain to the low wave-number domain and

$$a(\mu) = \pi \frac{\mu \Gamma\left(\dfrac{5}{6\mu} \right)}{\Gamma\left(\dfrac{1}{2\mu} \right) \Gamma\left(\dfrac{1}{3\mu} \right)} , \tag{55}$$

a dimensionless constant, which for $\mu = \mu_L$ is determined by the normalization condition

$$\sigma_L^2 = \int_{-\infty}^{\infty} F_L(k)\,dk .$$ (56)

If the turbulence were isotropic,

$$F_T(k) = F_V(k) = \tfrac{1}{2}\{F_L(k) - kF'_L(k)\} = \frac{l_L \sigma_L^2}{2\pi} \frac{1 + \frac{8}{3}\left(\frac{l_L k}{a(\mu_L)}\right)^{2\mu_L}}{\left\{1 + \left(\frac{l_L k}{a(\mu_L)}\right)^{2\mu_L}\right\}^{(5/6\mu_L)+1}} .$$

We postulate that $F_T(k)$ and $F_V(k)$ have the same functional dependence on k, except with the parameters (l_L, σ_L^2, μ_L) replaced by (l_T, σ_T^2, μ_T) and (l_V, σ_V^2, μ_V), respectively. In this way we secure that the isotropic spectra are contained as a special case. We thus have

$$F_T(k) = \frac{l_T \sigma_T^2}{2\pi} \frac{1 + \frac{8}{3}\left(\frac{l_T k}{a(\mu_T)}\right)^{2\mu_T}}{\left\{1 + \left(\frac{l_T k}{a(\mu_T)}\right)^{2\mu_T}\right\}^{(5/6\mu_T)+1}}$$ (57)

and

$$F_V(k) = \frac{l_V \sigma_V^2}{2\pi} \frac{1 + \frac{8}{3}\left(\frac{l_V k}{a(\mu_V)}\right)^{2\mu_V}}{\left\{1 + \left(\frac{l_V k}{a(\mu_V)}\right)^{2\mu_V}\right\}^{(5/6\mu_V)+1}} .$$ (58)

The length scales l_L, l_T and l_V specify the wavelengths at which the transitions from local isotropy to the energy-containing regions of the spectra at longer wavelengths take place, while the μ-parameters govern the curvature of the spectra across the transition. Larger μ means a more abrupt transition.

For local isotropy ($k \gg \max(1/l_L, 1/l_T, 1/l_V)$), we get

$$\frac{3}{4} F_T(k) = \frac{3}{4} F_V(k) = F_L(k) = \frac{9\alpha}{55} \epsilon^{2/3} k^{-5/3} ,$$ (59)

so that

$$\frac{\sigma_L^3}{l_L} a^{5/2}(\mu_L) = \frac{\sigma_T^3}{l_T} a^{5/2}(\mu_T) = \frac{\sigma_V^3}{l_V} a^{5/2}(\mu_V) = \left(\pi \frac{9\alpha}{55}\right)^{3/2} \epsilon ,$$ (60)

where α is the Kolmogorov constant and ϵ the rate of dissipation of kinetic energy.

It has been experimental practice to determine the so-called peak wavelengths λ_L, λ_T and λ_V from the measured spectra. They are defined as the wavelengths for which $kF_L(k)$, $kF_T(k)$ and $kF_V(k)$ have their maxima. It is easily shown that

$$\lambda_L = \left\{\frac{2}{3}\right\}^{-1/2\mu_L} \frac{2\pi}{a(\mu_L)} l_L ,$$

$$\lambda_T = \left\{\frac{5}{3}\sqrt{\mu_T^2 + \frac{6}{5}\mu_T + 1} - \left(\frac{5}{3}\mu_T + 1\right)\right\}^{1/2\mu_T} \frac{2\pi}{a(\mu_T)} l_T , \tag{62}$$

and

$$\lambda_V = \left\{\frac{5}{3}\sqrt{\mu_V^2 + \frac{6}{5}\mu_V + 1} - \left(\frac{5}{3}\mu_V + 1\right)\right\}^{1/2\mu_V} \frac{2\pi}{a(\mu_V)} l_V . \tag{63}$$

Using (60)–(63), we get

$$h(\mu_L) = 2\pi \left(\frac{55}{9\alpha}\right)^{3/2} \frac{\sigma_L^3}{\lambda_L \epsilon} , \tag{64}$$

$$j(\mu_T) = 2\pi \left(\frac{55}{9\alpha}\right)^{3/2} \frac{\sigma_T^3}{\lambda_T \epsilon} , \tag{65}$$

and

$$j(\mu_V) = 2\pi \left(\frac{55}{9\alpha}\right)^{3/2} \frac{\sigma_V^3}{\lambda_V \epsilon} , \tag{66}$$

where

$$h(\mu) = \sqrt{\frac{\left[\dfrac{\pi}{a(\mu)}\right]^3}{\left\{\dfrac{2}{3}\right\}^{1/\mu}}} \tag{67}$$

and

$$j(\mu) = \sqrt{\frac{\left[\dfrac{\pi}{a(\mu)}\right]^3}{\left\{\frac{5}{3}\sqrt{\mu^2 + \frac{6}{5}\mu + 1} - (\frac{5}{3}\mu + 1)\right\}^{1/\mu}}} . \tag{68}$$

Figure 1 shows the two functions $h(\mu)$ and $j(\mu)$. We see that when μ increases, in particular beyond unity, the solutions to $h(\mu) = constant$ and $j(\mu) = constant$ become less accurate; in fact, it can be shown that both functions have a maximum so that there are cases with dual solutions.

In the following two subsections, we shall demonstrate how to determine the parameters μ_L, μ_T and μ_V from bulk parameters σ_L^2, σ_T^2, σ_V^2, λ_L, λ_T, λ_V and ϵ reported in the literature. Later we shall give examples of how to fit μ_L, μ_T and μ_V directly to measured spectra.

There is an interesting and useful by-product of this model; namely, the

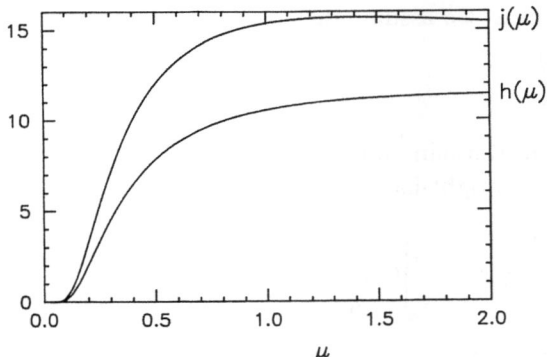

Fig. 1.   The functions $h(\mu)$ and $j(\mu)$.

formulations proposed here provide a convenient method for determining the integral length scales l_L, l_T and l_V of the turbulent velocity field. (More correctly, the integral length scales of the transverse and the vertical velocity components are $l_T/2$ and $l_V/2$, respectively.) This is done by substituting μ_L, μ_T and μ_V in (61)–(63). Lenschow and Stankov (1986) discuss methods for determining integral scales of first- and second-order moments in the CBL. The method we suggest corresponds to extrapolation to zero wave number with a more general velocity spectrum than that suggested by Lenschow and Stankov (1986).

3.1. APPLICATION TO THE NEUTRAL SURFACE LAYER

Kaimal *et al.* (1972) suggest a set of expressions for $F_L(k)$, $F_T(k)$ and $F_V(k)$ in the neutral atmospheric surface layer, from which we can derive λ_L, λ_T and λ_V in terms of the measurement altitude z:

$$\begin{Bmatrix} \lambda_L \\ \lambda_T \\ \lambda_V \end{Bmatrix} = \begin{Bmatrix} 22z \\ 6.33z \\ 2.13z \end{Bmatrix}. \tag{69}$$

The same spectra give the variances

$$\begin{Bmatrix} \sigma_L^2 \\ \sigma_T^2 \\ \sigma_V^2 \end{Bmatrix} = \begin{Bmatrix} 4.77u_*^2 \\ 2.68u_*^2 \\ 1.46u_*^2 \end{Bmatrix}, \tag{70}$$

where

$$u_*^2 = -\lim_{z \to 0}(\langle wu \rangle) \tag{71}$$

is the surface stress.

In the neutral case, we assume that

$$\epsilon = \frac{u_*^3}{\kappa z}, \tag{72}$$

where κ is the von Kármán constant.

Substituting in the right-hand sides of (64)–(66), we get

$$\begin{Bmatrix} h(\mu_L) \\ j(\mu_T) \\ j(\mu_V) \end{Bmatrix} = \kappa \alpha^{-3/2} \begin{Bmatrix} 44.95 \\ 65.79 \\ 78.62 \end{Bmatrix}. \tag{73}$$

From this, we see that the constant $\kappa \alpha^{-3/2}$ needs to be specified. According to Frenzen and Hart (1983), the so-called K- von K product $\alpha_1 \kappa^{4/3} \equiv \frac{18}{55} \alpha \kappa^{4/3}$ is equal to 0.160. Frenzen (1988, private communication) later re-analyzed his data and found that the K- von K product is probably somewhat larger that this, and can be as large as 0.170. We use $\alpha_1 \kappa^{4/3} = 0.165$ and $\kappa = 0.4$ (Zhang et al., 1988). Substituting into (73), we get

$$\begin{Bmatrix} h(\mu_L) \\ j(\mu_T) \\ j(\mu_V) \end{Bmatrix} = \begin{Bmatrix} 8.04 \\ 11.76 \\ 14.05 \end{Bmatrix}. \tag{74}$$

Solving (74) for μ_L, μ_T and μ_V, we get

$$\begin{Bmatrix} \mu_L \\ \mu_T \\ \mu_V \end{Bmatrix} = \begin{Bmatrix} 0.52 \\ 0.49 \\ 0.68 \end{Bmatrix}. \tag{75}$$

From (61)–(63) we have

$$\begin{Bmatrix} l_L \\ l_T \\ l_V \end{Bmatrix} = \begin{Bmatrix} 5.24 z \\ 2.52 z \\ 0.59 z \end{Bmatrix}, \tag{76}$$

which, together with (70) and (76), completely determine the one-dimensional velocity spectra (53), (57) and (58) of the neutral surface layer. These constitute an alternative representation to that of Kaimal et al. (1972) of generally accepted data.

Figure 2 shows the three spectra using the present formulation and that due to Kaimal et al. (1972). The difference between the two formulations is well within experimental uncertainty.

3.2. APPLICATION TO THE CONVECTIVE BOUNDARY LAYER

In the CBL it is also possible to obtain from the literature the bulk parameters entering the right-hand sides of (64)–(66). It is convenient to use as scaling

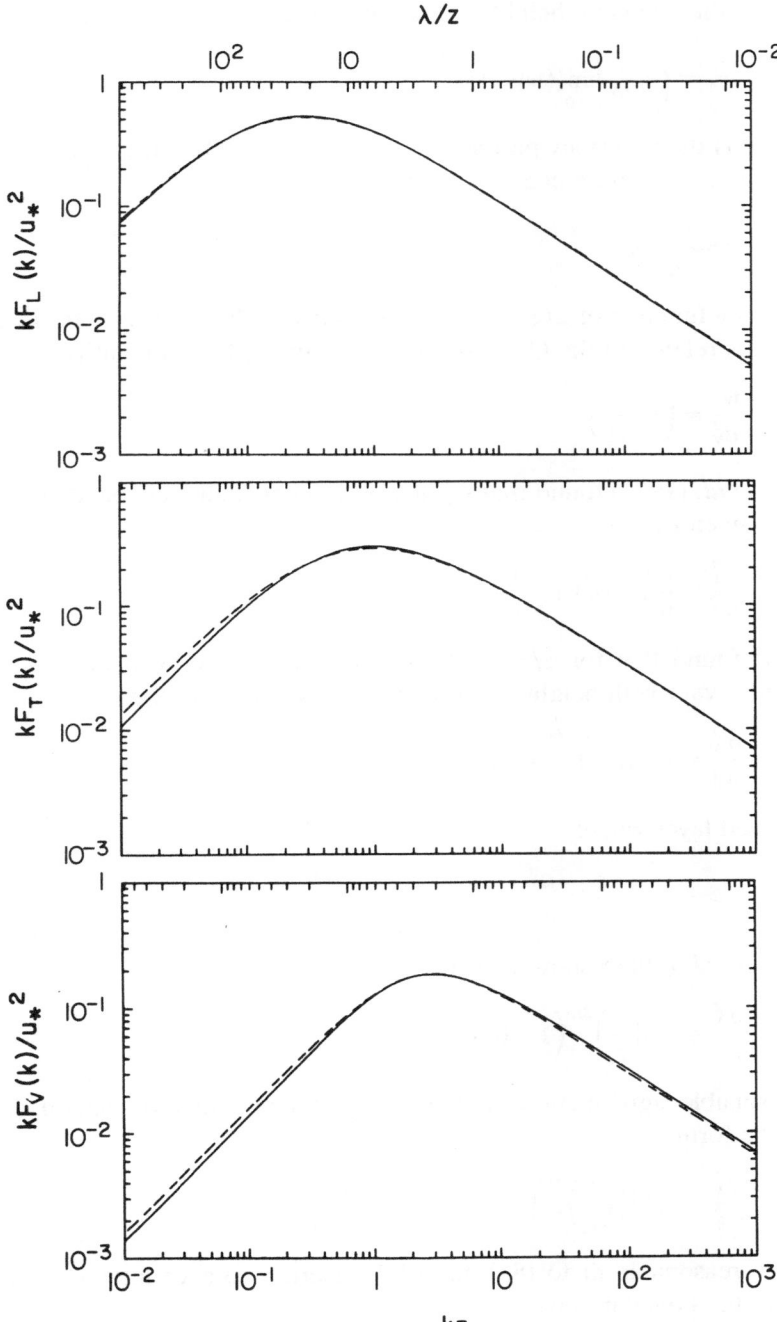

Fig. 2. Spectra, normalized by u_*^2, in the neutral surface layer for the formulation of Kaimal *et al.* (1972) (dashed lines) and for the formulation given by (53), (57) and (58). The wave number k and the wavelength $\lambda = 2\pi/k$ are normalized by means of the height z.

parameters the inversion height z_i and the convective velocity scale

$$w_* = \left\{ z_i \frac{g}{\Theta} \lim_{z \to 0} (\langle w\Theta' \rangle) \right\}^{1/3}, \tag{77}$$

where g/Θ is the buoyancy parameter and $\lim_{z \to 0}(\langle w\Theta' \rangle)$ is the surface heat flux. The rate of dissipation can now be written

$$\epsilon = \frac{w_*^3}{z_i} \psi_\epsilon, \tag{78}$$

where ψ_ϵ is a function of z/z_i and the ratio between the surface stress u_*^2 and w_*^2. This ratio is related to the Obukhov length L and z_i by the identity

$$\frac{w_*^2}{u_*^2} = \left(-\frac{z_i}{\kappa L} \right)^{2/3}. \tag{79}$$

Panofsky *et al.* (1977) found that σ_L^2 and σ_T^2 are equal and independent of height. They are given by

$$\frac{\sigma_L^2}{w_*^2} = \frac{\sigma_T^2}{w_*^2} = 0.3 + 4 \frac{u_*^2}{w_*^2}. \tag{80}$$

They also found that for $z/z_i < 0.1$, the variance of the vertical velocity component does vary with height and is approximated quite well by

$$\frac{\sigma_V^2}{w_*^2} = 1.44 \left(\frac{z}{z_i} \right)^{2/3} + 1.6 \frac{u_*^2}{w_*^2}. \tag{81}$$

In the mixed layer where

$$z \gg -L \text{ or } \frac{z}{z_i} \gg \frac{u_*^3}{w_*^3}, \tag{82}$$

Lenschow *et al.* (1980) showed that

$$\frac{\sigma_V^2}{w_*^2} = 1.8 \left(\frac{z}{z_i} \right)^{2/3} \left(1 - 0.8 \frac{z}{z_i} \right)^2 \tag{83}$$

is in reasonable agreement with both aircraft and tethered balloon data. We suggest the form

$$\frac{\sigma_V^2}{w_*^2} = 1.44 \left\{ \left(\frac{z}{z_i} \right)^{2/3} \left(1 - 0.7 \frac{z}{z_i} \right)^2 + \frac{10}{9} \frac{u_*^2}{w_*^2} \right\}, \tag{84}$$

which is a reasonable fit to (81) and (83), in order to account for the effect of stability in the boundary layer.

The peak wavelengths λ_L, λ_T and λ_V have been estimated experimentally by both Kaimal *et al.* (1976) and Caughey and Palmer (1979). They agree that for $0.01 \leq z/z_i \leq 1$

$$\frac{\lambda_L}{z_i} = \frac{\lambda_T}{z_i} = 1.5 \tag{85}$$

are somewhat uncertain, but reasonable estimates.

There is also considerable experimental scatter in λ_V. Caughey and Palmer (1979) found that for $0.1 \leqslant z/z_i \leqslant 1$,

$$\frac{\lambda_V}{z_i} = 1.8 \left\{ 1 - \exp\left(-4\frac{z}{z_i}\right) - 0.0003 \exp\left(8\frac{z}{z_i}\right) \right\} \tag{86}$$

is a good fit to their own Ashchurch data as well as the Minnesota data (Kaimal *et al.*, 1976).

According to (78), we must finally specify ψ_ϵ. Højstrup (1982) found that for $0.1 \leqslant z/z_i \leqslant 0.8$, the Minnesota data are consistent with

$$\psi_\epsilon = \left\{ 0.75 + 1.84\left(\frac{z}{z_i}\right)^{-2/3} \left(1 - \frac{z}{z_i}\right)^2 \frac{u_*^2}{w_*^2} \right\}^{3/2}. \tag{87}$$

We can now rewrite the three equations (64)–(66) for the determination of μ_L, μ_T and μ_V as follows:

$$h(\mu_L) = H\left(\frac{z}{z_i}, \frac{u_*}{w_*}\right), \tag{88}$$

$$j(\mu_T) = H\left(\frac{z}{z_i}, \frac{u_*}{w_*}\right) \tag{89}$$

and

$$j(\mu_V) = J\left(\frac{z}{z_i}, \frac{u_*}{w_*}\right), \tag{90}$$

where

$$H(s, t) = 28.3 \left\{ \frac{0.3 + 4t^2}{0.75 + 1.84 s^{-2/3}(1 - s)^2 t^2} \right\}^{3/2} \tag{91}$$

and

$$J(s, t) = \frac{40.8}{1 - e^{-4s} - 0.0003\, e^{8s}} \left\{ \frac{s^{2/3}(1 - 0.7s)^2 + \frac{10}{9}t^2}{0.75 + 1.84 s^{-2/3}(1 - s)^2 t^2} \right\}^{3/2}. \tag{92}$$

According to this derivation, we expect (91) and (92) to be valid for $0.1 \leqslant s \leqslant 0.8$ and $t \geqslant 0$.

Figure 3 shows the profiles of μ_L, μ_T and μ_V, and the integral length scales l_L, l_T and l_V. We note that a small amount of mechanical production changes the profiles significantly. The spectra at $z/z_i = 0.1$, 0.5 and 0.9, and $u_*/w_* = 0.00$ (extremely convective; no production of turbulence energy by shear) and 0.15 (typical daytime continental boundary layer) are shown in Figure 4. We see that

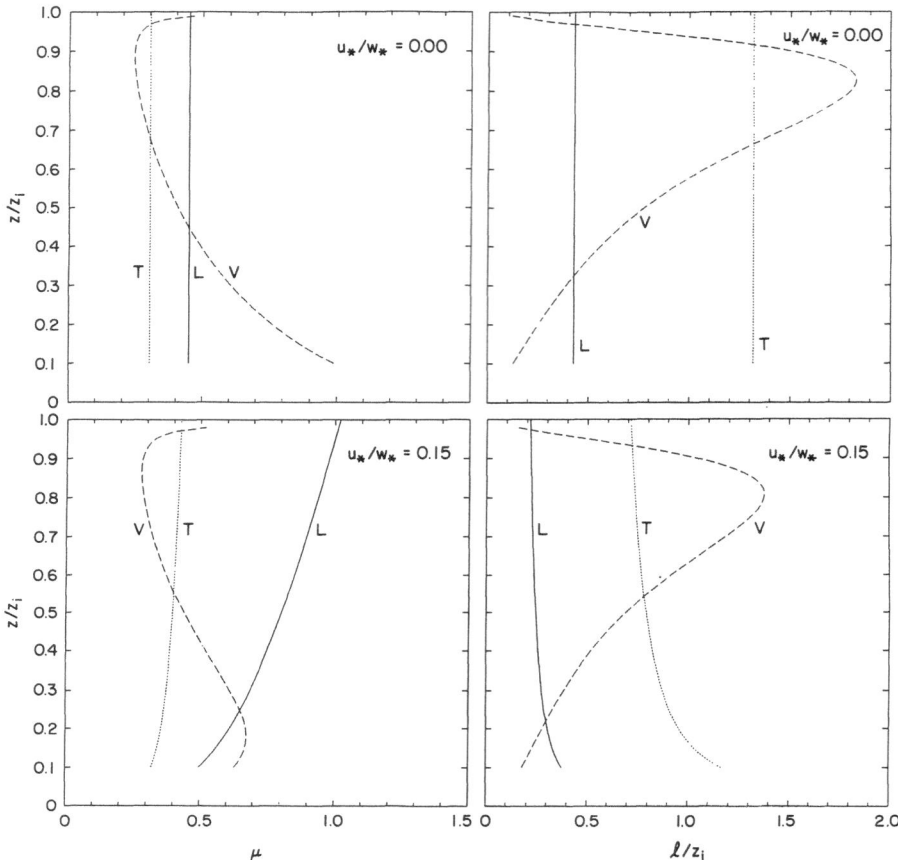

Fig. 3. Profiles of (μ_L, μ_T, μ_V) and (l_L, l_T, l_V) for the extremely convective boundary layer $u_*/w_* = 0.00$ $(-z_i/L = \infty)$ and for $u_*/w_* = 0.15$ $(-z_i/L = 120)$.

when there is mechanical production, all the length scales decrease; at the same time, the horizontal spectra become more pointed and the vertical spectrum flatter. This is in qualitative agreement with the observation that the length scales grow with decreasing stability (Lenschow *et al.*, 1988) and with the fact that for $u_*/w_* = 0$, buoyant turbulent energy production enters exclusively through the vertical velocity component and that mechanical turbulent energy production, when present ($u_*/w_* \neq 0$), enters through the longitudinal velocity component (e.g., Businger, 1982).

By inspecting the height variation of the vertical-velocity spectrum and comparing with Figure 3, we observe how the larger value of μ_V at the lowest level corresponds to a more pointed spectrum with a more abrupt transition. Furthermore, this transition moves to higher wave numbers as we approach the surface. This is consistent with the fact that buoyant turbulent energy production

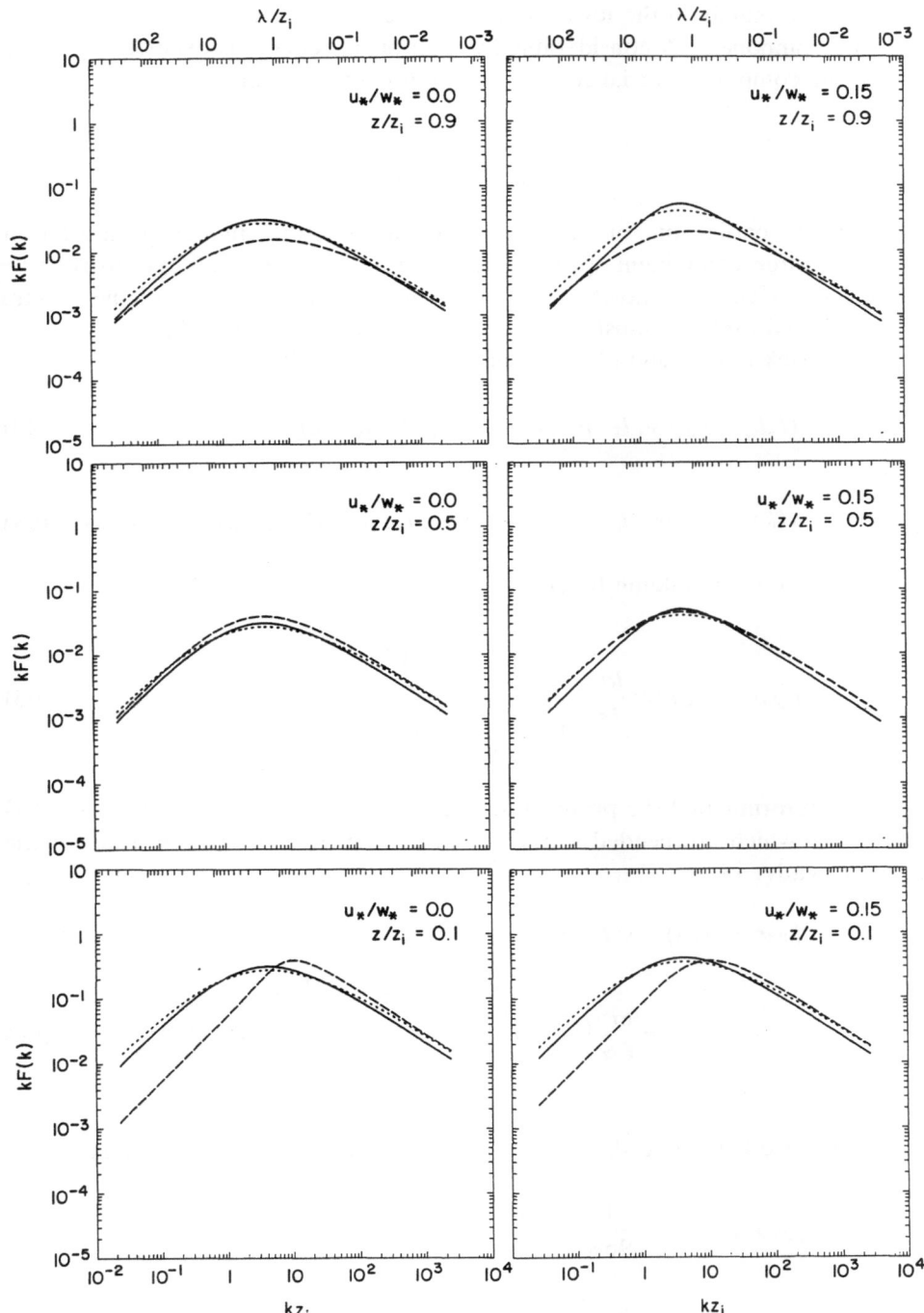

Fig. 4. Spectra in the convective boundary layer at the altitudes $z/z_i = 0.1$, 0.5 and 0.9 for $u_*/w_* = 0.00$ and 0.15.

is generated mainly in the lower part of the boundary layer through the vertical velocity component. A considerable fraction of this energy is exported upwards, where the component variances becomes approximately equal.

4. Solutions

As discussed earlier, one of our objectives is to obtain the model spectral tensor from the three component spectra, measured at a single point. We proceed by solving for $\alpha(s)$, $\beta(s)$ and $A_2 + A_3$, given by (45)–(47), in terms of the spectra (53), (57) and (58). We must therefore first evaluate $f'''(s)$ and $g'''(s)$.

Going back to (22) and (23), we see that we may write

$$H(k) = F_0(\sigma_T, l_T, \mu_T; k) - F_0(\sigma_V, l_V, \mu_V; k) \tag{93}$$

and

$$J(k) = 2F_0(\sigma_L, l_L, \mu_L; k) - F_0(\sigma_T, l_T, \mu_T; k) - F_0(\sigma_V, l_V, \mu_V; k), \tag{94}$$

using the common building block

$$F_0(\sigma, l, \mu; k) = \frac{l\sigma^2}{2\pi} \frac{1 + \dfrac{8}{3}\left(\dfrac{lk}{a(\mu)}\right)^{2\mu}}{\left\{1 + \left(\dfrac{lk}{a(\mu)}\right)^{2\mu}\right\}^{(5/6\mu)+1}}. \tag{95}$$

We have reformulated the problem according to (26)–(28) and we see that $f'''(s)$ and $g'''(s)$, which are needed in the solutions, can be expressed in terms of the third derivative of

$$f_0(\sigma, l, \mu; s) = s^2 F_0(\sigma, l, \mu; s^{-1/2})$$

$$= \frac{l\sigma^2}{6\pi}\left(\frac{a(\mu)}{l}\right)^{5/3} \frac{8 + 3\left(\dfrac{a^2(\mu)}{l^2}s\right)^{\mu}}{\left\{1 + \left(\dfrac{a^2(\mu)}{l^2}s\right)^{\mu}\right\}^{(5/6\mu)+1}} s^{17/6}. \tag{96}$$

To do this, we have used the results by Kristensen et al. (1983) to obtain

$$f_0'''(\sigma, l, \mu; s) = \frac{1}{96\pi}\frac{\sigma^2 a^2(\mu)}{l}\left(\frac{a^2(\mu)}{l^2}s\right)^{-1/6}$$

$$\times \sum_{n=1}^{4} \frac{C_n(\mu)}{\left\{1 + \left(\dfrac{a^2(\mu)}{l^2}s\right)^{\mu}\right\}^{(5/6\mu)+n}}, \tag{97}$$

where

$$
\left.\begin{aligned}
C_1(\mu) &= 40(1-\mu)(1-2\mu)(2-\mu) \\
C_2(\mu) &= \tfrac{140}{3}(1-\mu)(1-2\mu)(5+6\mu) \\
C_3(\mu) &= \tfrac{10}{9}(5+6\mu)(5+12\mu)(7-12\mu) \\
C_4(\mu) &= \tfrac{10}{27}(5+6\mu)(5+12\mu)(5+18\mu)
\end{aligned}\right\}. \tag{98}
$$

We now have

$$
f'''(s) = f_0'''(\sigma_T, l_T, \mu_T; s) - f_0'''(\sigma_V, l_V, \mu_V; s), \tag{99}
$$

$$
g'''(s) = 2f_0'''(\sigma_L, l_L, \mu_L; s) - f_0'''(\sigma_T, l_T, \mu_T; s) - f_0'''(\sigma_V, l_V, \mu_V; s). \tag{100}
$$

According to (45), (46), (99) and (100), we need to evaluate integrals of the form

$$
\gamma_0(q, \sigma, l, \mu; s) = qs^q \int_0^s t^{1-q} f_0'''(\sigma, l, \mu; t)\, dt. \tag{101}
$$

Substituting (97), (101) can be written

$$
\gamma_0(q, \sigma, l, \mu; s) = \frac{1}{96\pi} \frac{\sigma^2 l^3}{a^2(\mu)} q\eta^q \sum_{n=1}^4 C_n(\mu) \int_0^\eta \frac{t^{5/6-q}}{(1+t^\mu)^{(5/6\mu)+n}}\, dt, \tag{102}
$$

where

$$
\eta = \frac{a^2(\mu)}{l^2} s. \tag{103}
$$

The integral can be evaluated in terms of the incomplete beta function

$$
B_z(a, b) \equiv \int_0^z s^{a-1}(1-s)^{b-1}\, ds. \tag{104}
$$

We get

$$
\gamma_0(q, \sigma, l, \mu; s) = \frac{1}{96\pi} \frac{q}{\mu} \frac{\sigma^2 l^3}{a^2(\mu)}
$$

$$
\times \eta^q \sum_{n=1}^4 C_n(\mu) B_{\eta^\mu/(1+\eta^\mu)}\left(\frac{11}{6\mu} - \frac{q}{\mu}, n + \frac{q-1}{\mu}\right). \tag{105}
$$

We can now state the solutions in the following way:

$$
\alpha(s) = \gamma_0\left(+\frac{1}{\sqrt{2}}, \sigma_T, l_T, \mu_T; s\right) - \gamma_0\left(+\frac{1}{\sqrt{2}}, \sigma_V, l_V, \mu_V; s\right)
$$

$$
+ \gamma_0\left(-\frac{1}{\sqrt{2}}, \sigma_T, l_T, \mu_T; s\right) - \gamma_0\left(-\frac{1}{\sqrt{2}}, \sigma_V, l_V, \mu_V; s\right) \tag{106}
$$

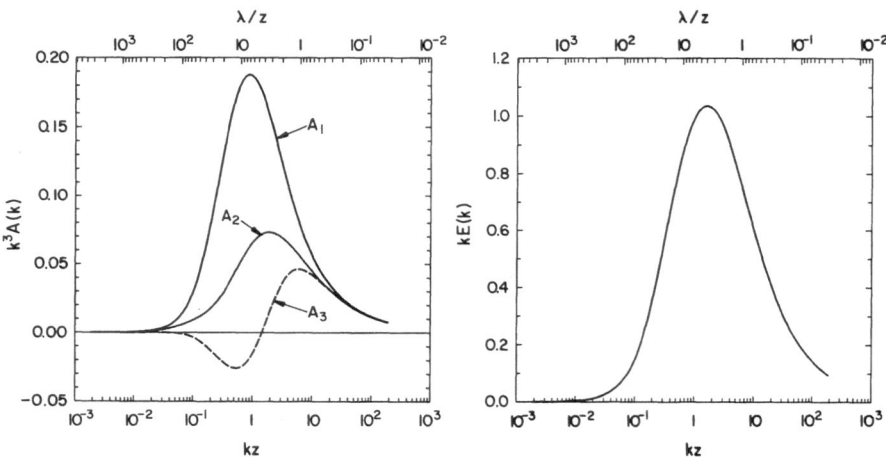

Fig. 5. $A_1(k)$, $A_2(k)$, $A_3(k)$ and the energy spectrum $E(k)$ in the neutral surface layer.

and

$$\beta(s) = 2\gamma_0(\tfrac{1}{3}, \sigma_L, l_L, \mu_L; s) - \gamma_0(\tfrac{1}{3}, \sigma_T, l_T, \mu_T; s) - \gamma_0(\tfrac{1}{3}, \sigma_V, l_V, \mu_V; s).$$

(107)

As we have seen, Equations (28) and (47) can now be used to determine $A_1(k)$, $A_2(k)$ and $A_3(k)$. In other words, by fitting expressions of the form (53), (57) and (58) to measured, one-point velocity spectra, (106) and (107) can be used to find an analytical expression for our assumed non-isotropic spectral tensor (5).

4.1. THE SPECTRAL TENSOR IN THE NEUTRAL SURFACE LAYER

The three functions $A_1(k)$, $A_2(k)$ and $A_3(k)$ have been determined for the neutral surface layer. They are displayed in Figure 5. As we see, $A_3(k)$ is not positive everywhere, which means that the tensor $\Phi_{ij}(\mathbf{k})$ is not positive semidefinite, as must be required. One consequence is that Schwarz's inequality (9) does not hold so that we can get coherences greater than one. This will be demonstrated later. One reason for this behaviour of the real surface layer may be lack of homogeneity in the vertical direction. The solution scheme we have given above does not automatically guarantee that $\Phi_{ij}(\mathbf{k})$ becomes positive semidefinite. The energy spectrum, given by (13), is also shown in Figure 5.

4.2. THE SPECTRAL TENSOR IN THE CONVECTIVE BOUNDARY LAYER

Figures 6a and 6b show $A_1(k)$, $A_2(k)$ and $A_3(k)$ at different levels in the CBL for completely convective conditions $u_*/w_* = 0$ and for a typical daytime boundary layer $u_*/w_* = 0.15$. We see that in some of these cases, namely for $z/z_i = 0.1$ and therefore closest to the surface layer, $A_3(k)$ is negative for the smallest values of k. However, it is not nearly as pronounced as is the case in the surface layer and

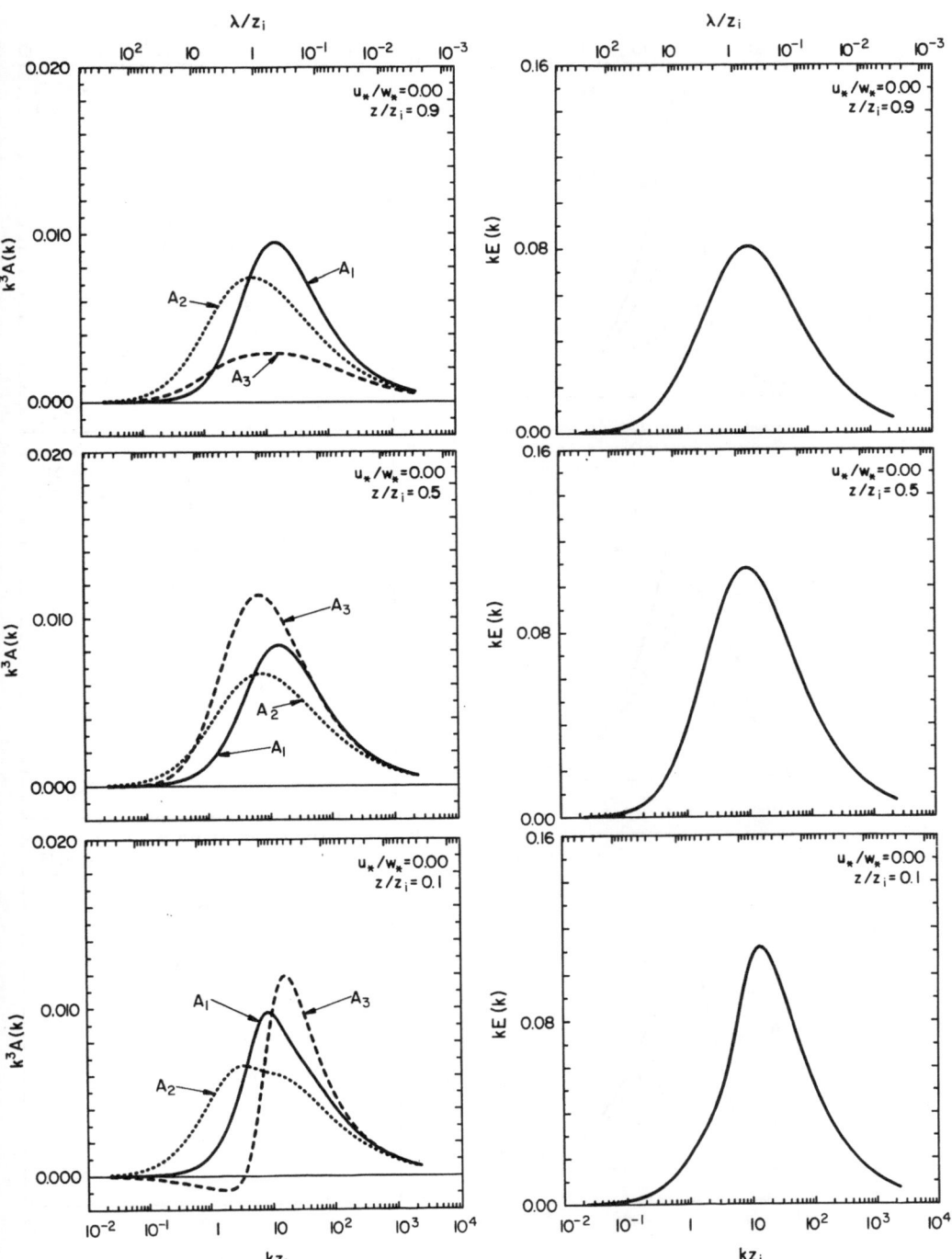

Fig. 6a. $A_1(k)$, $A_2(k)$, $A_3(k)$ and $E(k)$ in the extremely convective boundary layer ($u_*/w_* = 0$) at the altitudes $z/z_i = 0.1$, 0.5 and 0.9.

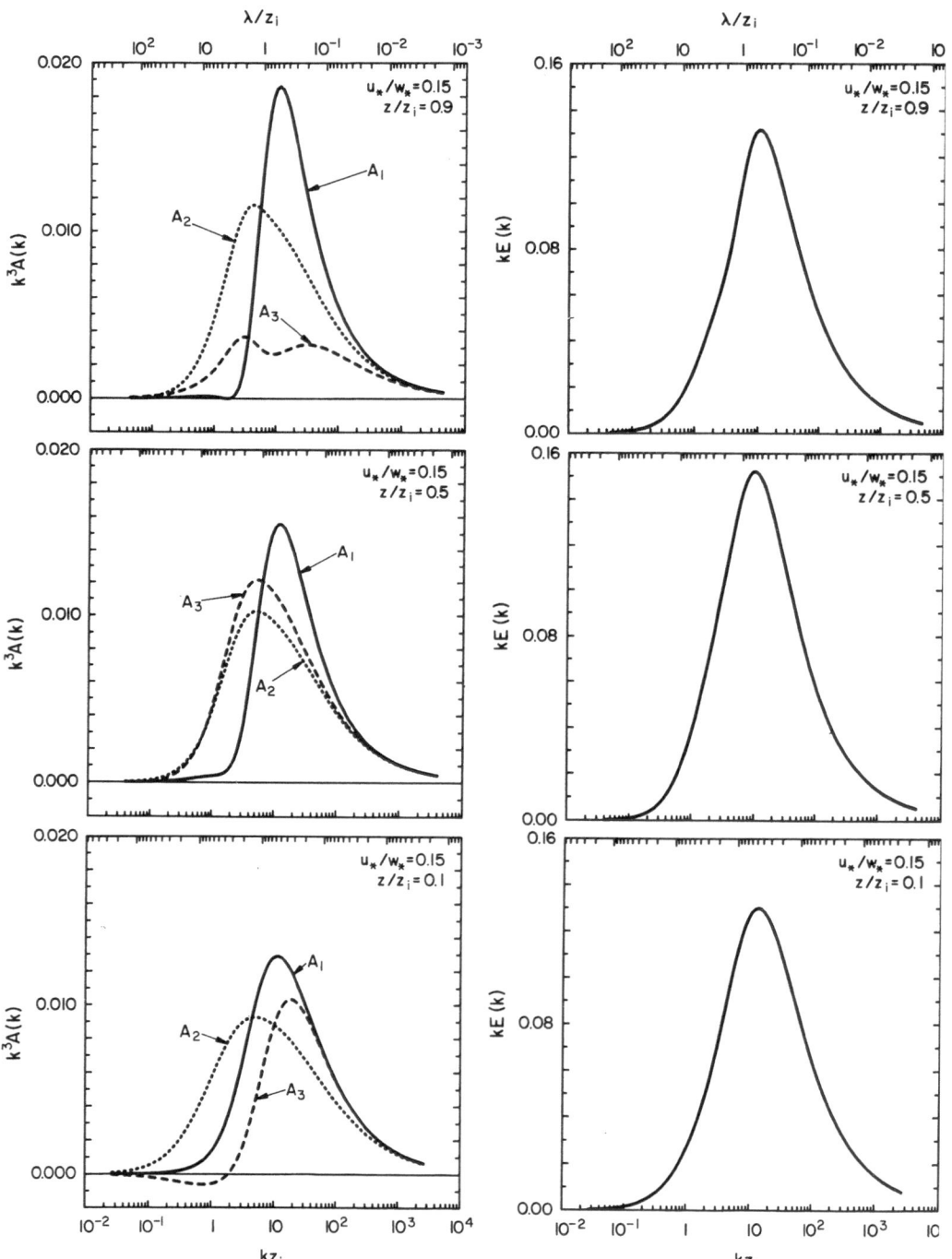

Fig. 6b.   Same as Figure 6a, but with $u_*/w_* = 0.15$.

we believe that our approach to determine the spectral tensor in the CBL is useful.

5. Applications of the Spectral tensor

One of the important applications of the spectral tensor is to investigate the spatial structure of turbulence. A useful variable for doing this is the spatial coherence function. Therefore, we show how to obtain this function from the spectral tensor. We derive the spectral coherences of the velocity components for a displacement

$$\mathbf{D} = D \cos \theta \mathbf{i}_2 + D \sin \theta \mathbf{i}_3 \tag{108}$$

perpendicular to the mean-wind $U\mathbf{i}_1$ and for an angle θ with the horizon. For the space-time auto-covariance $R_{ij}(\mathbf{D}, \tau)$ between velocity components u_i and u_j, Taylor's hypothesis can be formulated as

$$R_{ij}(\mathbf{D}, \tau) = R_{ij}(-U\tau\mathbf{i}_1 + \mathbf{D}, 0). \tag{109}$$

In order to compute the coherences, we must first derive the cross-spectra

$$\chi_{ij}(\mathbf{D}, \omega) = \frac{1}{2\pi} \int_{-\infty}^{\infty} R_{ij}(\mathbf{D}, \tau) e^{-i\omega\tau} d\tau. \tag{110}$$

Since

$$R_{ij}(\mathbf{r}, 0) = \oint \Phi_{ij}(\mathbf{k}) e^{i\mathbf{k} \cdot \mathbf{r}} d^3 k, \tag{111}$$

we can rewrite (110) in terms of the double integral

$$\chi_{ij}(\mathbf{D}, \omega) = \frac{1}{U} \int_{-\infty}^{\infty} e^{ik_2 D \cos \theta} dk_2 \int_{-\infty}^{\infty} e^{ik_3 D \sin \theta} dk_3$$

$$\times \Phi_{ij}\left(-\frac{\omega}{U}\mathbf{i}_1 + k_2\mathbf{i}_2 + k_3\mathbf{i}_3\right) \tag{112}$$

or, using the transformation (17),

$$\chi_{ij}(\mathbf{D}, \omega) = \frac{1}{U} \int_{0}^{\infty} \kappa \, d\kappa \int_{0}^{2\pi} e^{iD\kappa \cos(\Theta - \theta)} d\Theta$$

$$\times \Phi_{ij}\left(-\frac{\omega}{U}\mathbf{i}_1 + \kappa \cos \Theta \mathbf{i}_2 + \kappa \sin \Theta \mathbf{i}_3\right). \tag{113}$$

The coherence is then defined as (Kristensen and Jensen, 1979)

$$\text{coh}_{ij}(\mathbf{D}, \omega) = \frac{|\chi_{ij}(\mathbf{D}, \omega)|^2}{\chi_{ij}(\mathbf{0}, \omega)\chi_{ij}(\mathbf{0}, \omega)} . \tag{114}$$

We must now evaluate the six different elements of the symmetric tensor $\Phi_{ij}(-(\omega/U)\mathbf{i}_1 + \kappa \cos \Theta \mathbf{i}_2 + \kappa \sin \Theta \mathbf{i}_3)$. Using the short-hand notation

$$k = \sqrt{\left(-\frac{\omega}{U}\right)^2 + \kappa^2} \tag{115}$$

for convenience,

$$\Phi_{11}\left(-\frac{\omega}{U}\mathbf{i}_1 + \kappa \cos \Theta \mathbf{i}_2 + \kappa \sin \Theta \mathbf{i}_3\right)$$

$$= \frac{1}{k^4}\left\{A_1(k)\kappa^4 + A_2(k)\left(\frac{\omega}{U}\right)^2 \kappa^2 \cos^2 \Theta + A_3(k)\left(\frac{\omega}{U}\right)^2 \kappa^2 \sin^2 \Theta\right\}, \tag{116}$$

$$\Phi_{22}\left(-\frac{\omega}{U}\mathbf{i}_1 + \kappa \cos \Theta \mathbf{i}_2 + \kappa \sin \Theta \mathbf{i}_3\right)$$

$$= \frac{1}{k^4}\left\{A_1(k)\left(\frac{\omega}{U}\right)^2 \kappa^2 \cos^2 \Theta + A_2(k)\left(\left(\frac{\omega}{U}\right)^2 + \kappa^2 \sin^2 \Theta\right)^2\right.$$

$$\left. + A_3(k)\kappa^4 \cos^2 \Theta \sin^2 \Theta\right\}, \tag{117}$$

$$\Phi_{33}\left(-\frac{\omega}{U}\mathbf{i}_1 + \kappa \cos \Theta \mathbf{i}_2 + \kappa \sin \Theta \mathbf{i}_3\right)$$

$$= \frac{1}{k^4}\left\{A_1(k)\left(\frac{\omega}{U}\right)^2 \kappa^2 \sin^2 \Theta + A_2(k)\kappa^4 \cos^2 \Theta \sin^2 \Theta\right.$$

$$\left. + A_3(k)\left(\left(\frac{\omega}{U}\right)^2 + \kappa^2 \cos^2 \Theta\right)^2\right\}, \tag{118}$$

$$\Phi_{23}\left(-\frac{\omega}{U}\mathbf{i}_1 + \kappa \cos \Theta \mathbf{i}_2 + \kappa \sin \Theta \mathbf{i}_3\right)$$

$$= \frac{\kappa^2 \cos \Theta \sin \Theta}{k^4}\left\{A_1(k)\left(\frac{\omega}{U}\right)^2 - A_2(k)\left(\left(\frac{\omega}{U}\right)^2 + \kappa^2 \sin^2 \Theta\right)\right.$$

$$\left. - A_3(k)\left(\left(\frac{\omega}{U}\right)^2 + \kappa^2 \cos^2 \Theta\right)\right\}, \tag{119}$$

$$\Phi_{31}\left(-\frac{\omega}{U}\mathbf{i}_1 + \kappa \cos\Theta\mathbf{i}_2 + \kappa \sin\Theta\mathbf{i}_3\right)$$

$$= \frac{\frac{\omega}{U}\kappa \sin\Theta}{k^4}\left\{A_1(k)\kappa^2 - A_2(k)\kappa^2 \cos^2\Theta\right.$$

$$\left. + A_3(k)\left(\left(\frac{\omega}{U}\right)^2 + \kappa^2 \cos^2\Theta\right)\right\}, \tag{120}$$

and

$$\Phi_{12}\left(-\frac{\omega}{U}\mathbf{i}_1 + \kappa \cos\Theta\mathbf{i}_2 + \kappa \sin\Theta\mathbf{i}_3\right)$$

$$= \frac{\frac{\omega}{U}\kappa \cos\Theta}{k^4}\left\{A_1(k)\kappa^2 + A_2(k)\left(\left(\frac{\omega}{U}\right)^2 + \kappa^2 \sin^2\Theta\right)\right.$$

$$\left. - A_3(k)\kappa^2 \sin^2\Theta\right\}. \tag{121}$$

In order to facilitate the integrations over Θ, we introduce the definition

$$S_n(x) = \frac{J_n(x)}{2x^n}, \tag{122}$$

where $J_n(x)$ is the Bessel Function of the first kind. Using the relations

$$x^2 S_{n+1}(x) = 2nS_n(x) - S_{n-1}(x) \tag{123}$$

and

$$S'_n(x) = -xS_{n+1}(x), \tag{124}$$

we get the following list of useful integrals:

$$\int_0^{2\pi} e^{ix\cos\Theta}\, d\Theta = 4\pi S_0(x) \tag{125}$$

$$\int_0^{2\pi} e^{ix\cos\Theta}\cos^2\Theta\, d\Theta = 4\pi\{S_0(x) - S_1(x)\} \tag{126}$$

$$\int_0^{2\pi} e^{ix\cos\Theta}\sin^2\Theta\, d\Theta = 4\pi S_1(x) \tag{127}$$

$$\int_0^{2\pi} e^{ix\cos\Theta} \cos^4\Theta \, d\Theta = 4\pi \frac{x^2-3}{x^2} \{S_0(x) - 2S_1(x)\} \tag{128}$$

$$\int_0^{2\pi} e^{ix\cos\Theta} \sin^4\Theta \, d\Theta = -\frac{12\pi}{x^2} \{S_0(x) - 2S_1(x)\} \tag{129}$$

$$\int_0^{2\pi} e^{ix\cos\Theta} \cos^2\Theta \sin^2\Theta \, d\Theta = \frac{4\pi}{x^2} \{3S_0(x) + (x^2-6)S_1(x)\} \tag{130}$$

$$\int_0^{2\pi} e^{ix\cos\Theta} \cos\Theta \, d\Theta = i4\pi x S_1(x) \tag{131}$$

$$\int_0^{2\pi} e^{ix\cos\Theta} \cos\Theta \sin^2\Theta \, d\Theta = -i\frac{4\pi}{x} \{S_0(x) - 2S_1(x)\} . \tag{132}$$

These integrals are used to evaluate the longitudinal and horizontal displacement cross-spectra in the next two sections. The data with which we want to compare our theory are either coherences of the longitidunal velocity component ($i = j = 1$) or coherences with horizontal displacement only ($\theta = 0$).

5.1. LONGITUDINAL-DISPLACEMENT CROSS-SPECTRUM

We rewrite (113) for the longitudinal component

$$\chi_{11}(\mathbf{D}, \omega) = \frac{1}{U} \int_0^\infty \kappa \, d\kappa \int_0^{2\pi} e^{iD\kappa\cos\Theta} \, d\Theta$$

$$\times \Phi_{11}\left(-\frac{\omega}{U}\mathbf{i}_1 + \kappa\cos(\Theta + \theta)\mathbf{i}_2 + \kappa\sin(\Theta + \theta)\mathbf{i}_3\right), \tag{133}$$

using the fact that the integrand is 2π periodic in Θ. The longitudinal-displacement velocity cross-spectrum is

$$\chi_{11}(\mathbf{D}, \omega) = \frac{4\pi}{U} \int_0^\infty \frac{\kappa^3}{k^4} \, d\kappa \Big\{ A_1(k)\kappa^2 S_0(D\kappa)$$

$$+ A_2(k)\left(\frac{\omega}{U}\right)^2 [S_0(D\kappa)\cos^2\theta - S_1(D\kappa)\cos(2\theta)]$$

$$+ A_3(k)\left(\frac{\omega}{U}\right)^2 [S_0(D\kappa)\sin^2\theta + S_1(D\kappa)\cos(2\theta)] \Big\}. \tag{134}$$

5.2. Horizontal-Displacement Cross-Spectra

As mentioned before, there are six ways to combine velocity components in cross-spectra. When $\theta = 0$, two of these are identically zero due to the identity

$$\int_0^{2\pi} e^{ix \cos \Theta} G(\cos \Theta) \sin \Theta \, d\Theta = 0 . \tag{135}$$

Inspecting (119) and (120), we find in view of (135)

$$\chi_{23}(D\mathbf{i}_2, \omega) = \chi_{31}(D\mathbf{i}_2, \omega) = 0 . \tag{136}$$

The four non-zero horizontal-displacement cross-spectra are

$$\chi_{11}(D\mathbf{i}_2, \omega) = \frac{4\pi}{U} \int_0^\infty \frac{\kappa^3}{k^4} d\kappa$$

$$\times \left\{ A_1(k)\kappa^2 S_0(D\kappa) + A_2(k)\left(\frac{\omega}{U}\right)^2 [S_0(D\kappa) - S_1(D\kappa)] \right.$$

$$\left. + A_3(k)\left(\frac{\omega}{U}\right)^2 S_1(D\kappa) \right\} , \tag{137}$$

$$\chi_{22}(D\mathbf{i}_2, \omega) = \frac{4\pi}{U} \int_0^\infty \frac{\kappa}{k^4} dk$$

$$\times \left\{ A_1(k)\left(\frac{\omega}{U}\right)^2 \kappa^2 [S_0(D\kappa) - S_1(D\kappa)] + A_2(k)\left[\left\{\left(\frac{\omega}{U}\right)^4 - 3\left(\frac{\kappa}{D}\right)^2\right\} S_0(D\kappa) \right. \right.$$

$$\left. + 2\left(\frac{\kappa}{D}\right)^2 \left\{\left(\frac{\omega}{U}\right)^2 D^2 + 3\right\} S_1(D\kappa)\right]$$

$$\left. + A_3(k)\left(\frac{\kappa}{D}\right)^2 [3 S_0(D\kappa) + \{(D\kappa)^2 - 6\} S_1(D\kappa)] \right\} , \tag{138}$$

$$\chi_{33}(D\mathbf{i}_2, \omega) = \frac{4\pi}{U} \int_0^\infty \frac{\kappa}{k^4} d\kappa$$

$$\times \left\{ A_1(k)\left(\frac{\omega}{U}\right)^2 \kappa^2 S_1(D\kappa) + A_2(k)\left(\frac{\kappa}{D}\right)^2 [3 S_0(D\kappa) + \{(D\kappa)^2 - 6\} S_1(D\kappa)] \right.$$

$$\left. + A_3(k)\left[\left\{k^4 - 3\left(\frac{\kappa}{D}\right)^2\right\} S_0(D\kappa) - 2\left\{k^2\kappa^2 - 3\left(\frac{\kappa}{D}\right)^2\right\} S_1(D\kappa)\right] \right\} , \tag{139}$$

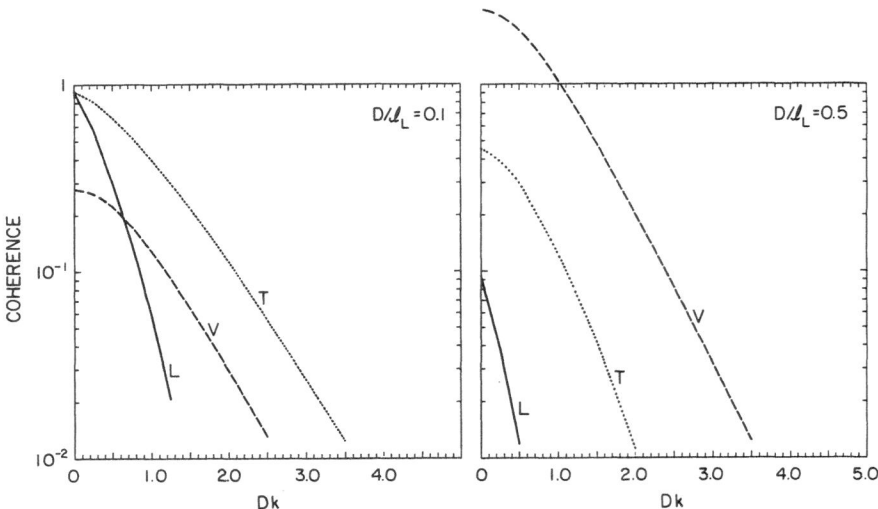

Fig. 7.   Coherences in the neutral surface layer. They are calculated on the basis of the A-functions shown in Figure 5. We see that, as a consequence of $A_3(k)$ not being positive for all values of the wave number, the calculated coherences can exceed one.

$$\chi_{12}(D\mathbf{i}_2, \omega) = i \frac{4\pi}{U} \frac{\omega}{U} \int\limits_0^\infty \frac{\kappa^2}{k^4} \, d\kappa$$

$$\times \left\{ A_1(k)D\kappa^3 S_1(D\kappa) + A_2(k)\left[-\frac{\kappa}{D} S_0(D\kappa) + \left\{ \left(\frac{\omega}{U}\right)^2 D\kappa + 2\frac{\kappa}{D} \right\} S_1(D\kappa) \right] \right.$$

$$\left. + A_3(k)\frac{\kappa}{D}[S_0(D\kappa) - 2S_1(D\kappa)] \right\}.   \tag{140}$$

5.2.1. *Spectral Coherence in the Neutral Surface Layer*

We have calculated the spectral coherences based on the solutions for $A_1(k)$, $A_2(k)$ and $A_3(k)$ found in Subsection 4.1. Since $A_3(k)$ is negative for small values of k, we expect that the coherences are in error. Figure 7 shows the coherences for two different displacements, $D/l_L = 0.1$ and $D/l_L = 0.5$. For the smaller value of D only the positive, i.e., "the locally isotropic" portion of $A_3(k)$ contributes significantly. However, when D is closer to l_L, eddies outside the locally isotropic domain come into play and negative values of $A_3(k)$ contribute more to the coherence. As pointed out in Section 2, Schwarz's inequality will then not hold, and coherences less than unity cannot be guaranteed. This, in fact, is what Figure 7 shows for the vertical velocity component when $D/l_L = 0.5$.

5.2.2. *Spectral Coherence in the Convective Boundary Layer*

The functions $A_1(k)$, $A_2(k)$ and $A_3(k)$ for the CBL, shown in Subsection 4.2, have been used to compute coherences for $D/l_L = 0.1$ and $D/l_L = 0.5$. The results

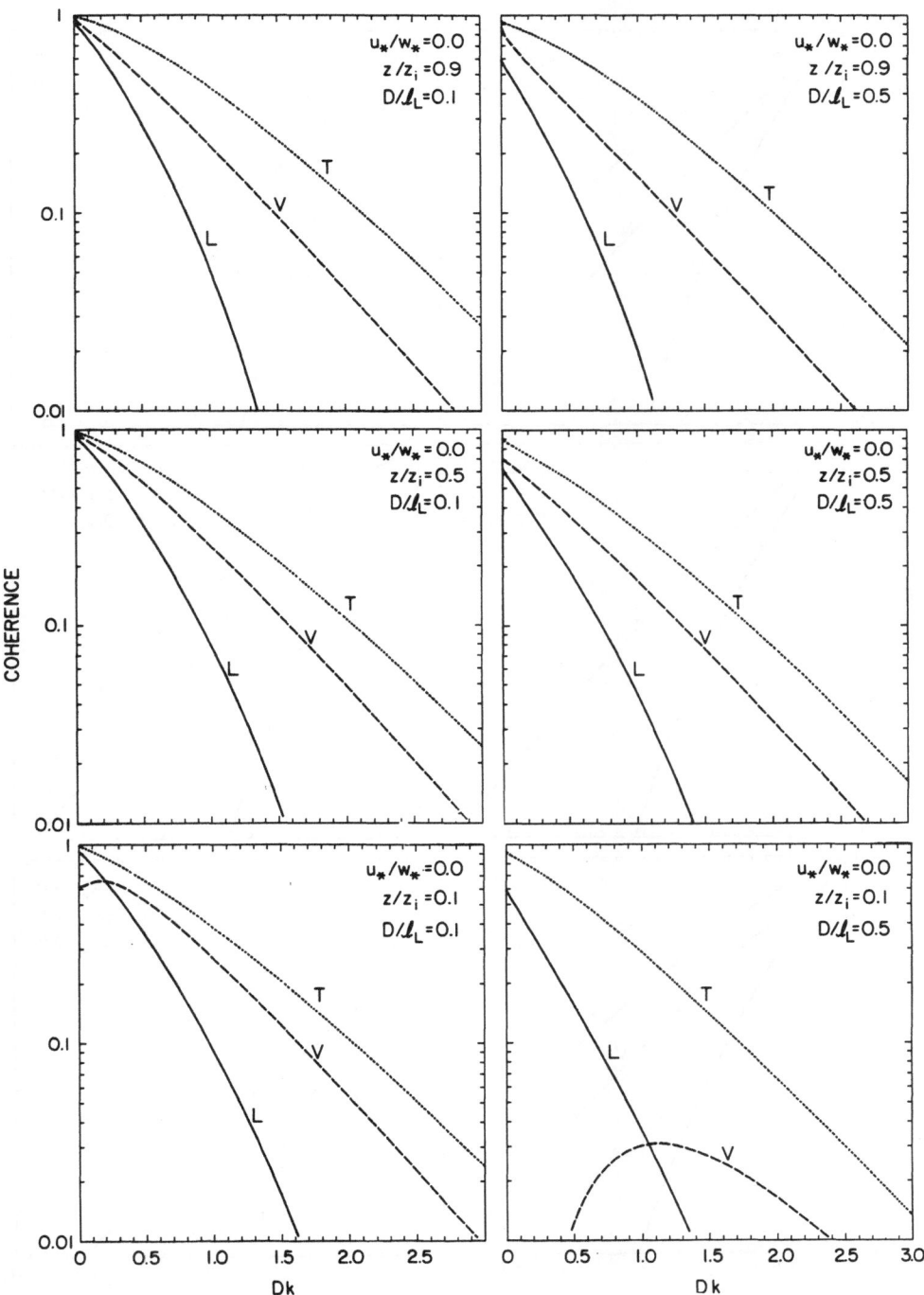

Fig. 8a. The coherences in the extremely convective boundary layer ($u_*/w_* = 0$), calculated on basis of the A-functions shown in Figure 6a.

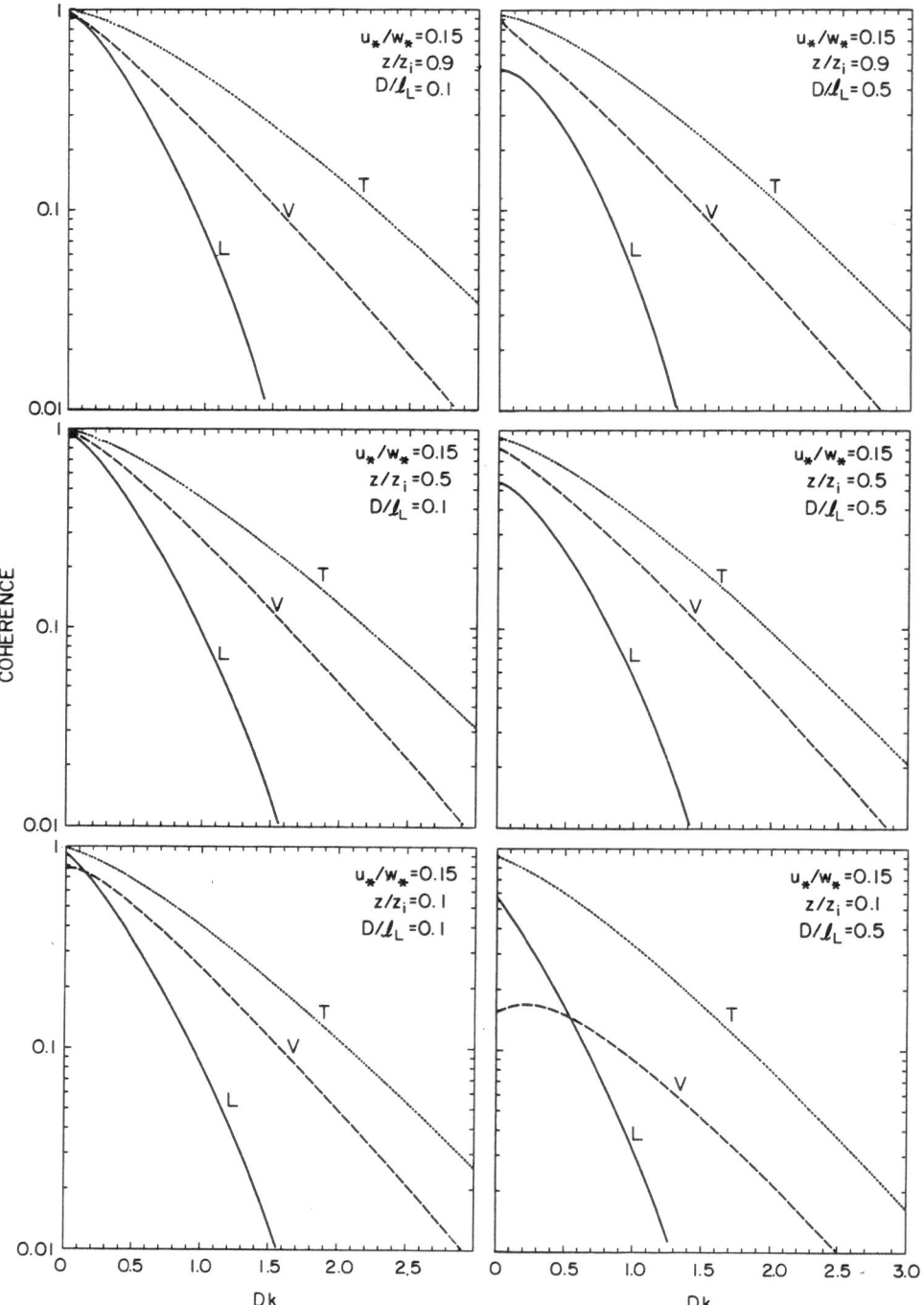

Fig. 8b. The coherences in the convective boundary layer with $u_*/w_* = 0.15$, calculated on basis of the A-functions shown in Figure 6b.

are shown in Figures 8a and 8b. Since $A_3(k)$ has considerably less negative area than is the case for the neutral surface layer, we do not see any cases where the coherence is greater than one. We note that when the displacement is not much smaller than the scale of the turbulence, the coherence goes to a value less than unity for $k \to 0$, as qualitatively predicted by Kristensen and Jensen (1979).

6. Experimental Validation

So far we have only discussed the application of the spectral tensor to the convective atmospheric boundary layer and the neutral surface layer assuming, without verification, that the model is accurate enough to provide realistic results. We therefore compare our predicted results, in the form of coherences, with experimentally determined coherences. We shall use data from two sources: the Lammefjord Experiment (LAMEX) (Courtney, 1987; 1988) and the Dual Aircraft Formation Flights Experiment (DAFFEX) (Lenschow and Kristensen, 1988).

It is difficult to find other experimental data in the literature suitable for evaluation of the spectral tensor as outlined here and, at the same time, containing experimentally determined spectral coherences between turbulent velocity components measured with several displacements under different experimental conditions. There is, however, another technique for obtaining observational data to validate our model. The NCAR Electra aircraft measures the three velocity components at the nose as well as the airspeed at the left wing tip 14.8 m from the longitudinal axis of the airplane. By flying the Electra at various heights above the ground, we can obtain the spectral tensor and the longitudinal-velocity coherence for different values of D/l_L. We have calculated these quantities for several flight legs from the Dynamic and Chemistry of the Marine Stratocumulus (DYCOMS) (Lenschow et al. (1988) and First ISCCP (International Satellite Cloud Climatology Project) Regional Experiment (FIRE) (Albrecht et al., 1988) which were carried out in the stratiform-capped boundary layer off the California coast. Unfortunately, in the CBL, the flight levels are not low enough to give values of D/l_L which are sufficiently large to provide coherences that differ from those predicted on the basis of local isotropy. Therefore, we do not discuss these results further. Since l_L is smaller in the stably stratified boundary layer, it may be possible to obtain large enough values of D/l_L there to test our model for the spectral tensor.

One way to use the data to validate our model is to use the experimentally determined component spectra $F_L(k)$, $F_T(k)$ and $F_V(k)$ to determine the three functions $A_1(k)$, $A_2(k)$ and $A_3(k)$ and consequently the spectral tensor $\Phi_{ik}(\mathbf{k})$. With $\Phi_{ik}(\mathbf{k})$, it is possible to predict the coherences and compare them with the directly measured coherences.

6.1. Lamex

Three masts, shown in Figure 9, were erected in Lammefjord, an area of land reclaimed from a shallow, flat-bottomed fjord on the island of Zealand in

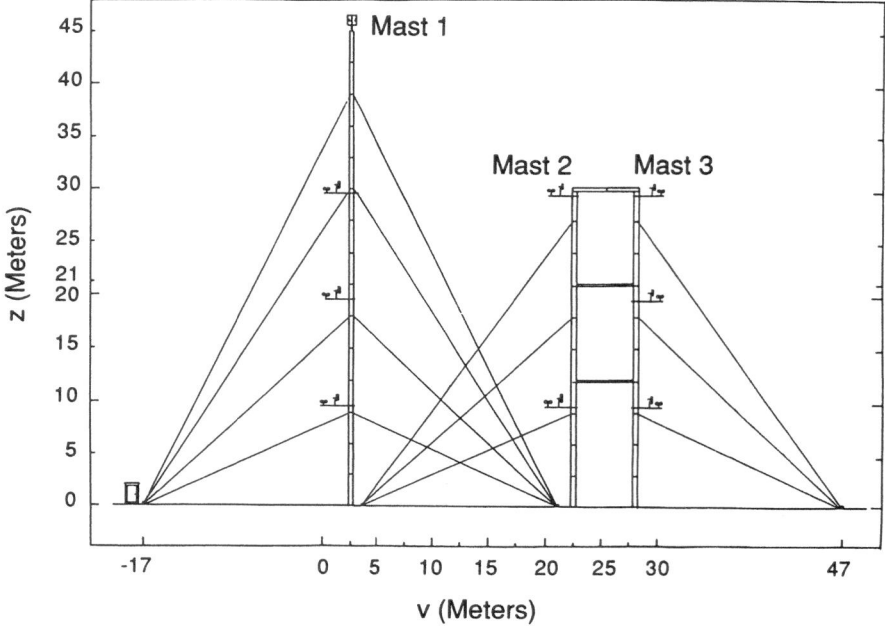

Fig. 9. The three masts at the Lammefjord site in Denmark. Wind vanes and cup anemometers are mounted at 10 and 30 m on all three masts and at 20 m on masts 1 and 3. A three-dimensional sonic anemometer is mounted on the top of mast 1 (45 m). The lateral displacements between instruments at the same height are 10, 20 and 30 m.

Denmark, in order to study the spectral coherence between longitudinal velocity components at points displaced over a plane normal to the prevailing wind, corresponding to the rotor plane of a wind turbine. The fetch upwind of the prevailing wind direction was 2500 to 3000 m, where the level did not vary more than 1 m. Identical cup anemometers with distance constants of 1.7 m and wind vanes with distance constants of 0.7 m and damping ratios of 0.5 were mounted at 10 and 30 m heights on all three masts and at 20 m on the two outer masts, mast 1 and mast 3. The lateral displacements between instruments at the same height were 10, 20 and 30 m. A three-component sonic anemometer was mounted on the top of the tallest mast, mast 1.

The fundamental sampling rate was $16\,\mathrm{s}^{-1}$, but the data were averaged and recorded every second to reduce spectral aliasing. The data were recorded continuously and stored permanently on a 1 Gbyte removable optical disk, a so-called WORM (Write Once Read Many). One disk had capacity for 34 days of data.

The measurements were started about the beginning of June 1987 and terminated at the end of June 1988. The system worked essentially uninterrupted with a data recovery of about 80%. Courtney (1988) has given a detailed description of the experiment.

A period of 50 hours of data, starting late August 1, 1987, was chosen. Figure 10 shows that the mean wind direction was very constant in the entire period and that the mean wind speed at 10 m dropped linearly with time from about 13 m s^{-1} to about 9 m s^{-1} over the period. The angle between the mean wind direction and the mast array was about 53°.

The entire time series was divided into 800 consecutive realizations. For each realization, we calculated for each pair of signals, the two auto-spectra and the cross-spectrum and normalized them by means of the variances of the realization. The arithmetic mean of the mean wind speeds at the two positions for each period was used to convert frequency to wave number, assuming Taylor's hypothesis. In this way we eliminated the effect of trend in the wind speed. We then computed the mean of the auto- and cross-spectra from the 800 realizations. The coherence is finally determined as the absolute square of the mean cross-spectrum divided by the two auto-spectra. The phase is computed from the cross-spectrum.

Figure 11a shows the experimental coherences and phases for three different values of D/l_L, where D is the lateral displacement. We note that the phases change linearly with Dk. This of course corresponds to the angle the mean wind direction forms with the plane of the mast array. The lateral displacements have been corrected accordingly by multiplying the actual horizontal distances by $\cos(37°)$. The solid lines in the coherence plots are the predicted coherences. They are determined on the basis of the "generic" spectra and the A-functions shown in Figures 2 and 5, pertaining to the neutral surface layer. In principle, we could have used the three-component sonic anemometer at 45 m height on the top of mast 1 for spectral measurements. However, due to wake-effects from the transducers, which still need to be evaluated to establish correction procedures, we decided not to use the sonic anemometer.

According to Figure 11a, the coherence with the smallest value of D/l_L is well predicted. For the larger values of D/l_L, we predict too large a coherence. We find, however, that the prediction of the intersections with the ordinate axis ($Dk = 0$) is consistent with the experimental findings.

Figure 11b shows the coherences for three different values of D/l_L, where D now is a vertical displacement. The solid line is again the prediction, based on the A-functions, shown in Figure 5. Here we overestimate all the coherences, while the intersection with the ordinate axis seems again to be consistent with the experimental data.

Pielke and Panofsky (1970) concluded on the basis of data with considerable scatter that the phase is essentially a linear function of Dk and that the slope s would be equal to D/z. If we ignore the curvature of our measured phases, we find indeed

$$s \approx 5 \times \frac{D}{l_L} \approx \frac{D}{z}, \qquad (141)$$

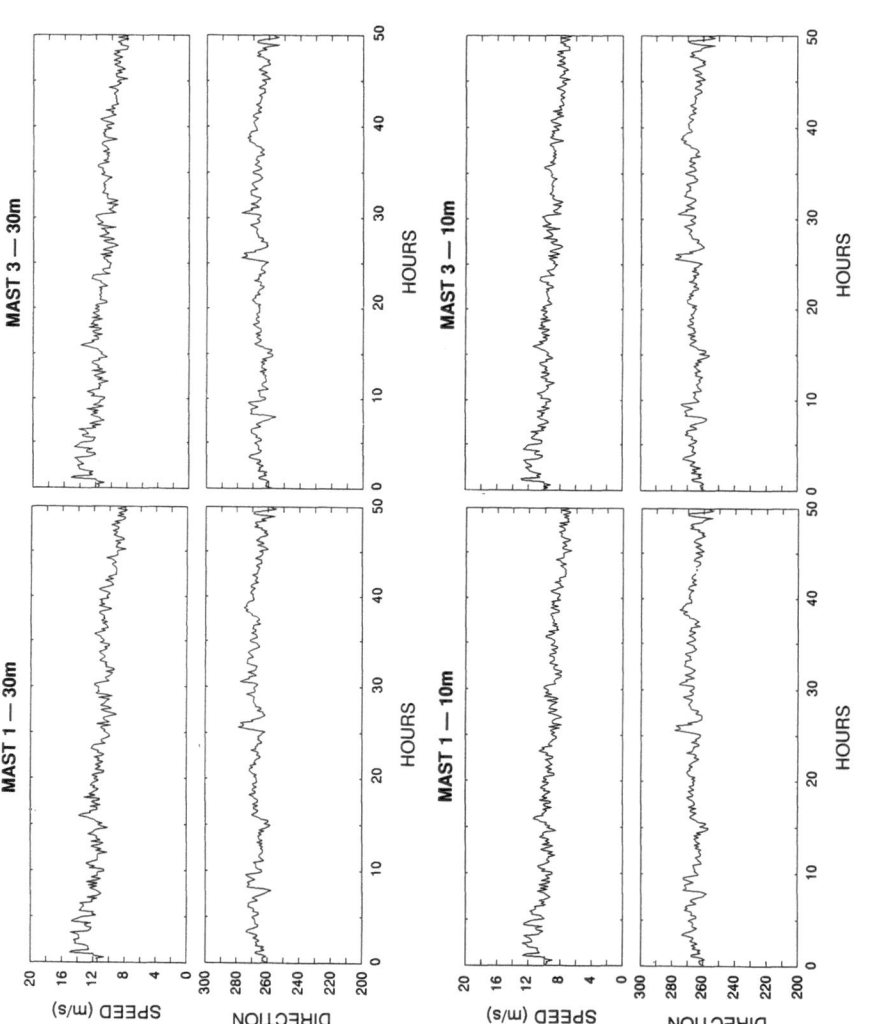

Fig. 10. Wind speed and direction at 10 and 30 m height for a 50 hr period on masts 1 and 3.

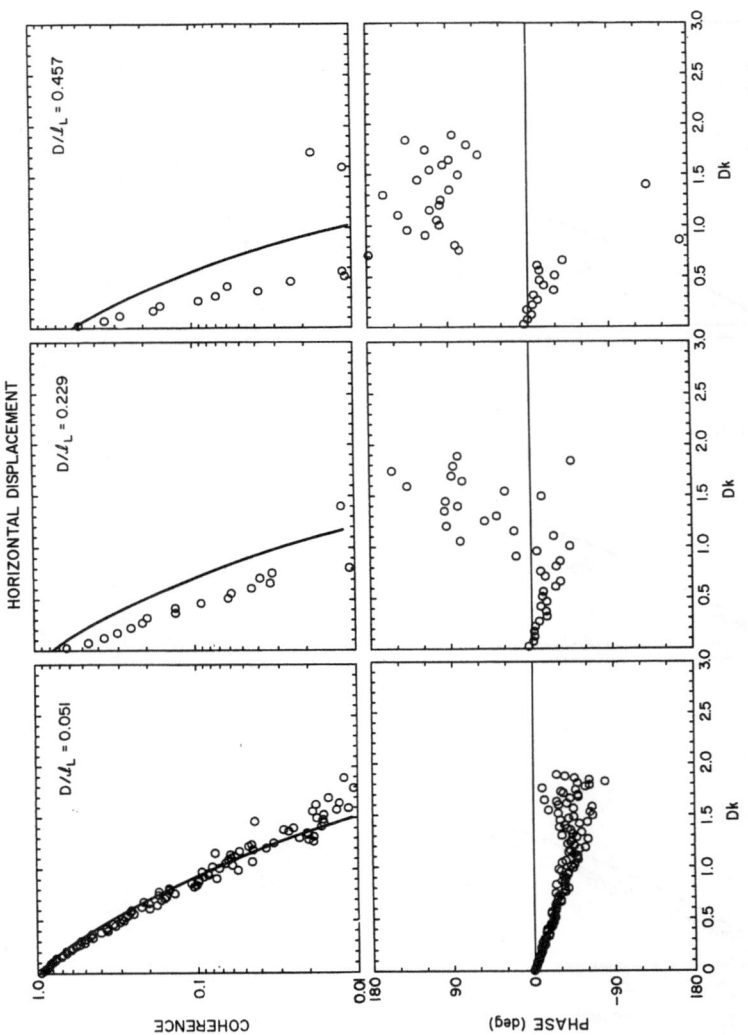

Fig. 11a. Directly measured LAMEX longitudinal-velocity coherences and phases with horizontal displacements for three values of D/l_L. The solid lines are the coherences predicted on basis of the A-functions shown in Figure 5, pertaining to the neutral surface layer.

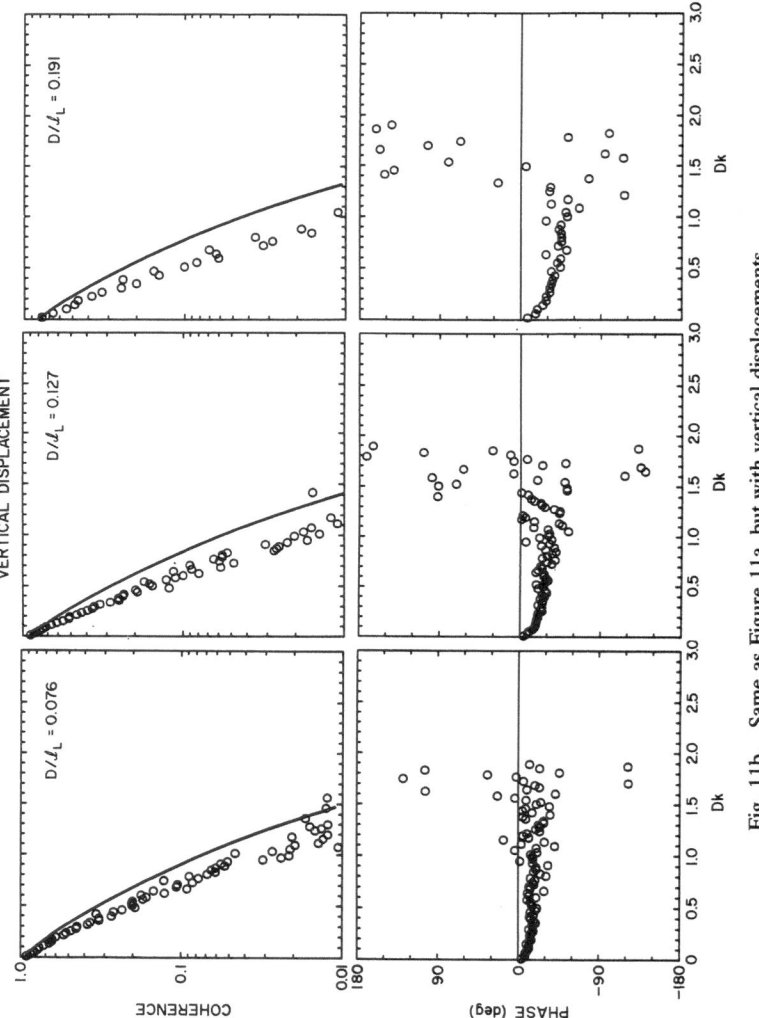

Fig. 11b.   Same as Figure 11a, but with vertical displacements.

where we have used (76). We note, however, that the slope for $Dk < 0.1$ is greater than this, which implies that at long wavelengths, the eddy structures have more tilt than for $Dk > 0.1$.

6.2. DAFFEX

In September 1979, a series of six flights, called the Dual Aircraft Formation Flight Experiment (DAFFEX), were carried out with the two NCAR Queen Air aircraft in the CBL over eastern Colorado. Each aircraft was instrumented with a fixed-vane gust probe mounted at the tip of a nose boom for measuring the velocity of the air with respect to the aircraft, and an Inertial Navigation System (INS) for measuring the velocity and attitude angles of the airplane in an earth-based coordinate system. As discussed by Lenschow (1986), the velocity of the air with respect to the earth is then computed from these measurements. The two aircraft were flown in formation, mostly with lateral separation but sometimes with vertical separation. The pilots attempted to hold a constant displacement (for all three components of the displacement vector) between the two aircraft during each flight leg, with lateral displacements varying from about 15 m to 1.6 km. Most of the legs were about 360 s duration and the airplane true airspeeds were between 70 and 80 m s^{-1}. The legs were flown at various levels within the boundary layer above a minimum height of about 30 m above ground.

Relative position information for each aircraft at 10 s^{-1} was obtained by integrating the airplane velocity components computed from the INS on each aircraft. The relative displacement vector between the aircraft was then obtained from their difference. The displacement vector was measured absolutely by analyzing frames of film taken with a time-lapse camera at a rate of 0.5 s^{-1}. The camera was mounted normal to the longitudinal axis in the horizontal plane of one aircraft, pointed towards the other. The horizon was used as a reference for the vertical displacement. The mean and trend of the displacement components obtained from the INS differences were replaced by those obtained from the film analysis, since the INS has excellent time resolution and short-term accuracy, but relatively poor long-term accuracy. The technique, as well as its accuracy and limitations, is described and discussed by Lenschow and Kristensen (1988).

We have analysed about 20 flight legs, out of which we have selected six cases for presentation here. When calculating the coherences directly from the data, we computed raw spectral estimates without subdividing the records, as we did in the analysis of the LAMEX data. The raw spectral estimates were smoothed by averaging over bins consisting of 8 raw estimates. This procedure is equivalent to dividing the time series into 8 realizations and averaging the spectra after calculating them for each realization (Kristensen and Kirkegaard, 1986). The auto-spectra were in this case determined by first calculating the average raw auto-spectra from the two aircraft. The three parameters μ, l and σ^2 for the component spectra $F_L(k)$, $F_T(k)$ and $F_V(k)$ were then determined by a standard iterative least-squares fit procedure using *MINPACK-1* (Moré *et al.*, 1980). This means that, in contrast to what we did in the LAMEX analysis, we are here actually testing the method we are proposing: using single-point streamwise component spectra to derive the spectral tensor $\Phi_{ij}(\mathbf{k})$, defined by the three A-functions.

Figure 12a shows the results of the analysis of a flight leg of duration 390 s, starting 1030 LST on 17 September. The airspeed was $U_a = 74$ m s^{-1} so the total length of the leg was 28860 m, and the heading was 272°. The inversion height z_i and the measurement height z were 277 m and 170 m, respectively, and the mean lateral displacement was $D = 26$ m. The least-squares fit resulted in the following spectral parameters:

$$\begin{Bmatrix} \mu_L \\ \mu_T \\ \mu_V \end{Bmatrix} = \begin{Bmatrix} 0.497 \\ 0.460 \\ 1.352 \end{Bmatrix},$$

$$\begin{Bmatrix} l_L \\ l_T \\ l_V \end{Bmatrix} = \begin{Bmatrix} 150.1 \text{ m} \\ 173.8 \text{ m} \\ 96.1 \text{ m} \end{Bmatrix}$$

and

$$\begin{Bmatrix} \sigma_L^2 \\ \sigma_T^2 \\ \sigma_V^2 \end{Bmatrix} = \begin{Bmatrix} 0.397 \text{ m}^2 \text{ s}^{-2} \\ 0.436 \text{ m}^2 \text{ s}^{-2} \\ 0.647 \text{ m}^2 \text{ s}^{-2} \end{Bmatrix}.$$

In the upper three frames of Figure 12a we see, from left to right, the spectra, the A-functions and the energy spectrum. The lower six frames show the coherences and the phases for the three velocity components: the longitudinal (L), the transverse (T) and the vertical (V). In this case we have $D/l_L = 0.17$, which is a relatively small value. Indeed, the coherences are close to one for $Dk = 0$. However, the vertical-velocity coherence has a maximum for $Dk \approx 0.2$, shown by the prediction and supported by the directly determined coherence.

Figure 12b shows another example. The record starts at 1104 LST on 14 September and lasts 360 s. The other directly determined parameters are $U_a = 73$ m s^{-1}, $z_i = 1389$ m, $z = 358$ m, $D = 81$ m and the heading = 89°. We derive from the spectra:

$$\begin{Bmatrix} \mu_L \\ \mu_T \\ \mu_V \end{Bmatrix} = \begin{Bmatrix} 0.688 \\ 0.567 \\ 1.220 \end{Bmatrix},$$

$$\begin{Bmatrix} l_L \\ l_T \\ l_V \end{Bmatrix} = \begin{Bmatrix} 199.5 \text{ m} \\ 315.8 \text{ m} \\ 346.0 \text{ m} \end{Bmatrix}$$

and

$$\begin{Bmatrix} \sigma_L^2 \\ \sigma_T^2 \\ \sigma_V^2 \end{Bmatrix} = \begin{Bmatrix} 0.723 \text{ m}^2 \text{ s}^{-2} \\ 0.928 \text{ m}^2 \text{ s}^{-2} \\ 1.710 \text{ m}^2 \text{ s}^{-2} \end{Bmatrix}.$$

Here we have $D/l_L = 0.41$, so we expect a smaller longitudinal-velocity coherence than in the first case. In particular, the coherence for $Dk = 0$ should be smaller as indeed seems to be the case.

Figure 13 shows component spectra and coherences for vertical velocity components in four other cases that we have analysed.

In all the coherences there is a lot of scatter, in contrast to the LAMEX case. The reason is that the number of degrees of freedom in this case is 8, whereas it

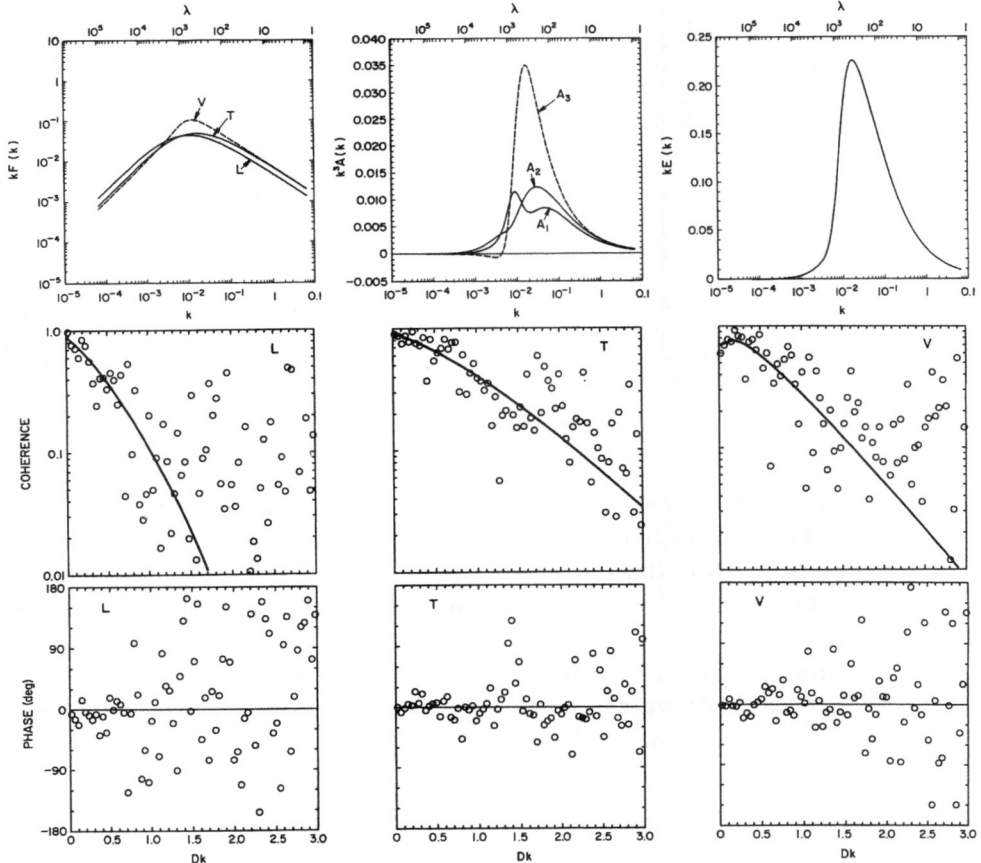

Fig. 12a.   Results of velocity data analysis from a flight leg of DAFFEX. The three spectra, the three A-functions and the energy spectrum are shown in the top row. Below are the coherences and phases for the three velocity components, measured and predicted on the basis of the measured spectra.

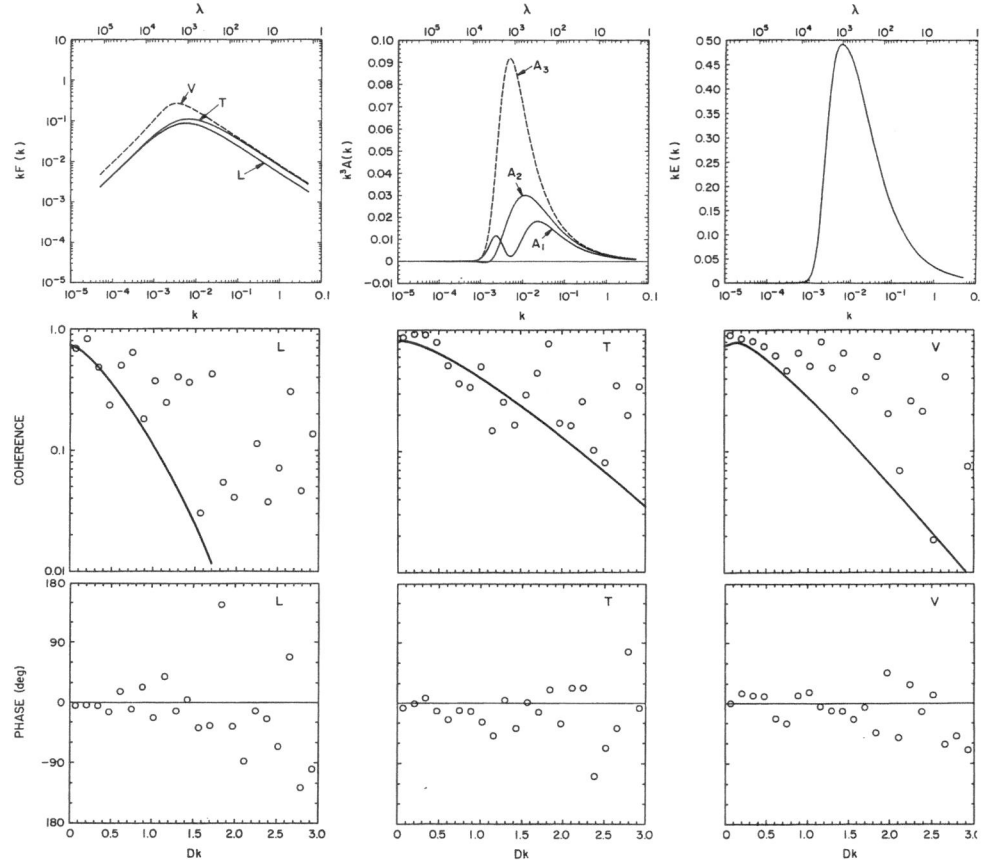

Fig. 12b. Same as Figure 12a.

was 800 in the LAMEX analysis. Fewer degrees of freedom cause more scatter and also systematic overestimation of the coherences, as discussed by Kristensen and Kirkegaard (1986). Indeed, in all cases analysed, there seems to be a tendency for the measured values in the coherence plots to exceed the predicted curves.

It is perhaps too strong to state that the measured coherences support the predictions, but at least we can conclude that they do not seem to contradict them, when we take the statistical scatter into account.

7. Conclusions

We have developed a kinematic model of turbulence which allows us to estimate the spectral tensor from the three one-dimensional spectra. The component spectra have the property that they approach local isotropy at high wave

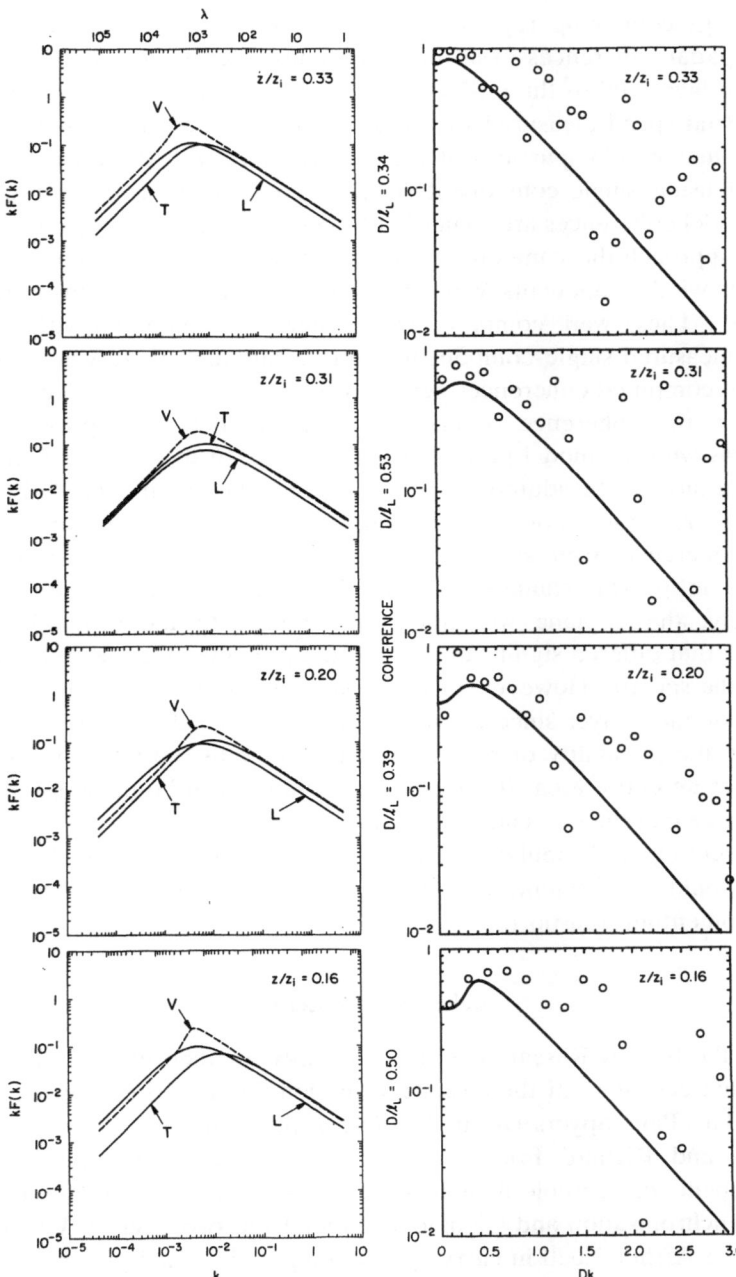

Fig. 13. Results of velocity data analysis from flight legs with four different values of z/z_i and D/l_L. To the left, the spectra and to the right, the measured and predicted lateral coherences of the vertical velocity components.

numbers, but may have differing length scales and curvature at low wavenumbers in the energy-containing region of the spectra. We have used the model to estimate spatial coherences from measured one-dimensional spectra in the neutral surface layer and in the CBL. Coherences were calculated for a 50-hr time series of wind speed measured by a set of vertically and horizontally displaced cup anemometers. Comparisons of these coherences with those calculated from our model using single-component spectra reported in the literature indicate that the model coherences are somewhat larger than the measured ones, although they both approach the same limiting values at small wave numbers. Coherences for all three wind components were obtained from two aircraft flying in formation in the CBL. These were compared with coherences computed with our model using the measured single-component spectra. Although there was considerable scatter, the computed coherences generally agreed with the measured ones.

Thus far, the coherence observations are insufficient to provide detailed comparisons with the model predictions. Therefore, further observational studies would be desirable. In addition to further direct measurements of the velocity components from towers or formation aircraft, other possibilities exist. We have obtained coherences from single aircraft measurements of airspeed on the nose and the wingtip. Unfortunately, the displacement was not large enough (or alternatively, the airplane was not close enough to the ground) to obtain coherences that differed significantly from that predicted for the locally isotropic region of the spectra. However, this approach may well be feasible in the stably stratified boundary layer since the length scales are smaller. Doppler radars and lidars offer the possibility of remote measurements of the radial velocity component over an entire area. If the spatial resolution can be made small enough, the coherence can then be calculated.

The present model formulation deals only with a homogeneous atmosphere. A next step would be to extend it to include a boundary layer with shear, i.e., with vertical momentum transport.

Acknowledgements

We thank the NCAR Research Aviation Facility for their enthusiastic efforts in planning and carrying out the aircraft formation flights and processing the data. In particular, Paul Spyers-Duran facilitated the planning and operations for DAFFEX, and Richard Friesen and Robert Lackman (Scientific Computor Division) spent considerable time developing the required techniques for dealing with the synchronization and alignment of data from two aircraft. We appreciate the efforts of Arthur Isbell in carrying out computations and plotting the results, and Chandran Kaimal in reviewing the manuscript. The Lammefjord Experiment (LAMEX) was funded by the European Commission under its Wind Research Programme. Many people at Risø have contributed to the project. We wish to express particular thanks to Gunnar Dalsgaard for designing, planning and coordinating the project so successfully.

References

Albrecht, B. A., Randall, D. A., and Nicholls, S.: 1988, 'Observations of Marine Stratocumulus During FIRE', *Bull. Amer. Meteorol. Soc.* **69**, 618–626.

Businger, J. A.: 1982, 'Equations and Concepts', in F. T. M. Nieuwstadt and H. van Dop (eds.), *Atmospheric Turbulence and Air Pollution Modelling*, D. Reidel Publ. Co., Dordrecht, Holland, pp. 1–36.

Courtney, M. S.: 1987, 'The Lammefjord Experiment', *Proceedings of the American Wind Energy Association Annual Conference*, October, San Francisco, CA.

Courtney, M. S.: 1988, 'An Atmospheric Turbulence Data Set for Wind Turbine Research', *Proceedings of the British Wind Energy Annual Conference*, April, London, UK.

Caughey, S. J. and Palmer, S. G.: 1979, 'Some Aspects of Turbulence Structure through the Depth of the Convective Boundary Layer', *Quart. J. Roy. Meteorol. Soc.* **105**, 811–827.

Frenzen, P. and Hart, L. H.: 1983, 'A Further Note on the Kolmogorov–von Kármán Product and the Values of the Constants', *Proc. Sixth Symp. on Turbulence and Diffusion*, March 22–25, Boston, MA, 24–27.

Højstrup, J.: 1982, 'Velocity Spectra in the Unstable Boundary Layer', *J. Atmos. Sci.* **39**, 2239–2248.

Kaimal, J. C., Wyngaard, J. C., Izumi, Y. and Coté, O. R.: 1972, 'Spectral Characteristics of Surface Layer Turbulence', *Quart. J. Roy. Meteorol. Soc.* **98**, 563–589.

Kaimal, J. C., Wyngaard, J. C., Haugen, D. A., Coté, O. R., Izumi, Y., Caughey, S. J., and Readings, C. J.: 1976, 'Turbulence Structure in the Convective Boundary Layer', *J. Atmos. Sci.* **33**, 2152–2169.

Kristensen, L. and Jensen, N. O.: 1979, 'Lateral Coherence in Isotropic Turbulence and in the Natural Wind', *Boundary-Layer Meteorol.* **17**, 353–373.

Kristensen, L., Kirkegaard, P. and Lenschow, D. H.: 1983, 'Squashed Atmospheric Turbulence', Report Risø-R-478, Risø National Laboratory, Roskilde, Denmark, 79 pp.

Kristensen, L. and Kirkegaard, P.: 1986, 'Sampling Problems with Spectral Coherence', Report Risø-R-526, Risø National Laboratory, Roskilde, Denmark, 63 pp.

Lenschow, D. H.: 1986, 'Aircraft Measurements in the Boundary Layer', in D. H. Lenschow (ed.), *Probing the Atmospheric Boundary Layer*, Amer. Meteor. Soc., Boston, MA, 39–55.

Lenschow, D. H. and Kristensen, L.: 1989, 'Applications of Dual Aircraft Formation Flights', *J. Atmos. Oceanic Technol.*, in press.

Lenschow, D. H., Paluch, I. R., Bandy, A. R., Pearson, Jr. R., Kawa, S. R., Weaver, C. J., Heubert, B. J., Kay, J. G., Thornton, D. C., and Driedger, III, A. R.: 1988, 'Dynamics and Chemistry of Marine Stratocumulus (DYCOMS) Experiment', *Bull. Amer. Meteorol. Soc.* **69**, 1058–1067.

Lenschow, D. H. and Stankov, B. B.: 1986, 'Length Scales in the Convective Boundary Layer', *J. Atmos. Sci.* **43**, 1198–1209.

Lenschow, D. H., Zhang, S. F., and Stankov, B. B.: 1988, 'The Stably Stratified Boundary Layer over the Great Plains: II. Horizontal Variations and Spectra', *Boundary-Layer Meteorol.* **42**, 123–135.

Lumley, J. L. and Panofsky, H. A.: 1964, *Atmospheric Turbulence*, Interscience Monographs and Texts in Physics and Astronomy, John Wiley & Sons, New York, xi + 239 pp.

Moré, J. J., Garbow, B. S., and Hillstrom, K. E.: 1980, 'User Guide for MINPACK-1', Report ANL-80-74, Argonne National Laboratory, Argonne, IL, 261 pp.

Panofsky, H. A., Tennekes, H., Lenschow, D. H., and Wyngaard, J. C.: 1977, 'The Characteristics of Turbulent Velocity Components in the Surface Layer under Convective Conditions', *Boundary-Layer Meteorol.* **11**, 355–361.

Panofsky, H. A. and Dutton, J. A.: 1984, *Atmospheric Turbulence*, John Wiley & Sons, New York, xix + 397 pp.

Pielke, R. A. and Panofsky, H. A.: 1970, 'Turbulence Characteristics Along Several Towers', *Boundary-Layer Meteorol.* **1**, 115–130.

Zhang, S. F., Oncley, S. P., and Businger, J. A.: 1988, 'A Critical Evaluation of the von Kármán Constant from a New Atmospheric Surface Layer Experiment', *Proc. Eighth Symp. on Turbulence and Diffusion*, April 25–29, San Diego, CA, 148–150.

ON THE TEMPERATURE SPECTRUM IN THE CONVECTIVE BOUNDARY LAYER

ZBIGNIEW SORBJAN

Department of Geological and Geophysical Sciences, The University of Wisconsin-Milwaukee, Milwaukee, WI 53201, U.S.A.

(Received in final form 20 September, 1988)

Abstract. A model for the temperature spectrum in the convective boundary layer is presented. The model is developed by using local similarity parameterization of the mixed layer. The model is compared with an idealized temperature spectrum obtained during the Minnesota experiment and exhibits behavior very similar to that observed in the atmosphere.

1. Introduction

The behavior of temperature and velocity spectra in the convective boundary layer (CBL) is well known mostly due to the extensive experimental efforts in Kansas in 1968 (Kaimal *et al.*, 1972) and Minnesota in 1973 (Kaimal *et al.*, 1976). As a result of these two experiments, some quantitative spectral models have been developed. In the surface layer, velocity spectra were modeled by Kaimal (1972, 1978) and Højstrup (1981), and temperature spectra by Kaimal (1971, 1978) and Panofsky (1978). Højstrup (1982) developed a model for velocity spectra in the lower half of the convective boundary layer. Spectral models for temperature in the mixed layer have been lacking because, unlike velocity spectra, they cannot be conveniently generalized within the framework of mixed-layer similarity.

Experimental data (Kaimal *et al.*, 1976) indicate that the spectrum of temperature exhibits run-to-run variations in the low-frequency variance introduced by entrainment effects in the upper half of the boundary layer (Figure 1). At the low-frequency end, all curves converge into a single one which sometimes (but not always) extends upward in response to diurnal or advective trends in temperature (Kaimal, 1982). As illustrated by arrows in Figure 1, at inertial subrange frequencies, the spectral intensity decreases steadily with height to $0.5 z_i$, stays at a low value between 0.5 and $0.7 z_i$, and starts to rise again above $0.7 z_i$.

This peculiarity of the temperature spectrum can be explained in terms of local similarity in the CBL (Sorbjan, 1988). According to local similarity hypotheses, the turbulent heat flux and any statistical moment involving vertical velocity, temperature, pressure and concentration of passive species can be decomposed into two components: non-penetrative (hereafter denoted by *b*) and residual (hereafter denoted by *t*). In addition, each of these components can be scaled by the appropriate set of local scales. The local scales are chosen to coincide in the surface layer with free-convection local scales.

ZBIGNIEW SORBJAN

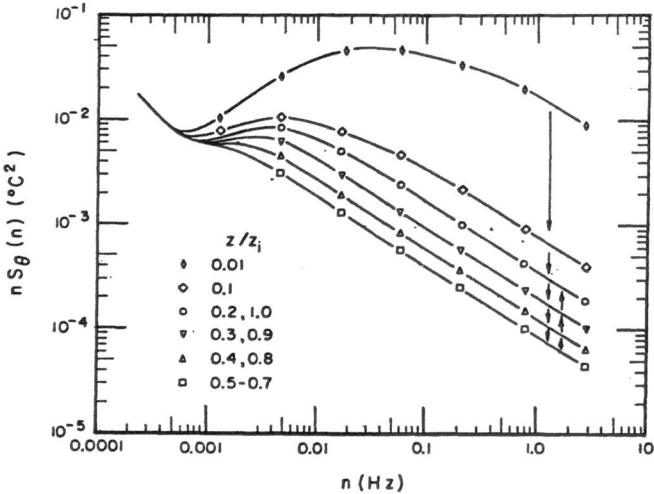

Fig. 1.   Idealized temperature spectrum in the convective boundary layer (after Kaimal *et al.*, 1976).

Local scales are displayed in Table I, where z_i is the height of the mixed layer – defined as a level where the heat flux is most negative, Δ is the depth of the interfacial layer between the mixed and the inversion layers (in this layer the heat flux increases from negative to zero), subscript "0" denotes the surface value and subscript "i" denotes the value at the top of the CBL, $\overline{w'\theta'}|_b$ and $\overline{w'\theta'}|_t$ are the "non-penetrative" and "residual" components of the heat flux, and finally, β is the buoyancy parameter.

Table I indicates that the heat flux is decomposed as follows: $\overline{w'\theta'} = \overline{w'\theta'}|_b + \overline{w'\theta'}|_t$. Similarly, for temperature variance, it can be obtained that (Sorbjan, 1988):

$$\overline{\theta'^2} = \overline{\theta'^2}|_b + \overline{\theta'^2}|_t \tag{1}$$

TABLE I

Definition of local scales in the convective boundary layer.

Parameter	b-scaling	t-scaling
Height	z	$Z = z_i - z + \Delta$
Velocity	$U_b = [\beta z \overline{w'\theta'}|_b]^{1/3}$	$U_t = [\beta Z \overline{w'\theta'}|_t]^{1/3}$
Temperature	$\Theta_b = -\dfrac{\overline{w'\theta'}|_b}{U_b}$	$\Theta_t = -\dfrac{\overline{w'\theta'}|_t}{U_t}$
Components of the heat flux	$\overline{w'\theta'}|_b = \overline{w'\theta'}|_0(1 - z/z_i)$	$\overline{w'\theta'}|_t = \overline{w'\theta'}|_i z/z_i$

and

$$\frac{\overline{\theta'^2}|_b}{\Theta_b^2} = c_{\theta b}$$

$$\frac{\overline{\theta'^2}|_t}{\Theta_t^2} = c_{\theta t},$$ (2)

where $c_{\theta b}$, $c_{\theta t}$ are coefficients independent of height. From Table I, it can also be found that:

$$\Theta_b = \Theta_*(z/z_i)^{-1/3}(1 - z/z_i)^{2/3}$$

$$\Theta_t = \Theta_* R^{2/3}(z/z_i)^{2/3}(1 - z/z_i + D)^{-1/3},$$ (3)

where Θ_* is the mixed-layer temperature scale, $D = \Delta/z_i$, and R is a ratio of the heat fluxes at the top and bottom of the boundary layer. It can be noticed that if $D \neq 0$, Θ_t is finite at the top of the mixed layer and as a consequence, $\overline{\theta'^2}$ is also finite on this level.

Equations (1), (2) and (3) depict a general rule of the local similarity parameterization in the mixed layer: a statistical moment of the turbulent field is expressed as a combination of two basic functions which are obtained in terms of dimensional analysis.

The curve described by (1)–(3) and plotted for $c_{\theta b} = 2$, $c_{\theta t} = 15$, $R = -0.2$, $D = 0.1$ is shown in Figure 2, together with the Minnesota data. The curve is in

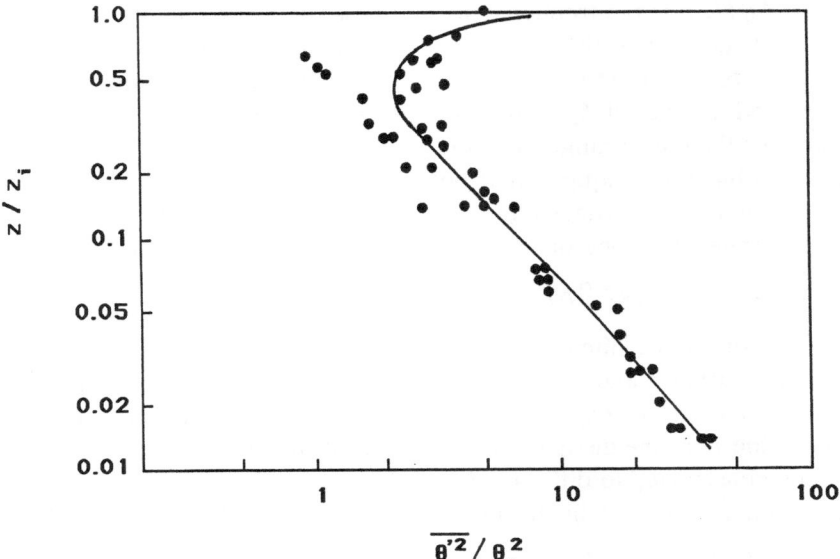

Fig. 2.   Temperature variance in the convective boundary layer (points, the Minnesota data, Kaimal et al., 1976; curve obtained from Equations (1)–(3)).

agreement with the empirical points. The points which depart from the rest in the middle of the mixed layer were probably obtained from runs with different values of parameters R and D. Unfortunately, these parameters cannot be evaluated from the Minnesota data.

In the next section, local similarity will be used to develop a model of the temperature spectrum in the CBL.

2. Temperature Spectra

2.1. MODEL

Following the local similarity approach, the logarithmic temperature spectrum can be split into two components:

$$nS_\theta(n) = nS_{\theta b}(n) + nS_{\theta t}(n) . \tag{4}$$

Each component when non-dimensionalized by the local scale is assumed to be a universal function of reduced frequencies only:

$$\frac{nS_{\theta b}(n)}{\Theta_b^2} = c_{\theta b} F_b(f_b)$$

$$\frac{nS_{\theta t}(n)}{\Theta_t^2} = c_{\theta t} F_t(f_t) , \tag{5}$$

where f_b and f_t are the reduced frequencies: $f_b \equiv f = nz/U$, and $f_t = n(z_i - z + \Delta)/U$, n is the frequency in s^{-1}, U is the wind velocity and $c_{\theta b}$ and $c_{\theta t}$ are constants describing the distribution of temperature variance.

Functions F_b and F_t in (5) can be found by imposing two constraints: one on the slope and a value of S_θ in the inertial subrange and one on the value of its integral over the entire range of frequencies.

The first constraint requires that both functions F have to be consistent with the '−2/3 law' in the inertial subrange of frequencies. Corrsin (1951) derived that in this subrange, the slope of the spectral curves is described by the expression:

$$nS_\theta(n) = \alpha_\theta \varepsilon^{-1/3} N k^{-2/3} , \tag{6}$$

where k is the wave number, $k = 2\pi n/U$, N and ε are dissipation rates for temperature variance and turbulent kinetic energy and α_θ is a universal constant, estimated from various experiments to be about 0.4 (e.g., Kaimal, 1978) (some authors define N as the dissipation rate of half of the temperature variance which causes the constant α_θ to differ by a factor of 2).

Dimensional analysis indicates that similarly to (6) in the inertial subrange:

$$nS_{\theta b}(n) = C_b \Theta_b^2 f^{-2/3}$$

$$nS_{\theta t}(n) = C_t \Theta_t^2 f_t^{-2/3} , \tag{7}$$

where C_b and C_t are coefficients. Moreover, for consistency of (7) and (6), it is required that the product $\varepsilon^{-1/3} N$ in (6) can be split into b- and t-components and scaled in terms of local scales:

$$\varepsilon^{-1/3} N = c_{pb} \frac{\Theta_b^2}{z^{2/3}} + c_{pt} \frac{\Theta_t^2}{(z_i - z + \Delta)^{2/3}}, \tag{8}$$

where c_{pb} and c_{pt} are coefficients.

Equations (4)–(8) lead to the conclusion that in the inertial subrange:

$$F_b = C_b^{-1} c_{\theta b} f^{-2/3}$$
$$F_t = C_t^{-1} c_{\theta t} f_t^{-2/3}, \tag{9}$$

where

$$C_b = (2\pi)^{-2/3} \alpha_\theta c_{pb}$$
$$C_t = (2\pi)^{-2/3} \alpha_\theta c_{pt}. \tag{9'}$$

Substituting (7) and (3) into (4) yields:

$$nS_\theta(n) = C_b \Theta_b^2 (f)^{-2/3} + C_t \Theta_t^2 (f_t)^{-2/3}$$

$$= \Theta_*^2 f_i^{-2/3} \left\{ C_b \frac{(1 - z/z_i)^{4/3}}{(z/z_i)^{4/3}} + C_t R^{4/3} \frac{(z/z_i)^{4/3}}{(1 - z/z_i + D)^{4/3}} \right\}. \tag{10}$$

This equation reflects the tendency depicted in Figure 1; for given $f_i = nz_i/U$ in the inertial subrange, the values of the function nS_θ initially decrease with increasing height and then increase when $z \to z_i$.

For the entire spectrum range, functions F_b and F_t in (4) will be assumed to have the following forms:

$$F_b(f) = \frac{f/f_{mb}}{(1 + 1.5 f/f_{mb})^{5/3}}$$

$$F_t(f) = \frac{f/f_{mt}}{(1 + 1.5 f/f_{mt})^{5/3}}, \tag{11}$$

where f_{mb} and f_{mt} are peak frequencies. Notice that the second formula formally should state that $F_t(f_t)$ is a function of the variable f_t/ft_{mt} (f_{tm} is a peak frequency for f_t). However, for convenience, f_t/ft_{mt} in (11) has been substituted by f/f_{mt}.

It can be verified through inspection that the integral from 0 to ∞ of the function $(1 + 1.5 f/f_m)^{-5/3}$ is equal to unity (it has to be remembered that $dn/n = d(f/f_m)/(f/f_m)$). Moreover, the high frequency slope is equal to $-2/3$, which is consistent with (9).

Comparing (11) with (9) shows that:

$$f_{mb} = C_{mb}$$

$$f_{mt} = C_{mt} \frac{z/z_i}{1 - z/z_i + D}, \tag{12}$$

where

$$C_{mb} = (1.5)^{5/2}(C_b/c_{\theta b})^{3/2}$$
$$C_{mt} = (1.5)^{5/2}(C_t/c_{\theta t})^{3/2} .$$

(12')

The peak wavelength $\lambda_m = z/f_m$ for both components can be obtained from (12) in the form:

$$\lambda_{mb}/z_i = C_{mb}^{-1} z/z_i$$
$$\lambda_{mt}/z_i = C_{mt}^{-1}(1 - z/z_i + D) .$$

(13)

The equations (13) predict a linear increase of λ_{mt}/z_i in the surface layer and a linear decrease of this quantity in the upper part of the mixed layer. This agrees with the observations. The Minnesota spectra (Kaimal et al., 1976) show that λ_{mt}/z_i increases linearly with height in the surface layer. Above the surface layer, λ_{mt}/z_i shows a tendency to increase slightly with height up to $0.5z_i$ and to decrease again above $0.7z_i$.

Equations (4)–(5), (11)–(12) and (3) describe a model of the temperature spectrum in the CBL. The model indicates that the temperature spectrum depends upon parameters R, D, Θ_* and z_i. The model equations require a specification of coefficients $c_{\theta b}$, $c_{\theta b}$, C_{mb}, C_{mb}. A proper choice of these coefficients is discussed in the next section.

2.2. EVALUATION OF THE MODEL

The model will be first examined in the surface layer. For $z/z_i \to 0$, the scale $\Theta_b^2 \to \kappa^{2/3} T_*^2(z/L)^{-2/3}$ (where κ is von Karman constant, L is Monin–Obukhov length and T_* is the surface-layer temperature scale), the scale $\Theta_t^2 \to 0$ and (10) takes the form:

$$nS_\theta(n)/T_*^2 = C_b\kappa^{2/3}(z/L)^{-2/3}f^{-2/3} .$$

(14)

The equation described by (14) is plotted in Figure 3. Points in this figure were obtained by Kaimal et al. (1972), using the Kansas data, for both $\kappa = 0.35$ and $f = 4$. The best fit of (14) with the empirical point is obtained for $C_b\kappa^{2/3} f^{-2/3} = 0.04$. This implies that $C_b = 0.2$. A similar value, $C_b = 0.195$, was obtained by Panofsky (1978).

For $C_b = 0.2$ and $c_{\theta b} = 2$, from (12') it can be evaluated that $C_{mb} = (1.5)^{5/2}(c_b/c_{\theta b})^{3/2} = 0.087$. This value is between $C_{mb} = 0.12$, obtained by Kaimal et al. in 1982, and $C_{mb} \approx 0.05$, suggested by Kaimal et al. in 1976. As reported by Kaimal et al. (1976), the last value was obtained from data characterized by a significant scatter of λ_{mt}/z_i, twice as large as the w-spectrum.

Moreover, from (9') it can be found that $c_{pb} = (2\pi)^{2/3} C_b/\alpha_\theta = 1.7$. For this value, Equation (8) is plotted versus the Minnesota data in Figure 4; here, it was assumed that $c_{pt} = 15.5$, $R = -0.2$ and $D = 0.1$. The figure indicates that the

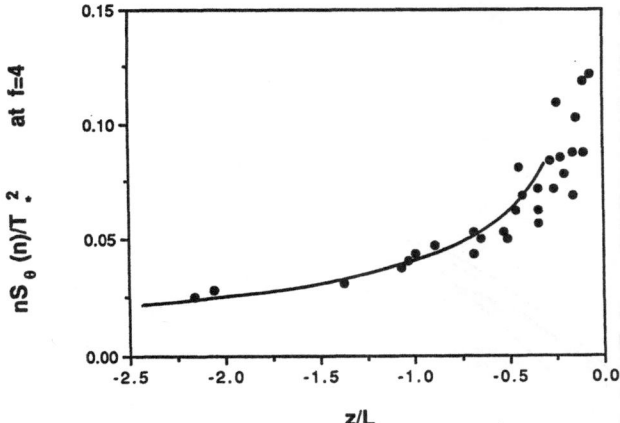

Fig. 3.   Temperature spectrum in the convective surface layer (points, the Kansas data, Kaimal *et al.*, 1972; curve, Equation (14)).

similarity curve obtained for $c_{pb} = 1.7$ slightly underestimates the observed values of the function $N\varepsilon^{-1/3}$ in the lower half of the mixed layer.

For $\alpha_\theta = 0.4$, $c_{pt} = 15.5$ and $c_{\theta t} = 15$, from (9'), it can be estimated that $C_t = (2\pi)^{-2/3}\alpha_\theta c_{pt} = 1.8$ and from (12') that $C_{mt} = (1.5)^{5/2}(C_t/c_{\theta t})^{3/2} = 0.11$. At the top of the mixed layer, $\lambda_{mt}/z_i = C_{mt}^{-1}D = 9.1D$. For $D = 0.1$, this value is 0.91 and is lower than the value 1.5 suggested by Kaimal *et al.* (1976).

The model equations (4)–(5), (11)–(12) and (3) are plotted in Figure 4 for

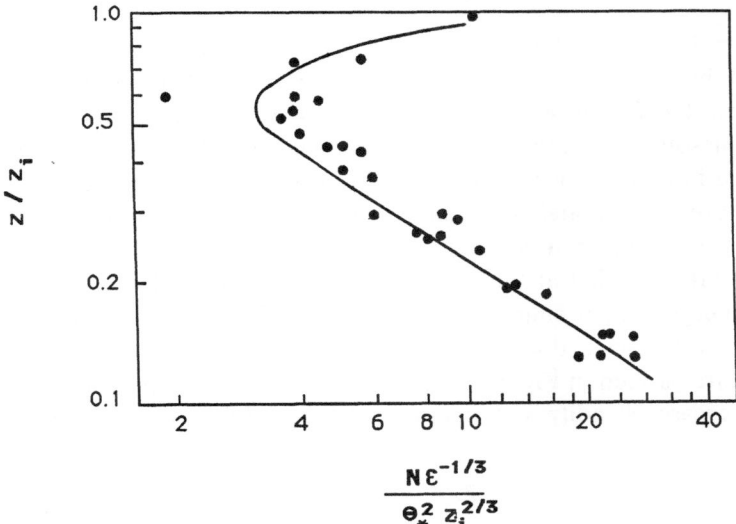

Fig. 4.   Function $N\varepsilon^{-1/3}$ in the convective boundary layer (points, the Minnesota data, Kaimal *et al.*, 1976; curve, Equation (8)).

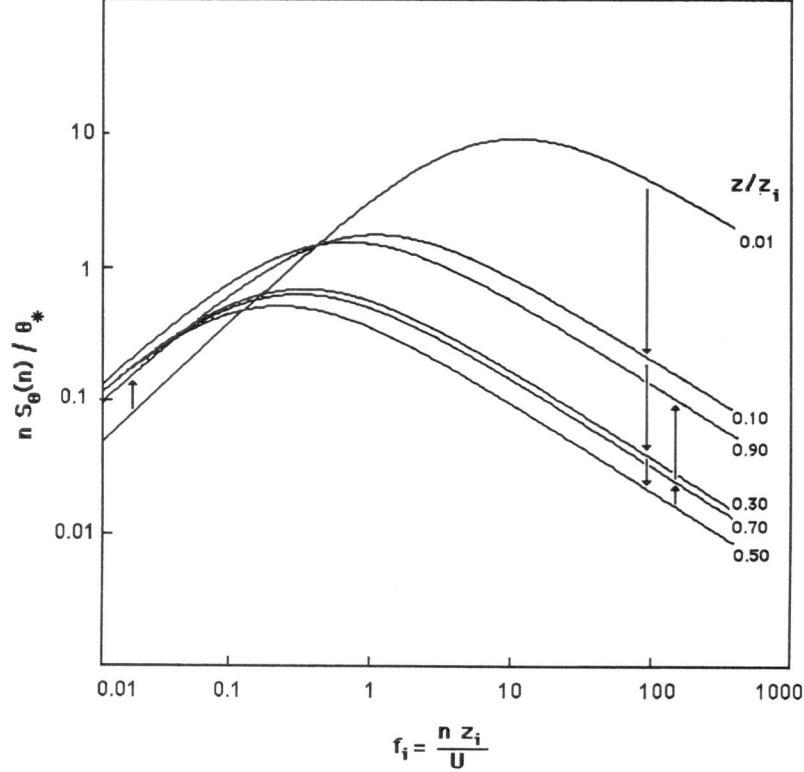

Fig. 5. Temperature spectrum obtained from the model.

$c_{\theta b} = 2$, $c_{\theta t} = 15$, $C_{mb} = 0.087$, $C_{mt} = 0.11$, and for parameters $R = -0.2$ and $D = 0.1$. It has to be stressed that all values of the adopted constants are approximate. Because the exact values of parameters R and D were not available for the Minnesota data, it was arbitrarily assumed that $R = -0.2$ and $D = 0.1$.

The curve in Figure 5 is shown in dimensionless coordinates. Idealized spectra in Figure 1 are dimensional. This makes the comparison of Figures 1 and 5 quite difficult. However, the general trend from Figure 1 is reproduced in Figure 5.

Figure 5 indicates that at the low-frequency end, above the surface layer, the curves converge into a single one as observed in the atmosphere. The low-frequency end, however, does not extend upward in response to the diurnal trend in temperature as seen in Figure 1. At inertial subrange frequencies, the spectral intensity decreases steadily with height to $0.5z_i$, and rises again above this level.

3. Conclusions

A model for the temperature spectrum in the CBL based on local similarity parameterization of the mixed layer has been presented. The model is designed to prescribe a proper slope and a value of the spectrum in the internal subrange and a proper value of the integral of the spectral function over the entire range of frequencies.

The spectral function obtained from the model agrees with the Kansas spectra in the surface layer and reproduces the tendency observed during the Minnesota experiment in the mixed layer. Values of the constants used in the model equations are approximate and require further consideration.

References

Corrsin, S.: 1951, 'On the Spectrum of Isotropic Temperature Fluctuations in an Isotropic Turbulence', *J. Appl. Phys.* **22**, 469–473.

Højstrup, J.: 1981, 'A Simple Model for Adjustment of Velocity Spectra in Unstable Conditions Downstream of an Abrupt Change in Roughness and Heat Flux', *Boundary-Layer Meteorology.* **21**, 341–356.

Højstrup, J.: 1981, 'Velocity Spectra in Unstable Planetary Boundary Layer', *J. Atmos. Sci.* **39**, 2239–2248.

Kaimal J. C., Wyngaard, J. C., Izumi, Y., and Coté, O. R.: 1972, 'Spectral Characteristics of Surface Layer Turbulence', *Quart. J. R. Meteorol. Soc.* **98**, 563–589.

Kaimal J. C., Wyngaard, J. C., Haugen, D. A., Coté, O. R., and Izumi, Y.: 1976, 'Turbulence Structure in the Convective Boundary Layer', *J. Atmos. Sci.* **33**, 2152–2169.

Kaimal, K. C.: 1978, 'Horizontal Velocity Spectra in a Unstable Surface Layer', *J. Atmos. Sci.* **35**, 18–24.

Kaimal, J. C., Eversole, R. A., Lenschow, D. H., Stankov, B. B., Kahn, P. H., and Businger, J. A.: 1982, 'Spectral Characteristics of the Convective Boundary Layer over Uneven Terrain', *J. Atmos. Sci.* **39**, 1098–1114.

Panofsky, H. A.: 1978, 'Matching in the Convective Planetary Boundary Layer', *J. Atmos. Sci.* **35**, 272–276.

Sorbjan, Z.: 1988, 'Local Similarity in the Convective Boundary Layer', *Boundary-Layer Meteorol.*, **45**, 237–250.

DEPENDENCE OF VELOCITY VARIANCE ON SAMPLING TIME*

GABRIELE C. WOLLENWEBER

Wildbadstr. 173, 5580 Traben-Trarbach, Federal Republic of Germany

and

HANS A. PANOFSKY

San Diego, CA, U.S.A. (Deceased 28 February 1988)

(Received 21 September, 1988)

Abstract. The dependence of the lateral and vertical velocity variances on sampling time have been investigated, based on recent spectral models. The results have been compared with observations at Vandenberg Air Force Base and at the Boulder Atmospheric Observatory.

For sampling times of 1 s to 1 hr, the variation of lateral variances is much smaller in neutral than in unstable conditions. In a stable atmosphere, the variances show a stronger increase with sampling time than predicted by the model. The vertical velocity variances do not increase significantly between 1 and 60 min. Again in stable conditions, the measurements show a more pronounced increase of variances as compared to the calculations, indicating significant energy between periods of a few to 20 min. This additional energy may be ascribed to gravity waves.

The models cannot be applied in cases with trends in the data. The model tends to underestimate the dependence of variances on sampling time in cases when systems, such as longitudinal vortices or gravity waves exist.

1. Introduction

To determine dispersion of gaseous or finely dispersed material released into the atmosphere, models are often used that rely on σ_y, σ_z, the lateral and vertical dispersion parameters, respectively. The variations of σ_y and σ_z with stability and distance to the source have been determined by field observations with limited sampling time τ. The popular Pasquill–Gifford scheme, for example, or modifications by Briggs (1973), are based on measurements of about 5 min sampling time or less.

In many applications, the same values have been applied to problems of a time scale of 30 min or even longer. In pollution problems in cities, for example, one-hour average concentrations are the basic quantities required. Actually one-hour concentrations may be significantly lower than averages over shorter periods. Further, in case of accidental release of very hazardous material, maximum concentrations have to be estimated for shorter periods. In general, it would be useful to be able to find a relationship between concentrations and sampling time.

To describe the widening of a plume as a function of sampling time, a power law (Hanna *et al.*, 1982) is often used.

* This work was undertaken while the authors were associated with the Scripps Institution of Oceanography, La Jolla, California.

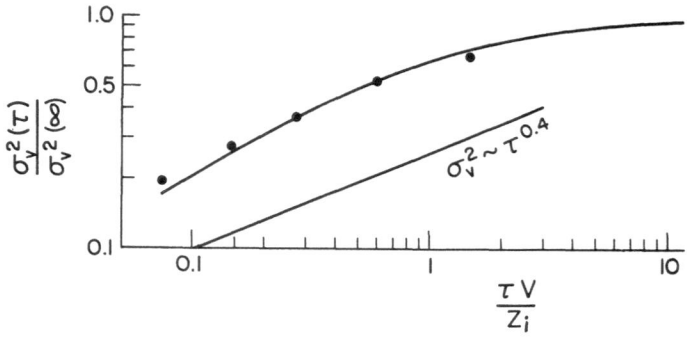

Fig. 1. Ratio of wind direction variances as a function of sampling time τ (solid line – approximation, dots – measurements, straight line – power law).

$$\sigma_y = A\tau^p .$$

Here A is a constant and τ is the sampling time. The power p is usually taken as 0.2.

Gifford (1975) assumed the power to be constant. The standard deviations of horizontal and vertical velocity components grow with increasing sampling time τ. The main reason for this is that larger eddies are taken into account. The rate of growth depends on variables such as mean wind speed and stability. The power p should be small in near neutral conditions with little low frequency energy, and should be larger in unstable conditions. As a result, the power p cannot be expected to be constant. In addition to this, the variances should become independent of sampling time and the power should tend to zero for sampling times reaching the spectral gap.

To illustrate how well the power law describes the dependence of variances on sampling time, Figure 1 shows measurements of σ_v^2 made by Buell (1985) at Vandenberg Air Force Base together with a power law corresponding to the popular $\sigma_v \approx \tau^{0.2}$ formula (for greater convenience, the power law is drawn with an offset). Also shown is an approximation of the filter function applied to the convective part of Højstrup's (1982) spectrum.

2. Theory

According to a modified form of Taylor's diffusion theorem (see, e.g., Pasquill and Smith, 1983), the dispersion parameters can be estimated in terms of wind direction fluctuations σ_θ and distance x from the source:

$$\sigma_y = \sigma_\theta \cdot x \cdot f(x) .$$

The function $f(x)$ decreases from unity as x increases. As shown in Pasquill and Smith (1983), there exist several models for $f(x)$, which depend to some extend on stability. Thus the effect of sampling time on σ_y is due largely to the effect of sampling time on σ_θ, at least for small x.

The width of the plume and the standard deviation of wind direction can be produced by eddies of different sizes. It is assumed that the variance of wind direction is the sum of variances caused by fluctuations with periods less than 1 hr and periods greater than 1 hr. Variances of periods exceeding 1 hr can be produced by mesoscale and large-scale systems. They may produce trends in time series of 1 hr and because of this, only observations with no significant trend will be analyzed here.

This investigation will concentrate on small-scale turbulence only, driven by wind shear and vertical temperature difference. The results should not be applied if systems exist with periods near 1 hr as for example in cases of longitudinal vortices. The models will also fail in cases when gravity waves contribute significantly to the variances. In those cases, the model will underestimate the dependence of variances on τ for longer sampling times.

It is possible to express the variance of wind direction as an integral over the spectral density $S_\theta(f)$ by Equation (1):

$$\sigma_\theta^2 = \int_0^\infty S_\theta(f)\, \mathrm{d}f, \tag{1}$$

where f is the frequency. If fluctuations are considered only over a finite sampling time, low frequencies will be partly excluded. To describe the effect on variances, a filter function has to be applied. The filter (Pasquill and Smith, 1983) depends on sampling time and frequency:

$$\sigma_\theta^2(\tau) = \int_0^\infty \left(1 - \frac{\sin^2 \pi f \tau}{(\pi f \tau)^2}\right) S_\theta(f)\, \mathrm{d}f. \tag{2}$$

Equation (2) gives approximately the same result as a sharp cutoff of the spectrum at a frequency of $f = \frac{1}{2\tau}$. The difference in integrating Equation (2) and using the approximation of a cutoff of the spectral energy will be shown later.

To study the influence of sampling time on variances, spectral models proposed by Højstrup (1981, 1982) are used to integrate Equation (2) numerically. They are valid for unstable and neutral conditions and can be applied to the surface layer. They were designed for the lateral velocity component and show good agreement with the Kansas data. Højstrup (1981, 1982) assumed that there are two contributions to spectral density and that there is only weak interaction between them. So each part can be modeled separately. The low frequency part is buoyancy produced and does not explicitly depend on height above the surface but on the height of the lowest inversion z_i. This term is essentially the same as in Kaimal's (1978) model. The high frequency part is produced by wind shear. It is characterized by the observation height z and the mean wind speed. If the height increases, the spectrum is shifted towards lower frequencies.

In contrast to Kaimal's model, this spectrum has a continuous transition from unstable to neutral conditions and a single valued neutral case. Both spectra show the inertial subrange with a $-\frac{5}{3}$ slope. Højstrup's model will be used in the surface layer, where $z \ll z_i$:

$$\frac{fS_v(f)}{u_*^2} = \frac{0.95 n_i}{(1+2n_i)^{5/3}}\left(\frac{z_i}{-L}\right)^{2/3} + \frac{17n}{(1+9.5n)^{5/3}}. \tag{3}$$

Here u_* is the friction velocity, $n_i = fz_i/V$, $n = fz/V$, L is the Monin–Obukhov length and V is the mean velocity. To be able to use Højstrup's spectrum, it has to be assumed that the variances of wind direction can be estimated by variances of the lateral velocity component. With the spectral density given by Equation (3), it is possible to integrate Equation (2) numerically. Two parameters a and a_i are introduced to represent the product of sampling time and mean velocity, divided by characteristic lengths z and z_i:

$$a = V\tau/z, \qquad a_i = V\tau/z_i.$$

To obtain the relative change of variance, the equation is normalized by the variance for a sampling time of infinity:

$$\frac{\sigma_v^2(\tau)}{\sigma_v^2(\infty)} \equiv \frac{\sigma_\theta^2(\tau)}{\sigma_\theta^2(\infty)} = \frac{3.8F(a)+(z/-L)^{2/3}G(a_i)}{3.8+(z_i/-L)^{2/3}}. \tag{4}$$

As a result of this integration, the variances of the lateral velocity component will be explicit functions of z, z_i, L, wind speed and the non-dimensional sampling times a and a_i, respectively.

In Equation (4), the influence of two different physical effects can be seen. The function $G(a_i)$ is the ratio of variances for pure convective turbulence. $G(a_i)$ is characterized by the height of the well mixed layer and does not depend on height as long as the mean wind speed does not change with height. In neutral conditions, Equation (4) is controlled by $F(a)$ only. $F(a)$ represents the pure mechanical case. Figure 2 shows the ratio of variances for the two limiting cases as functions of a and a_i, respectively.

If it is assumed that $V\tau$ is constant, an increase in observation height z results in a decrease in a. As can be seen from Figure 2a, the mechanical contribution increases rapidly with decreasing height. There are only slight changes for values of $a > 100$ resulting in variances that are up to 4% smaller than for greater values of a. With a mean wind speed of $V = 10$ m/s and a height of $z = 20$ m, the sampling time has to be 3–4 min. But for a height of $z = 100$ m, the variance has to be determined for at least 16–17 min to show the same small difference compared to longer sampling times.

Figure 2b shows the increase of the ratio of variances for decreasing mixing-layer depth in pure convective conditions (again $V\tau$ is kept constant). For

$\sigma_\theta^2(\tau)/\sigma_\theta^2(\infty)$ as function of a

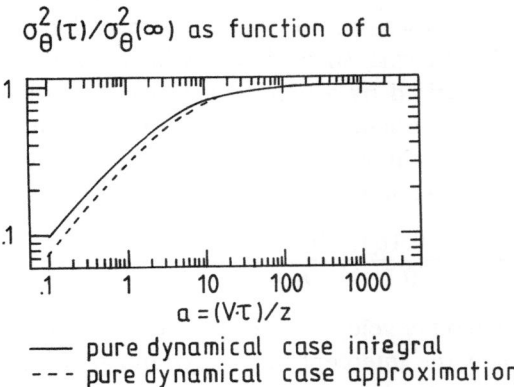

— pure dynamical case integral
- - - pure dynamical case approximation

Fig. 2a.   Dynamical part of Equation (4).

$\sigma_\theta^2(\tau)/\sigma_\theta^2(\infty)$ as function of a_i

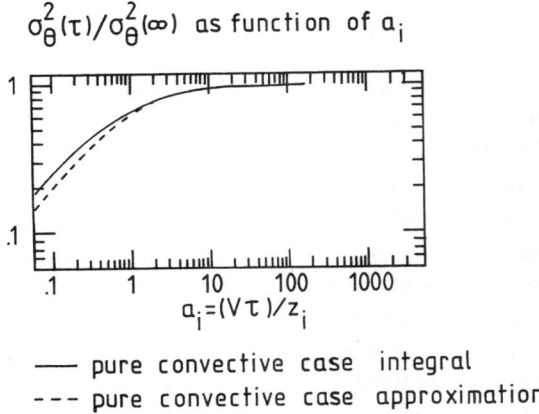

— pure convective case integral
- - - pure convective case approximation

Fig. 2b.   Convective part of Equation (4).

$a_i > 10$, the ratio is up to 4% smaller than for longer sampling times. As a consequence, the variances for $z_i = 1000$ m have to be determined over periods that are 2 times longer than for $z_i = 500$ m under the assumption that the mean wind speed is the same in both cases.

These results have great practical implications, since in the case of an accidental release of a toxic substance, it is necessary to know what sampling times will be long enough to determine the dispersion parameters with sufficient accuracy.

In Figure 2a and 2b, the dashed lines show the results when the filter function is approximated by a sharp cutoff of the spectrum at $f = \frac{1}{2\tau}$. For $a > 10$ and $a_i > 3$, the difference between the integration and the approximation is small. The difference increases for smaller values of a and a_i, respectively, and then cannot be disregarded.

To include stable cases, we adopted a lateral velocity spectrum suggested by Kaimal *et al.* (1972), who showed that with appropriate scaling, the velocity spectra are characterized by a family of curves according to z/L. Normalizing the spectra with the variances rather than the friction velocity and by using a modified frequency scale n/n_0, all spectra exhibit one universal shape that is approximated by Equation (5):

$$\frac{fS_v(f)}{\sigma_v^2} = \frac{0.164(n/n_0)}{1 + 0.164(n/n_0)^{5/3}}.$$ (5)

Following a method developed by Moraes (1988), we can determine the scaling frequency n_0 (n_0 is the intercept of the extrapolated inertial subrange spectrum with the $fS_v(f)/\sigma_v^2 = 1$ line):

$$n_0 = \beta_s(1 + 3.7z/D)$$

$$\beta_s = 0.045(1 + 0.628h/L)^{0.9}/(1 + 0.422h/L)$$ (6)

$$D = (1 - z/h)^{5/4}L$$

where h is the height of the stable boundary layer.

The spectral density given by Equation (5) is used to integrate Equation (2) numerically. Figure 3 shows the results of the integration as a function of a, with $h/L = 13.3$. There are only slight changes in the ratio of variances with a, reflecting the fact that the energy-containing eddies are rather small under stable stratification as compared to neutral or unstable conditions.

Under stable conditions, boundary-layer turbulence is less understood than its neutral or unstable counterparts. In some cases, the stable boundary layer is influenced by processes in addition to turbulence, e.g., gravity waves or horizontal vortices as described by Zhou and Panofsky (1983) or Le Mone (1973). These processes result in significant variations in wind direction and speed in periods of the order of 10 to 30 min. They greatly affect the energy spectrum and are not addressed by the stable model.

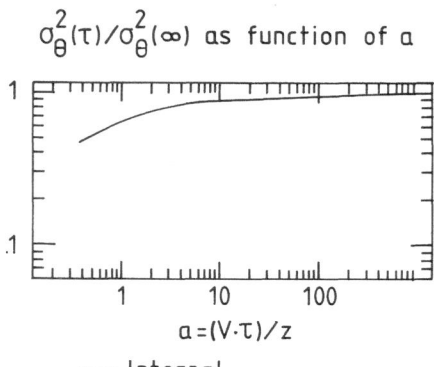

$\sigma_\theta^2(\tau)/\sigma_\theta^2(\infty)$ as function of a

$a = (V \cdot \tau)/z$

—— Integral

Fig. 3.   Ratio of wind direction variances according to Equation 5 and Equation 6.

To study the influence of sampling time on vertical plume spread, an approach analogous to Equation (2) is used. A spectrum of vertical velocity variances was chosen from several models suggested in Panofsky and Dutton (1984). This spectrum is easy to integrate and fits the observations well:

$$\frac{fS_w(f)}{u_*^2} = 2n \left(1 + 1.5 \frac{n}{n_m}\right)^{-5/3} \quad \text{where} \quad \left(n_m \equiv \frac{f_m z}{V}\right) \tag{7}$$

where n_m is the peak frequency where $fS_w(f)$ has its maximum. The peak frequency depends on stability and is about 0.5 in neutral and stable conditions and 0.2 in unstable conditions. In Panofsky and Dutton (1984), interpolation formulas for n_m as a function of z/L can be found.

The model chosen implies that the peak frequency but not the shape of the spectrum depends on stability. This is a simplification because vertical velocity spectra widen as z/L decreases. The distance to the ground limits the low frequency part of the vertical velocity spectrum. So the effect of sampling time on the vertical velocity variances is expected to be small. That gives us reason to believe that it is justified to use a rather simple model instead of a more complicated one as suggested, e.g., by Højstrup (1982).

Equation (2) has to be integrated numerically with the vertical velocity spectrum instead of the lateral velocity spectrum. The parameter $a = V\tau/z$ is introduced and the resulting equation is divided by the variance for a sampling time of $\tau = \infty$:

$$\frac{\sigma_w^2(\tau)}{\sigma_w^2(\infty)} = \phi\left(\frac{z}{V\tau n_m}\right). \tag{8}$$

In Figure 4, the ratio of variances is shown as a function of $a \cdot n_m$. The variances do not increase much for $a \cdot n_m > 12$ and are up to 4% smaller than variances for $\tau = \infty$. For smaller values of $a \cdot n_m$, the variances decrease. The

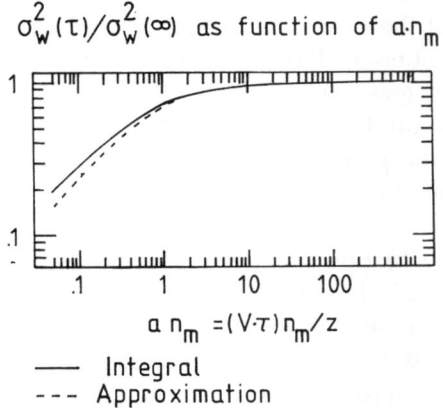

$$\sigma_w^2(\tau)/\sigma_w^2(\infty) \quad \text{as function of } a \cdot n_m$$

$$a\, n_m = (V \cdot \tau)\, n_m / z$$

—— Integral
--- Approximation

Fig. 4.   Ratio of vertical velocity variances according to Equation (8).

dashed curves show the results of the approximated filter function. Good agreement can be found for $a \cdot n_m > 2$.

3. Comparison with Observations

To compare the results of these models with measurements under varying meteorological conditions, data sets of Vandenberg AFB and the Boulder Atmospheric Observatory were used.

At Vandenberg AFB, data from one of the meteorological towers were evaluated. The tower has a uniform fetch over chaparral of at least 1 km for onshore winds. Wind direction and wind velocity fluctuations were gathered with a Geotech anemometer and a wind vane at a height of 16 m. The Monin–Obukhov length could not be determined accurately, but measurements of z_i were available. Therefore, computations were made for a number of reasonable estimates of z_i/L based on wind speed, time of day and an estimated vertical temperature profile. The Vandenberg observations include standard deviations for sampling periods from 15 s to 1 hr.

At Boulder, all three velocity components were measured at heights of 10 to 300 m. The Monin–Obukhov length was determined by eddy correlation techniques. Mixing-layer depth was determined in a few cases but could be estimated from additional measurements in the other cases. The terrain is not completely flat, and there are a few houses in the vicinity. In most cases, data taken from the two lowest measurement levels were compared with the calculations. The Boulder data included standard deviations for sampling periods of 1 min to 1 hr.

The observations represent the variances of wind direction as a function of sampling time normalized by the variance of the longest sampling period of 1 hr. The model results are normalized by variance of a sampling period of $\tau = \infty$. In the examples shown, the differences between a sampling time of 1 hr and infinity are very small. Since vertical velocity components were only measured at Boulder, a comparison with model results is limited to this measurement site.

In Figure 5, results of the model are compared with measurements at Vandenberg AFB in July around noon. The mixing-layer height is not very great, so that the variances do not increase much with τ. The variation of variance with τ is much larger than in neutral conditions. With reasonable choices of z_i/L, the model fits well. Since the graphs are bilogarithmic, a simple power law should appear as a straight line, but these results show that a constant exponent is not appropriate.

In Figure 6, a stable situation is compared with the results of the neutral and stable models. As expected, the ratio of variances depends less on τ for the stable model as compared to the neutral model. But the measurements increase much faster than predicted by both models. Since the slope of the graph equals fS_θ for $f = \frac{1}{2\tau}$, additional energy must be contributed in periods between 1 and 20 min. Presumably there is a spectral peak that might be produced by gravity waves.

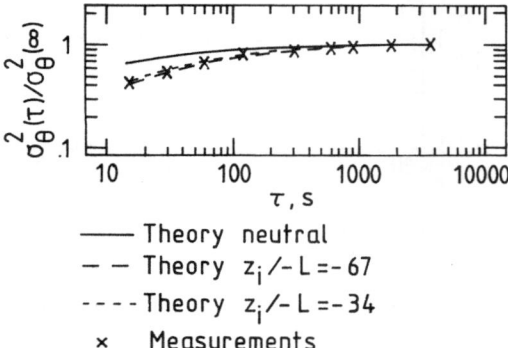

Fig. 5. Ratio of wind direction variances at Vandenberg, unstable case.

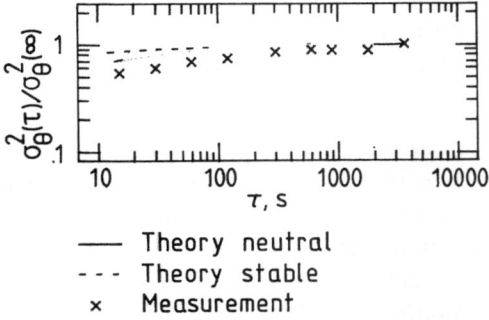

Fig. 6. Ratio of wind direction variances at Vandenberg, stable case.

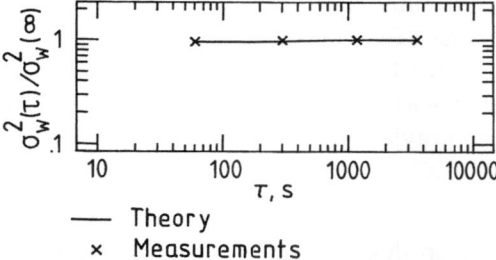

Fig. 7a. Ratio of vertical velocity variances at Boulder, level I, stable case.

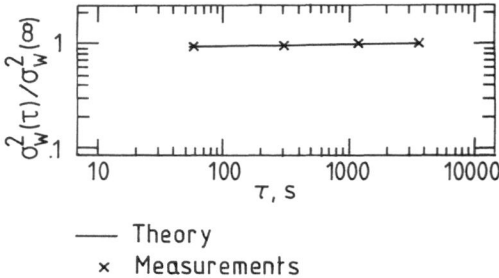

— Theory
× Measurements

Fig. 7b.   Ratio of vertical velocity variances at Boulder, level II, stable case.

Figure 7 shows the variation of vertical velocity variance for a nighttime situation. At the two lowest levels, the variances do not increase much with sampling time. Obviously there is not much low frequency energy. Further studies at other levels (not shown here) indicate considerable low frequency energy with periods of the order of minutes, reaching a maximum at heights of 100–150 m. This might be produced by gravity waves.

5. Conclusion

To study the dependence of the lateral and vertical dispersion parameters on sampling time, models have been derived that are based on Højstrup's (1981, 1982) spectral models for neutral and unstable air and on a spectral model by Kaimal et al. (1972) for stable cases, in a version revised by Moraes (1988). Models suggested by Panofsky and Dutton (1984) are used to determine vertical velocity variances. The models were compared with measurements. Models and observations agreed well as long as measurements that contained characteristic periods of about 1 hr or trends were omitted. In unstable conditions, the lateral variances change considerably faster than in neutral air particularly for small τ. For stable air, the variances increased faster than the models predicted. In this case, considerable energy was found at periods of the order of 10 min. Presumably this energy is contributed by gravity waves.

For vertical velocities, the dependence of variances on sampling time is generally weak at the two lowest measurement levels at Boulder. For a single stable case, substantial variation with τ was found between heights of 100 to 150 m, possibly due to gravity waves.

Acknowledgement

The authors would like to thank Gordon Schacher and Charles Skupniewicz for making the Vandenberg data available, and Chandran Kaimal and John Gaynor for the Boulder data and for some initial data processing. We are particularly grateful to Jürgen Richter of NOSC (US Navy) for general encouragement.

References

Briggs, G. A.: 1973, 'Diffusion Estimation for Small Emissions', ATDL Contribution File No. 79, Atmospheric Turbulence and Diffusion Laboratory.

Buell, R. J.: 1985, 'Mean Flow and Turbulence in Complex Terrain', Master thesis, Naval Postgraduate School, Monterey, CA.

Gifford, F. A.: 1975, 'Atmospheric Dispersion Models for Environmental Pollution Applications', in: *Lectures on Air Pollution and Environmental Impact Analysis*, Boston, MA, 29 Sept.–3 Oct. 1975, American Met. Soc., Boston, MA.

Hanna, S. R., Briggs, G. A., and Hosker, R. F.: 1982, 'Handbook on Atmospheric Diffusion', U.S. Dept. of Energy DOE/TIC 11223.

Højstrup, J.: 1981, 'A Simple Model for the Adjustment of Velocity Spectra in Unstable Conditions to an Abrupt Change in Roughness and Heat Flux', *Boundary-Layer Meteorol.* **21**, 341–356.

Højstrup, J.: 1982, 'Velocity Spectra in the Unstable Boundary Layer', *J. Atmos. Sci.*, **39**, 2239–2248.

Kaimal, J. C., Wyngaard, J. C., Izumi, Y., and Cote, O. R.: 1972, 'Spectral Characteristics of Surface Layer Turbulence', *Quart. J. R. Meteorol. Soc.*, **98**, 563–589.

Kaimal, J. C.: 1978, 'Horizontal Velocity Spectra in an Unstable Surface Layer', *J. Atmos. Sci.*, **35**, 18–23.

Le Mone, M. A.: 1973, 'The Structure and Dynamics of Horizontal Roll Vortices in the Planetary Boundary Layer', *J. Atmos. Sci.*, **30**, 1077–1091.

Moraes, O. L. L.: 1988, 'The Velocity Spectra in the Stable Atmospheric Boundary Layer', *Boundary-Layer Meteorol.* **43**, 223–230.

Panofsky, H. A. and Dutton, J. A.: 1984, 'Atmospheric Turbulence', Wiley-Interscience, New York.

Pasquill, F. and Smith, F. B.: 1983, 'Atmospheric Diffusion', Ellis Horwood Ltd., Halstead Press, Chichester, England.

Zhou, L. and Panofsky, H. A.: 1983, 'Wind Fluctuations in Stable Air at the Boulder Tower', *Boundary-Layer Meteorol.* **25**, 353–362.

WAVE DRAG IN THE PLANETARY BOUNDARY LAYER OVER COMPLEX TERRAIN*

GEORGE CHIMONAS

School of Geophysical Sciences, Georgia Institute of Technology, Atlanta, Georgia 30332-0420, U.S.A.

and

CARMEN J. NAPPO

Atmospheric Turbulence and Diffusion Division Air Resources Laboratory/NOAA, Oak Ridge, TN 37831-2456, U.S.A.

(Received in final form 27 September, 1988)

Abstract. The concepts of mountain-induced wave drag are applied to the smaller scale problem of the boundary layer over complex terrain. It is found that the Reynolds stress and surface drag caused by surface-generated waves can be at least as large as those conventionally associated with turbulence. Conditions in which wave effects are important are identified.

1. Introduction

The dominant features of the planetary boundary layer over flat terrain are now well understood and can be simulated with a fair degree of accuracy. But almost all concentrations of population and industry occur in regions of rather complex terrain, creating the need for reliable models of complex flow fields. This has been a major concern of boundary-layer research in the recent decade (Egan, 1984).

Research has largely developed along two lines: the development of techniques for determining the path of the fluid as it negotiates its three-dimensional obstacle course (Hunt and Richards, 1984; Cermak, 1984), and the development of better methods for computing diffusion within such a flow (Gifford, 1984; Wyngaard *et al.*, 1971; Berkowicz, 1984). While it is recognized that factors such as gravity-fed flows along sloping surfaces introduce new features into the dynamics, it is the general practice to model the boundary-layer profile and its characteristics with some local version of the models developed for flat terrain. This may well be valid in many situations, but there are conditions in which terrain features lead to characteristics that have no counterparts over a flat surface.

Such a situation arises when surface undulations create standing waves that interact with the flow. Surface-generated waves may find a critical level within the boundary layer, producing a coupling between the surface and an elevated region without modifying the intervening fluid. In these circumstances, flat-

---

* ATDD Contribution No. 88/5.

Boundary-Layer Meteorology **47**: 1–14, 1989.
© 1989 *Kluwer Academic Publishers.*

boundary-layer scalings which assume that the upper flow does not directly sense the surface (Nieuwstadt, 1984) require some revision, while the conventional turbulence theories provide only a portion of the Reynolds stress budget. Moreover, the critical-level interaction in the lee of a surface feature may produce horizontal eddies that enhance lateral diffusion.

In this paper we identify some of the conditions in which significant effects are likely. The analysis shows that modest undulations can introduce additional Reynolds stress and surface drag that are at least as large as those due to turbulence. The wave interaction is rather sensitive to the form of the flow profiles, so its parameterization may be cumbersome, and most wave effects will be realized in the upper parts of the boundary layer where parameterization and observation are presently quite poor.

2. Resolution of the Velocity Field and Reynolds Stress

At this time there is no rigorous way of distinguishing between the wave field and the turbulence field of the boundary layer. However, it is generally accepted that such a resolution is conceptually useful and will correctly distinguish important dynamic features through most of the disturbance spectrum. Accordingly, we write the velocity field V of the flow in terms of a mean component V_0, a wave component V_w, and a turbulent component V_t, in a Cartesian (x, y, z) representation

$$V = V_0 + V_w + V_t \tag{1a}$$

$$V_0 = (U, V, 0) \tag{1b}$$

$$V_w = (u, v, w) \tag{1c}$$

$$V_t = (u', v', w') . \tag{1d}$$

Gravity g acts in the negative z direction. It is assumed that the lower surface is horizontal in the mean, so that vertical velocities imposed by the terrain can be assigned to the wave field.

This analysis will be concerned entirely with the Reynolds stress, through which we shall estimate the importance of the wave field. Consider the component that involves the correlation of vertical and x-directed fluid motions,

$$\tau = -\langle \rho (U + u + u')(w + w') \rangle . \tag{2}$$

Brackets indicate some averaging procedure, and ρ is the fluid density. We make the usual boundary-layer approximation replacing ρ by a characteristic constant value $\bar{\rho}$, and write

$$\tau = -\bar{\rho} \langle u'w' \rangle - \bar{\rho} \langle uw \rangle . \tag{3}$$

The transition from (2) to (3) has omitted the wave-turbulence interaction terms

$\langle uw' + u'w \rangle$. Moreover, the dependence of $\langle u'w' \rangle$ on wave modification of the local stability will not be included. The theory of wave-overturning, saturation, and turbulence generation is based on this coupling (Hodges, 1967; Lindzen, 1967; Orlanski, 1972; Fritts, 1984), and other types of feedback interactions are also possible (Chimonas, 1972; Einaudi and Finnigan, 1981). However, these considerations are not special to flow over complex terrain, which is the concern of the present study.

Conventional boundary-layer studies have concentrated on the Reynolds stress resulting from the turbulent fluctuations. This stress may be parameterized in terms of the mean wind at some reference height, and a skin friction (or drag) coefficient C_D:

$$\tau = 0.5\rho C_D U^2 .\tag{4}$$

Sutton (1955) summarizes the development of this concept, and provides values for C_D derived under a variety of conditions. Values over land are all rather close to G. I. Taylor's early estimate of 0.005, found using a reference height of 30 m. Accepting this, we concentrate on estimating the surface-driven wave contribution to the stress, the second term on the right of (3).

3. The Wave Stress

A great deal of attention has been directed to the problem of atmospheric drag that arises from mountain-induced gravity waves (Smith, 1976; Bretherton, 1969). Flow over small undulations of the surface produces a fully analogous problem within the stable atmospheric boundary layer. Only the scaling is different. Figure 1 illustrates the situation. The mountain-wave problem has horizontal scales of 10's to 100's of kilometers, while for the boundary-layer problem, the scales are reduced to 100's or 1000's of meters. Moreover, the vertical displacements of the boundary layer will be a few tens of meters, miniscule when compared with mountain ranges. On the other hand, the boundary-layer drag acts within a shallow layer of fluid, so on a relative basis, the boundary-layer waves may be as important in their domain as the large-scale waves are in theirs. It will be shown that the drag associated with wave generation over slightly undulating terrain can be at least as large as the conventional turbulence-induced Reynolds stress discussed in the previous section.

The wave stress is computed with the linearized theory of periodic disturbances in an idealized mean flow. The governing equations are: Euler's equation for an inviscid fluid

$$\rho \frac{D\mathbf{V}}{Dt} = -\nabla p + \rho \mathbf{g} ;\tag{5}$$

the equation of continuity

$$(1/\rho)D\rho/Dt + \mathbf{\nabla} \cdot \mathbf{V} = 0 ;$$ (6)

and the adiabatic law, expressed here as conservation of potential temperature θ

$$D\theta/Dt = 0$$ (7)

where p is pressure and D/Dt is the derivative following a parcel of fluid

$$D/Dt = \frac{\partial}{\partial t} + \mathbf{V} \cdot \mathbf{\nabla} .$$ (8)

The potential temperature is related to density and pressure as

$$\theta = \text{constant} \cdot p^{1/\gamma}\rho^{-1}$$ (9)

where γ is the usual ratio of specific heats.

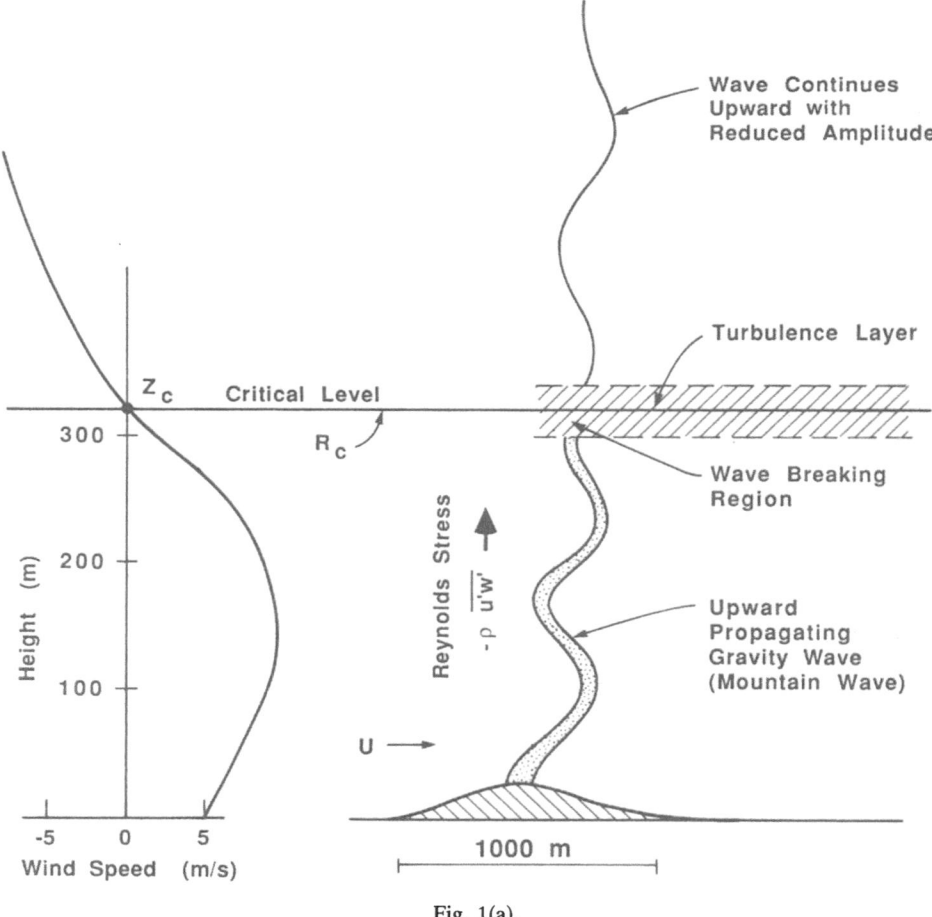

Fig. 1(a).

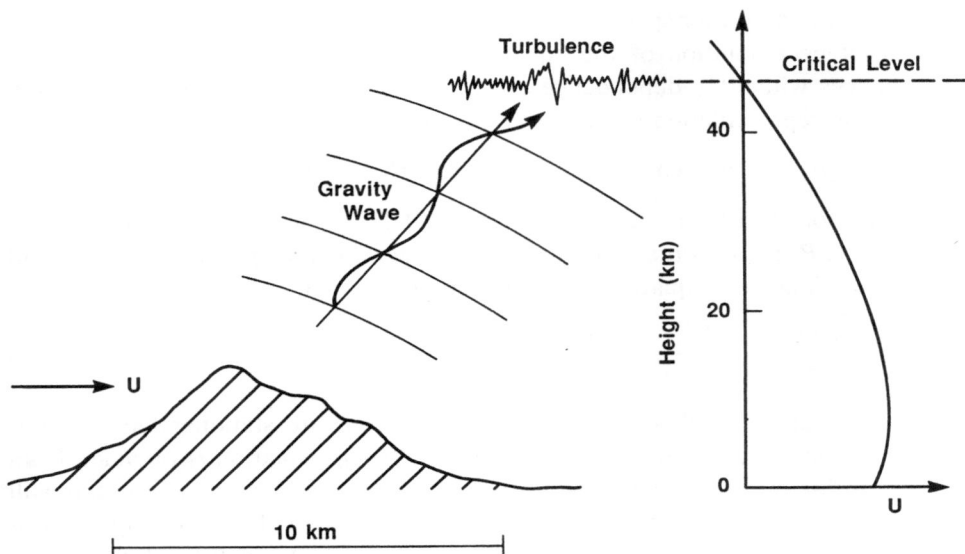

Fig. 1.   Gravity waves are generated by flow over surface formations and are reabsorbed by critical levels in overlying winds. (a) The scales of the boundary-layer waves that are the subject of the present study. (b) The scales of the well-known mountain wave.

Euler's equation is now linearized in disturbances about the mean state, and the terms governing the wave fields are equated. The mean density field $\rho_0(z)$ is replaced by the representative constant $\bar{\rho}$. The resulting equation is

$$\bar{\rho}\left(\frac{\partial}{\partial t}+\mathbf{V}_0\cdot\mathbf{\nabla}\right)\mathbf{V}_w + \bar{\rho}\mathbf{V}_w\cdot\mathbf{\nabla}\mathbf{V}_0 = -\mathbf{\nabla}p_w + \rho_w\mathbf{g}\,. \tag{10}$$

For low-phase-speed waves, the velocity that comes from particles swelling or shrinking to new densities (giving the first term in (6)) is much smaller than the wave velocity of accelerated particle motion. Thus, ignoring these small swelling/shrinking corrections, the continuity equation reduces to

$$\mathbf{\nabla}\cdot\mathbf{V}_w = 0\,. \tag{11}$$

In the same approximation, (7) provides the density perturbation field as

$$\left(\frac{\partial}{\partial t}+\mathbf{V}_0\cdot\mathbf{\nabla}\right)\rho_w = wN^2\bar{\rho}/g \tag{12}$$

where N is the mean Brunt-Vaisala frequency

$$N^2 = \frac{g}{\theta_0}\frac{d\theta_0}{dz}\,. \tag{13}$$

Equations (10)–(12) are linear in the wave fields and are homogeneous in time t and the horizontal coordinates (x, y). Any wave disturbance can therefore be represented as a Fourier spectrum in (x, y, t) space, and each spectral component must itself be a solution of the equations. In this work, only two-dimensional disturbances without y dependence are needed, and it is convenient to use the exponential representation for a harmonic component:

$$\{u_w, w_w, p_w, \rho_w\} = \{u(z), w(z), P(z), R(z)\} \exp[ik(x - ct)]. \tag{14}$$

When the forms (14) are substituted into (10)–(12), four equations in the four fields (u, w, P, R) are obtained. After some algebraic manipulations, these yield the Taylor–Goldstein equation governing the field $w(z)$:

$$\frac{d^2 w}{dz^2} + \left[\frac{N^2}{(c - U)^2} + \frac{d^2 U/dz^2}{(c - U)} - k^2 \right] w = 0. \tag{15}$$

The waves set up by flow over surface undulations are stationary in the conventional rest frame of the Earth, so their governing equation is obtained from (15) by setting c to zero. In this section, a model flow with U and N constant will be studied. This can be a poor representation in the stable boundary layer, but it suffices to provide general estimates. A more realistic flow will be studied later. For now, the surface-generated fields obey the simple limiting form of (15):

$$\frac{d^2 w}{dz^2} + m^2 w = 0, \tag{16}$$

where the constant m is given by

$$m^2 = [N^2/U^2 - k^2]. \tag{17}$$

The velocity fields (14) that satisfy all these restrictions, obey (11), and carry energy away from the surface (rather than downwards from a source in the high atmosphere) are thus

$$w = A \exp[i(kx + mz)] \tag{18}$$

and

$$u = -A(m/k) \exp[i(kx + mz)]. \tag{19}$$

Figure 2 shows the surface boundary condition – purely tangential flow – that determines the wave amplitude in terms of the surface slope dh/dx and the mean flow speed U that must be deflected over the undulation. This surface condition defines $A(k)$ of (18)–(19) for each Fourier k component. If the surface is an ideal sinusoid,

$$h = h_0 \exp(ikx), \tag{20}$$

the wave fields become

$$w = ikUh_0 \exp[i(kx + mz)] \tag{21}$$

$$u = -imUh_0 \exp[i(kx + mz)]. \tag{22}$$

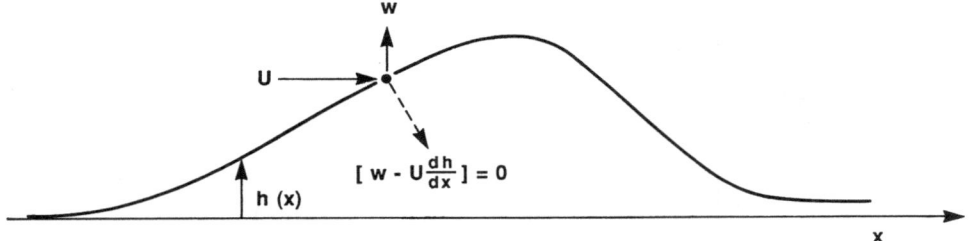

Fig. 2.   The lower boundary condition defines the wave disturbance so that there is no flow normal to the surface.

The wave contribution to the Reynolds stress, the final term in (3), is now computed to be

$$- \bar{\rho}\langle wu \rangle = 0.5\bar{\rho}(N^2/U^2 - k^2)^{1/2}k(Uh_0)^2 \tag{23}$$

if the square root is real, and zero otherwise.

This expression varies as the second power of the surface disturbance h_0, as must be expected since the stress is a bilinear function of forced wave fields. As a consequence, the wave stresses will become more important as the terrain features become more severe. There is no reason to expect similar behavior in the conventional turbulent contribution, which raises the possibility that the wave term could become overwhelming for all terrain that exceeds some degree of surface irregularity.

It is also apparent that this stress is a rather complex function of the velocity field, Figure 3. At low wind speeds, the stress (23) increases linearly with U. It maximizes to the value $0.25\bar{\rho}N^2h_0^2$, independent of the wavelength of the undulations, at the wind speed $N/k2^{1/2}$ and is zero for wind speeds greater than N/k. Thus, at low wind speeds the wave stress exceeds the turbulent stress, while at higher speeds the reverse is true.

For stable nocturnal conditions, N will typically be about $0.03\ \mathrm{s}^{-1}$, so the wave drag maximizes near wind speeds of $\lambda/300\ \mathrm{m\ s}^{-1}$, where λ is the wavelength of the undulations. That is, for undulations on a scale of a few hundred meters to a few kilometers, we can expect maximum wave drag at wind speeds of a few to $10\ \mathrm{m/s}$, which are quite typically observed values. At the wind speed that provides the maximum wave drag, the ratio of the turbulent drag represented by (4), to the wave drag represented by (23), is $C_D:(kh_0)^2$. As stated earlier, C_D is around 0.005, so for $kh_0 = 0.1$ the wave drag is twice the conventional drag. The value $kh_0 = 0.1$ represents quite gentle terrain (e.g., an undulation amplitude of 10 m with periodicity 600 m). To the extent that a simple calculation can tell us anything about the complex boundary layer, the conclusion must be that conditions are commonly met in which the wave drag is at least as large as the conventional turbulent drag.

For neutral or unstable stratification ($N \le 0$), the stress (23) is zero, so wave

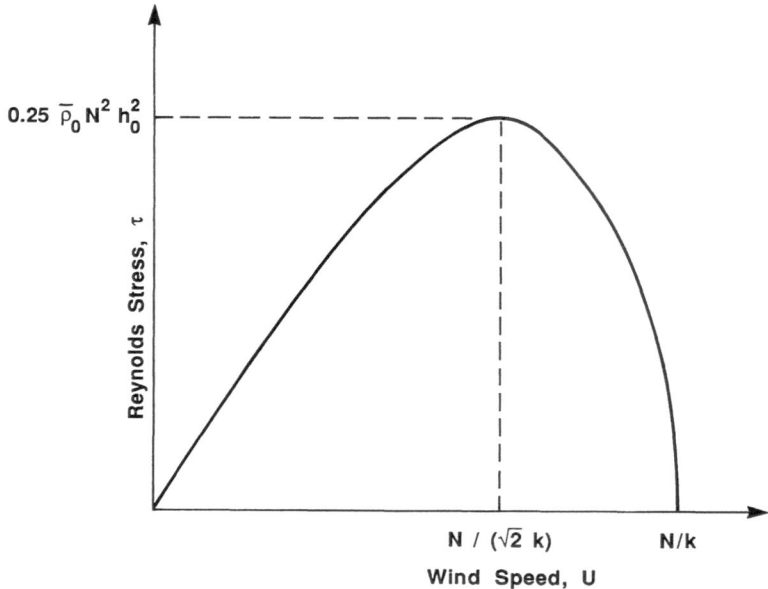

Fig. 3. Schematic of the wave-associated Reynolds stress for uniform flow over a sinusoidal surface. The initial rise of the drag is proportional to U, and the final part of the curve meets the abscissa at right angles. Formulas for the maximum are given in the text.

drag is a phenomenon confined to stably-stratified conditions. The wave stress is deposited in the wind field where the wave is absorbed. This may be a direct consequence of dissipation in the ambient turbulence, but research on waves in the mid- and upper-atmosphere suggests that the complicated processes of wave saturation and/or critical level absorption are more effective (Booker and Bretherton, 1967; Hodges, 1967; Fritts, 1984).

4. Directional Selectivity of Wave Drag

One aspect of wave drag is very different from conventional surface drag: whatever the wind direction, the wave drag acts in the direction of the un-dulations, while the conventional drag acts in the direction of the wind. This is illustrated in Figure 4. Only the wind component perpendicular to the ridges is subjected to the wave drag, and consequently the wind will tend to rotate towards a direction parallel to the ridges. In the real world there are few surfaces that approximate simple plane corrugations, but any real surface disturbance will have a non-isotropic Fourier representation causing some direction in x, y space to experience a maximum wave drag. The wind will tend to turn away from this direction. The wave drag favors flow along surface contours, strengthening the tendency of a stratified atmosphere to flow around an obstacle rather than over it.

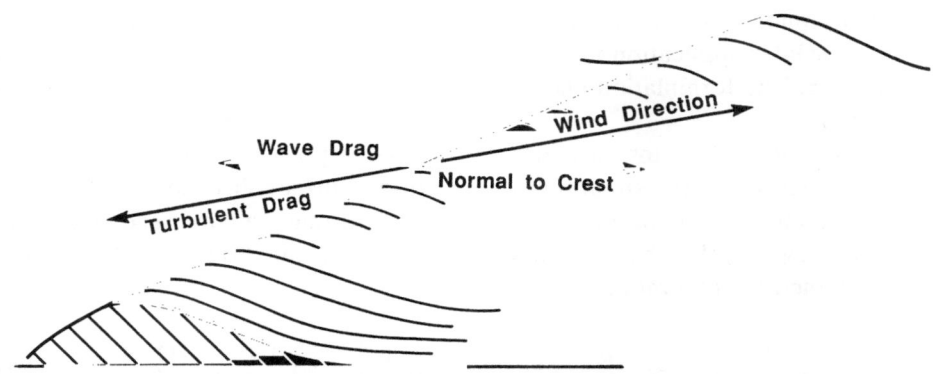

Fig. 4.   The directional preference introduced by wave drag. Conventional turbulent drag acts along
the wind direction, while wave drag is always normal to the surface contours.

5. Drag Over an Isolated Ridge

The sinusoidal surface may be regarded as the prototype for terrain with highly
organized undulations, as are found in wind-formed sand dunes or folded surface
regions. The prototype for an isolated feature is a single ridge, which will now be
considered.

In linearized theory, which is the normal starting point for mountain-wave
computations, the disturbance forced by a set of structures is the linear superposition of the disturbances forced by each individual structure considered
separately. Thus, the waves generated by a set of spaced ridges is the sum of the
disturbances associated with each ridge in isolation. Conversely, the disturbance
caused by an isolated surface feature can be computed by resolving the feature
into its Fourier representation and then summing the fields forced by each
sinusoidal component. That is, the ridge problem is just a superposition of
sinusoidal-surface responses, the formulation for which has already been given.

Let the ridge height $H(x)$ have a Fourier representation

$$H(x) = \frac{1}{2\pi} \int_{-\infty}^{\infty} dk\, h(k) \exp(ikx). \tag{24}$$

Then, as in Section (3), the wave fields associated with component $h(k) \exp(ikx)$
are

$$w(k, z) \exp(ikx) = ikUh(k)\hat{w}(k, z) \exp(ikx), \quad \hat{w}(k, 0) = 1 \tag{25}$$

and

$$u(k, z) \exp(ikx) = -Uh(k)\, d\hat{w}(k, z)/dz \exp(ikx) \tag{26}$$

where $\hat{w}(k, z)$, with normalization $\hat{w}(k, 0) = 1$, is the solution of the Taylor-

Goldstein equation, (15). This general form will be used to develop the stress formulas for later application to structured flow that does not allow a plane-wave solution in z. The formulation encompasses the plane-wave case, which will also be required.

Inverting the transforms provides the spatial representations of w and u. However, contrary to the situation of the periodic surface, it is now not possible to define a local value of the Reynolds stress through an average over one horizontal wavelength. The only simple stress-related function is the total ridge stress, obtained by integrating over all x:

$$-\bar{\rho} \int_{-\infty}^{\infty} wu \, dx = -\frac{\bar{\rho}}{\pi} \int_{-\infty}^{\infty} dk (Uh)^2 k \, \text{Im}\left[\hat{w}(k, z) \frac{d\hat{w}^*(k, z)}{dz} \right]. \qquad (27)$$

The asterisk represents complex conjugation, and Im[ ] selects the imaginary part of the expression within the braces.

Evaluation of this expression requires specification of the ridge shape, to determine $h(k)$, and specification of the flow, to determine $\hat{w}(k, z)$. The ridge will be taken to have a Gaussian shape,

$$H(x) = H_0 \exp[-(x/L)^2] \qquad (28)$$

so that

$$h(k) = H_0 L \pi^{1/2} \exp[-(kL/2)^2] . \qquad (29)$$

For the flow, we first re-use the uniform profile of stability and wind taken in Section (3). The stress (27) is then evaluated with (17)–(29) to provide

$$-\bar{\rho} \int_{-\infty}^{\infty} wu \, dx = \bar{\rho}(H_0 NL)^2 N/(U) \int_0^1 dy(1 - y)^{1/2} \exp[-y(LN/U)^2/2] \qquad (30)$$

if $N > 0$, and zero otherwise.

The integral can be expressed as a confluent hypergeometric function, but this does not make life any easier. The general behavior of (30) can be determined without much difficulty. In many ways, the stress given by (30) shares the characteristics of the stress over a uniform corrugation (see (23) et seq.). It is quadratic in the amplitude of the surface displacement, and is zero unless the flow is stably stratified.

Its behavior as a function of U is illustrated in Figure 5. Appropriate limits of the integral show that the stress is initially linear in U, and falls off as $1/U$ as the wind speed increases to satisfy $U/NL \gg 1$. This latter behavior is not as abrupt as the cut-off found for the corrugation, reflecting the continuum of wavenumbers contained in the ridge. Computation of the integral shows that the maximum stress occurs for a wind speed $U = 0.54NL$, at which speed the total drag

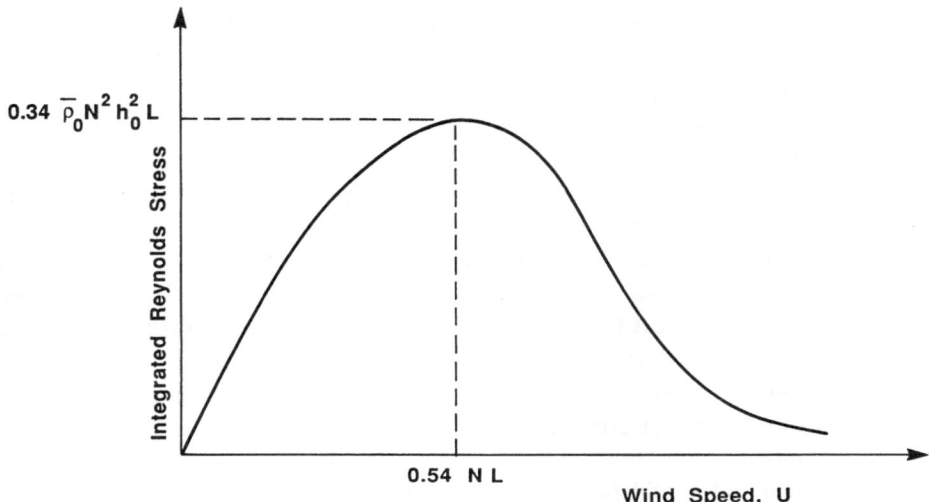

Fig. 5. Schematic of the wave-induced integrated Reynolds stress for uniform flow over an isolated ridge of height H_0 and half-width L. Formulas for the maximum are given in the text.

associated with the ridge is $0.34\bar{\rho}(NH_0)^2 L$. These values may be compared with the ones obtained for the corrugation. They are the same if the 'wavelength' of the ridge is taken to be 5 times its half-width L, and the integrated drag from the ridge is compared with the total stress over a length $1.4L$ of the corrugation.

Thus, a ridge and a corrugation with roughly similar physical scales generate about the same local wave stress above the disturbed surface. From a practical point of view, if terrain contains ridges, one needs to sum their individual contributions to the wave drag, and compare the result with the total turbulent drag over the same region.

6. The Influence of Wind Shear

The wave drag acts on the flow only where wave absorption occurs. While some absorption results from the general presence of turbulence, it is anticipated that the major part involves the non-linear process known as wave saturation (Hodges, 1967; Fritts, 1984). This occurs where the background wind speed is about the same as the wave phase speed. Since the wave is stationary, this happens if the mean flow U falls towards zero, and this will not always occur within the boundary layer. Indeed, the simpler boundary-layer winds found over flat regions are essentially monotonic with little turning, so they provide almost no possibility of saturation and the waves will escape into the overlying tropo-sphere. On the other hand, winds in or near complex terrain contain much more vertical structure, and reversals within the boundary layer are not unusual. Examples can be seen in the tower records obtained at the Boulder Atmospheric

Observatory (Gossard *et al.*, 1985; Hootman and Blumen, 1983). Drainage flows are often capped by a return flow that guarantees a zero wind about which stationary waves will saturate.

It is worth noting that such zeros of U will usually be in the mid- or upper-level boundary layer rather than in the region nearest the ground. They are only observed with the taller towers, or sensing systems specifically designed to get above the surface layer (as in Mahrt, 1985), and they affect the flow where models are presently least reliable. The wave stresses will therefore be significant in problems where the flow and mixing in the upper portion of the boundary layer over complex terrain must be accurately modeled. They probably have no direct influence on the lower part of the boundary layer.

Strong wind structure modifies wave propagation, which may change the Reynolds stress from that computed earlier for a uniform atmosphere. This will now be examined.

The Gaussian ridge will be used again, and the wind will be modeled with a hyperbolic-tangent profile that diminishes with height to approach, but not quite reach, zero. This allows an examination of the Reynolds-stress characteristics without the complexity of the singular critical-level problem. The critical-level problem will be presented in another paper devoted to the vorticity generated behind a three-dimensional surface projection.

The waves are governed by the Taylor–Goldstein Equation (15) with the phase velocity c set to zero

$$\frac{d^2 \hat{w}}{dz^2} + \left[\frac{N^2}{U^2} - \frac{1}{U} \frac{d^2 U}{dz^2} - k^2 \right] \hat{w} = 0 . \tag{31}$$

The boundary conditions on $\hat{w}(k, z)$ are that $\hat{w}(k, 0) = 1$, and $\hat{w}(k, \infty)$ is zero or corresponds to upward signal propagation (Hines, 1960). In previous sections, U and N have been set constant, so $\hat{w}$ was a simple exponential function, but in this section, the mean flow U is a function of z, so $\hat{w}$ is more complicated.

The Brunt–Vaisala frequency N is taken constant, and U is taken to be

$$2 U(z) = U_1 + U_2 + (U_2 - U_1) \text{Tanh}[(z - z_0)/s] . \tag{32}$$

The scale of wind variation, s, is chosen to be considerably smaller than the altitude of the wind inflexion, z_0, so U_1 is essentially the surface speed. The wind approaches the constant value U_2 in the upper half space where boundary conditions must be applied.

The integrated wave stress given by (27) must be evaluated with $\hat{w}$ (the solution of (31)) and $h(k)$ specified by (29). Equation (27) contains the factor $\text{Im}[\hat{w}(k, z) \, d\hat{w}^*(k, z)/dz]$, which has altitude z as an argument. However, it is known that the factor $\text{Im}[...]$ is constant except about a critical level (Miles 1961), so in this application the stress can be evaluated at any convenient height. The differential equation was solved with a standard integrating code. Setting $U_1 = U_2$ in the wind profile (32) gives the uniform flow, allowing verification of the numerical procedures by comparison with the analytic results obtained earlier. Numerical outputs for both the uniform and tangent-hyperbolic profiles are shown in Figures 6 and 7.

The main point that emerges is that an upward decrease of the wind field gives rise to greater Reynolds stress than is found if the wind holds constant. In the examples shown, the increase is around a factor two, strengthening the previous arguments that wave drag can be larger than conventional turbulent drag.

This increase is readily understood from wave properties. Waves come in two basic forms: internal waves, with oscillatory vertical profiles, and external waves, which change monotonically with height. The upward-propagating internal waves transfer energy and Reynolds stress in the vertical direction. But the evanescent external wave does not transfer either energy or Reynolds stress (although pairs of external waves can allow tunneling of energy and stress if there

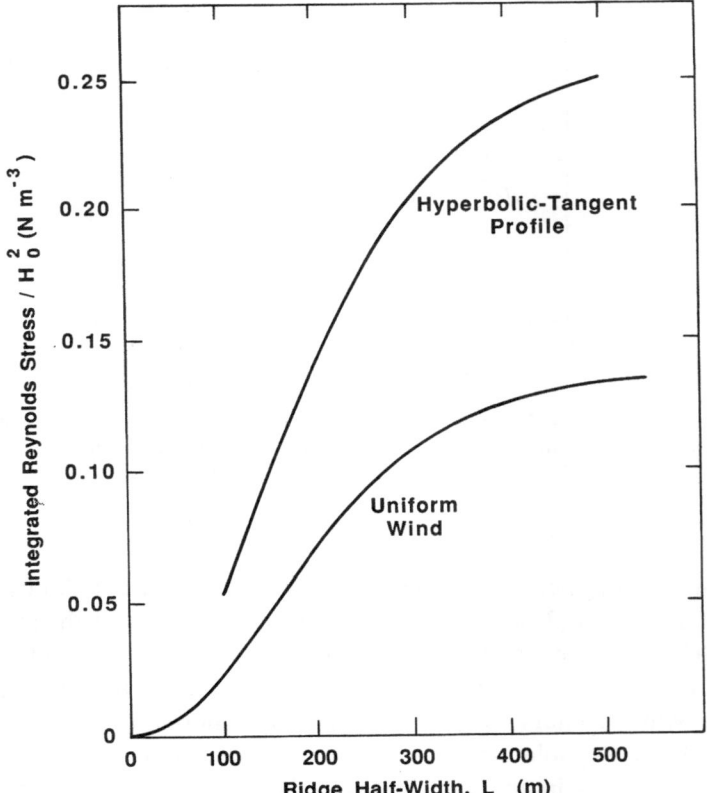

Fig. 6. Stress over a Gaussian ridge compared for the cases of a uniform wind and a hyperbolic-tangent wind profile. Stability is held constant at $N^2 = 10^{-3}\,\text{s}^{-2}$, and the ground-level wind is $4\,\text{m s}^{-1}$. The hyperbolic-tangent profile approaches $1\,\text{m s}^{-1}$ in the upper half-space, its inflection point is $200\,\text{m}$ above the surface, and its shear scale s is $50\,\text{m}$. The stress associated with the vertically-decreasing wind is about a factor 2 greater than that for the uniform wind.

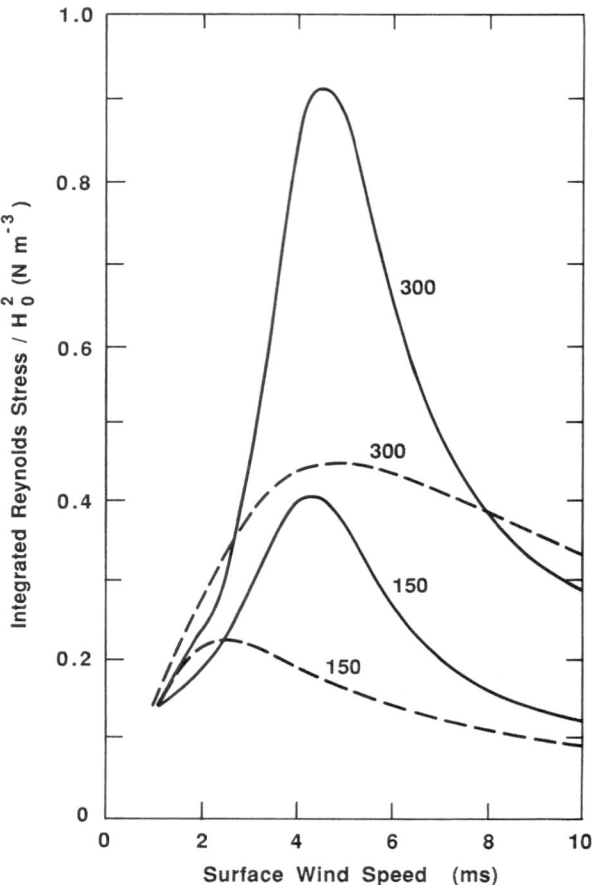

Fig. 7.   As in Figure 6, but now the abscissa is surface wind speed and only two values of the ridge
width L are considered. All other parameters have the values stated in Figure 6. Again, the
vertically-decreasing wind provides the greatest drag.

is an overlying region where an internal wave can carry it away). An example of
this difference is seen in (23), where the stress is non-zero only if the square root
is real, which is also the condition that the waves are internal. The Taylor–
Goldstein equation, (31), shows that in the upper half space, where U approaches
U_2, all waves with wavenumber less than N/U_2 are internal in the outer region,
and thus transfer a Reynold stress. As U_2 diminishes, more of the k-spectrum
falls in the internal domain and contributes to the stress. This is why the turning
back of the wind profile leads to an increase in the wave stress over the ridge.
Other features of Figures 6 and 7 suggest that, apart from the amplitudes, there is
little to distinguish the drag of the height-varying wind from its uniform counterpart. The peaks in Figure 7 have a different U dependence for the two cases,

but it is not clear that this feature is significant in practical applications where wind and temperature profiles are usually quite complex and poorly resolved.

7. Conclusions

Over complex terrain, the Reynolds stress associated with waves generated by surface variations may be considerably larger than the conventional Reynolds stress associated with turbulence. This does not require particularly rough terrain, although the wave effects do increase with the square of the surface displacement. The effect is much more dependent on the configuration of the boundary layer: the waves require stable stratification and a wind that falls to small values in the mid- or upper-boundary layer if any wave drag is to be effective there. The wave stresses probably only influence the higher parts of the boundary layer.

Acknowledgement

This work was supported in part by the National Science Foundation under grant NSF ATM-8519883.

References

Berkowitz, R.: 1984, 'Spectral Methods of Atmospheric Diffusion Modeling', *Boundary-Layer Meteorol.* **30**, 201–220.
Booker, J. R. and Bretherton, F. P.: 1967, 'The Critical Level for Internal Gravity Waves in a Shear Flow', *J. Fluid Mech.* **27**, 513–539.
Bretherton, F. P.: 1969, 'Momentum Transport by Gravity Waves', *Q.J.R. Meteorol. Soc.* **95**, 213–243.
Cermak, J. E.: 1984, 'Physical Modeling of Flow and Dispersion Over Complex Terrain', *Boundary-Layer Meteorol.* **30**, 261–292.
Chimonas, G.: 1972, 'The Stability of a Coupled Wave-Turbulence System in a Parallel Shear Flow', *Boundary-Layer Meteorol.* **2**, 444–452.
Egan, B. A.: 1984, 'Transport and Diffusion in Complex Terrain', *Boundary-Layer Meteorol.* **30**, 3–38.
Einaudi, F. and Finnigan, J. J.: 1981, 'The Interaction Between an Internal Gravity Wave and the Planetary Boundary Layer. Part 1: The Linear Analysis', *Q.J.R. Meteorol. Soc.* **107**, 793–806.
Fritts, D. C.: 1984, 'Gravity Wave Saturation in the Middle Atmosphere: A Review of Theory and Observations', *Rev. Geophys. Space Phys.* **22**, 275–308.
Gifford, F. A.: 1984, 'The Random Force Theory: Application to Meso- and Large-Scale Atmospheric Diffusion', *Boundary-Layer Meteorol.* **30**, 159–175.
Gossard, E. E., Gayner, J. E., Zamora, R. J., and Neff, W. D.: 1985, 'Fine Structure of Elevated Stable Layers Observed by Sounder and *in situ* Tower Sensors', *J. Atmos. Sci.* **42**, 2156–2169.
Hines, C. O.: 1960, 'Internal Gravity Waves at Ionospheric Heights', *Can. J. Phys.* **38**, 1441–1481.
Hootman, D. K. and Blumen, W.: 1983, 'Analysis of Nighttime Drainage Winds in Boulder, CO During 1980', *Mon. Weather Rev.* **111**, 1052–1061.
Hunt, J. C. R. and Richards, K. J.: 1984, 'Stratified Airflow Over One or Two Hills', *Boundary-Layer Meteorol.* **30**, 223–259.
Lindzen, R. S.: 1967, 'Thermally Driven Diurnal Tide in the Atmosphere', *Q.J.R. Meteorol. Soc.* **93**, 18–32.

Lindzen, R. S.: 1981, 'Turbulence and Stress Due to Gravity Wave and Tidal Breakdown', *J. Geophys. Res.* **86**, 9707–9714.

Mahrt, L.: 1985, 'Vertical Structure and Turbulence in the Very Stable Boundary Layer', *J. Atmos. Sci.* **42**, 2333–2349.

Miles, J. W.: 1961, 'On the Stability of Heterogeneous Shear Flow', *J. Fluid Mech.* **10**, 496–508.

Nieuwstadt, F. T. M.: 1984, 'Some Aspects of the Turbulent Stable Boundary Layer', *Boundary-Layer Meteorol.* **30**, 31–38.

Orlanski, I.: 1972, 'On the Breaking of Standing Internal Gravity Waves', *J. Fluid Mech.* **54**, 577–589.

Smith, R. B.: 1976, 'The Generation of Lee Waves by the Blue Ridge', *J. Atmos. Sci.* **33**, 507–519.

Sutton, O. G.: 1955, 'Micrometeorology', Robert E. Krieger Publishing Co., N.Y. Reprint 1977.

Wyngaard, J. C., Coté, O. R., and Izumi, Y.: 1971, 'Local Free Convection Similarity and the Budgets of Shear Stress and Heat Flux', *J. Atmos. Sci.* **28**, 1171–1182.

SPATIAL CORRELATION FUNCTIONS OF SURFACE-LAYER ATMOSPHERIC TURBULENCE IN NEUTRAL STRATIFICATION

B. A. KADER, A. M. YAGLOM, and S. L. ZUBKOVSKII

Institute of Atmospheric Physics, U.S.S.R. Academy of Sciences, Moscow, U.S.S.R.

(Received 30 September, 1988)

Abstract. The longitudinal (i.e., in the direction of the mean wind) spectra and cospectra of wind components and temperature fluctuations in the atmospheric surface layer during neutral conditions were carefully investigated by Kader (1984, 1987) for a broad range of wave numbers which included wavelengths far beyond the large-scale limit of the inertial subrange. At the same time, some direct measurements of spatial correlation functions of the longitudinal wind component and temperature were performed by Zubkovskii and Fedorov (1986) and Zubkovskii and Sushko (1987). Section 2 of the present paper gives a review of the available results on longitudinal spectra and cospectra of wind velocity and temperature fluctuations in neutral stratification and examines the consequences of these results related to the longitudinal autocorrelation and symmetrized cross-correlation functions of surface-layer turbulence. In Section 3 it is shown that the correlation equations of Section 2 agree satisfactorily with some recent measurements of the longitudinal correlation functions in the range of distances from 3 m to 100 m. Some measurements of the lateral correlation functions of atmospheric turbulence are also presented in Section 3. It is shown that these measurements lead to some predictions concerning the never-measured lateral space spectra of surface-layer turbulence.

1. Introduction

Most of the experimental investigations of turbulence within the atmospheric surface layer are concerned only with one-point statistical characteristics such as variances, covariances, and higher-order moments of turbulent fluctuations, or their time correlation functions and time spectra. Locally observed time correlation functions or frequency spectra are used sometimes for the approximate determination of the longitudinal (along the mean wind direction) space correlation functions and spectra with the aid of Taylor's frozen turbulence hypothesis. However, the degree of accuracy of Taylor's hypothesis for the meso-scale and large-scale disturbances is still uncertain and therefore the results obtained with its aid cannot be considered to be reliable. At the same time, the spatial structure of surface-layer turbulence is theoretically quite interesting and important for many applications. Therefore direct measurements of spatial turbulence characteristics and the analysis of these measurements are highly desirable.

Hans Panofsky (1962) was one of the first to study measured values of the space correlation functions of horizontal wind components in the surface layer over uniform flat ground. In subsequent years, Panofsky was also deeply concerned with the study of the spatial structure of surface-layer turbulence but his main attention was then focused on the analysis of frequency cross-spectra of turbulent fluctuations at two different space points (see, e.g., Chapter 9 of the book by Panofsky and Dutton, 1984, where many references to the original papers of Panofsky can be found). As for direct measurements of space correlation functions of surface-layer turbulence, most of the old measurements cover only a limited range of small distances either belonging to the inertial subrange or being only slightly longer than the large-scale limit of this subrange (see, e.g., quite typical papers by Koprov and Sokolov, 1973, and Volkov et al., 1975). However, Zubkovskii and Fedorov (1986) and Zubkovsky and Sushko (1987) have recently published some preliminary results of the direct measurements of space correlation functions for wind components and temperature, which cover a wide range of rather large distances outside the inertial subrange. The present paper is intended to analyse some data from a more complete collection of observations of such a type related to neutral thermal stratification and to compare these data with data on longitudinal spectra of surface-layer turbulence at such stratifications.

It is well known that the longitudinal spectra of the three wind components and temperature satisfy the Kolmogorov–Obukhov $-5/3$ power law in the inertial subrange of the wave numbers k (i.e., for $\eta^{-1} \gg k \gg z^{-1}$, where $\eta = (\nu^3/\epsilon)^{1/4}$ is the Kolmogorov length scale, ϵ – the mean energy dissipation rate, ν – the kinematic viscosity of air, and z – the height of the observation points above the ground); see, e.g., Monin and Yaglom (1975), Chapter, 8, and Yaglom (1981). Laboratory measurements of boundary-layer turbulence show that for $kz < 1$ (i.e., in the range of length scales larger than z), the same spectra often satisfy the -1 power law, i.e., they are proportional to k^{-1} (see, e.g., Perry and Abell, 1975, and Korotkov, 1976, where only the spectrum $E_{11}(k)$ of the longitudinal velocity component u is considered; the data confirming the validity of the -1 power law can also be found in some earlier works). Perry and Abell (1975) and Korotkov (1976) indicated that the -1 power law is a consequence of the assumption that the turbulence spectra at $kz < 1$ do not depend on the distance from the wall z and are fully determined by the friction velocity u_*; this assumption seems rather natural for the logarithmic sublayer of a turbulent boundary layer. Later Kader (1984, 1987) showed that the stated assumption is valid with good accuracy for all the longitudinal spectra and cospectra of velocity components and temperature fluctuations within the logarithmic sublayers of both laboratory turbulent boundary layers and air flows in the atmospheric surface layer. It was also discovered by Kader that the transition zone between the subranges, where the $-5/3$ power law and -1 power law are valid for the logarithmic sublayer spectra, is very narrow so that it is possible to assume that these two subranges adjoin each other. These facts supplemented by data of spectral measurements in the atmospheric surface layer permit one to obtain many important results concerning the longitudinal correlation functions of surface layer turbulence at neutral stratification. Section 2 is devoted to the consideration of these results while in Section 3, comparison of the results with direct measurements of the longitudinal correlation functions is considered.

In the conclusion it is noted that all available experimental data on spatial longitudinal spectra of turbulence are derived from locally measured time spectra with the aid of Taylor's hypothesis. Therefore a comparison of the correlation function results implied by these data with two-point measurements of space correlation functions can be considered as an indirect verification of the validity of Taylor's hypothesis for surface-layer turbulence.

2. Correlation Function Results Implied by Spectral Data

We shall assume as usual that the x-axis of the coordinate system is directed along the mean wind, the y-axis is perpendicular to the x-axis in the horizontal plane, and the z-axis is in the vertical direction, while the plane $z = 0$ coincides with the ground. Let $u_1 = u$, $u_2 = v$, $u_3 = w$, and $u_4 = \theta$ be fluctuating wind components along the coordinate axes and fluctuating temperature. The general spatial autocorrelation function of u_i fluctuations is

$$B_{ii}(\Delta x, \Delta y, \Delta z; z) = \langle u_i(x + \Delta x, y + \Delta y, z + \Delta z, t)u_i(x, y, z, t)\rangle, \qquad (1)$$

where t is time and the angular brackets denote the probabilistic averaging (which in practice is usually replaced by time averaging). The function (1) does not depend on t, x, and y since surface-layer turbulence is assumed to be stationary and horizontally homogeneous. As for the spatial cross-correlation functions, only the symmetrized form

$$B_{ij}(\Delta x, \Delta y, \Delta z; z) = \tfrac{1}{2}(\langle u_i(x + \Delta x, y + \Delta y, z + \Delta z, t)u_j(x, y, z, t)\rangle$$

$$+ \langle u_i(x, y, z, t)u_j(x + \Delta x, y + \Delta y, z + \Delta z, t)\rangle), \quad (1a)$$

will be considered. As a rule, only the longitudinal correlation functions $B_{ii}(\Delta x; z) = B_{ii}(\Delta x, 0, 0; z)$ and $B_{ij}(\Delta x; z) = B_{ij}(\Delta x, 0, 0; z)$ will be studied and instead of dimensional correlation functions B_{ii} and B_{ij}, the normalized (dimensionless) correlation functions

$$R_{ii}(\Delta x; z) = B_{ii}(\Delta x; z)/\sigma_i^2, \; \hat{R}_{ii}(\Delta x; z) = B_{ii}(\Delta x; z)/[u_*^{(i)}]^2, \qquad (2)$$

or

$$R_{ij}(\Delta x; z) = B_{ij}(\Delta x; z)/\sigma_i\sigma_j, \; \hat{R}_{ij}(\Delta x, z) = B_{ij}(\Delta x; z)/u_*^{(i)} u_*^{(j)}, \qquad (2a)$$

will be used. Here $\sigma_i = \langle u_i^2\rangle^{1/2}$ is the standard (root-mean-square) deviation of u_i while $u_*^{(1)} = u_*^{(2)} = u_*^{(3)} = u_* = \langle -uw\rangle^{1/2}$ and $u_*^{(4)} = \theta_* = Q/u_*$ (where $Q = \langle w\theta\rangle$ is the vertical temperature flux) are the typical scales of fluctuations u_i, $i = 1, 2, 3, 4$.

We shall consider only those points (x, y, z) of the atmospheric surface layer that belong to the logarithmic sublayer. In other words, it will be assumed that

$z \gg \max(h_0, \nu/u_*, \chi/u_*)$ (where h_0 is the mean height of the ground roughness and χ is the coefficient of thermal diffusivity) and that the influence of thermal stratification can be neglected at the height z. It follows from this that $\sigma_i/u_*^{(i)} = a_i$, where $i = 1, 2, 3, 4$ are universal constants. The measurements within the atmospheric surface layer show that apparently $a_1 \approx 2.7$, $a_2 \approx 2$, $a_3 \approx 1.25$, $a_4 \approx 2.9$; however, there is appreciable scatter in the data particularly in the values of a_4 (see, e.g., Monin and Yaglom, 1971; Yaglom, 1979; Kader and Yaglom, 1980; Panofsky and Dutton, 1983; and Kader, 1987, 1988). The values of the constants a_1, a_2, and a_3 resulting from measurements in laboratory logarithmic layers are only slightly smaller than atmospheric values, but laboratory measurements of $a_4 = \sigma_4/u_*^{(4)} = \sigma_\theta/\theta_*$ lead usually to a value which is close to 1.3, i.e., is much smaller than the atmospheric estimate of a_4. This striking discrepancy between the results of atmospheric and laboratory measurements of a_4 is apparently due to the influence of the always present thermal non-homogeneity of the ground. Because of non-homogeneity, σ_θ is far from being negligible even at practically neutral thermal stratification when Q is negligibly small, while for the laboratory flows $\sigma_\theta = 0$ at $Q = 0$.

Let us now denote by $E_{ii}(k; z)$, $i = 1, 2, 3, 4$, the one-dimensional space spectra of u_i in the mean wind direction and by $E_{ij}(k; z)$ the longitudinal cospectra (real parts of the cross-spectra) between u_i and u_j, where $i \neq j$, $i, j = 1, 2, 3, 4$. (The mirror symmetry of surface-layer turbulence with respect to the (x, z)-plane implies that $E_{12} = E_{23} = E_{24} = 0$ so that only cospectra $E_{13} = E_{uw}$, $E_{14} = E_{u\theta}$ and $E_{34} = E_{w\theta}$ must be considered). The autocorrelation function $B_{ii}(\Delta x; z)$ is the Fourier transform of the spectrum $E_{ii}(k; z)$:

$$B_{ii}(\Delta x; z) = \int_0^\infty \cos(k\Delta x) E_{ii}(k; z) \, dk , \tag{3}$$

while the Fourier transform of $E_{ij}(k; z)$, where $i \neq j$, is equal to the symmetrized cross-correlation function $B_{ij}(\Delta x; z)$:

$$B_{ij}(\Delta x; z) = \int_0^\infty \cos(k\Delta x) E_{ij}(k; z) \, dk . \tag{3a}$$

Dimensional analysis of the logarithmic sublayer implies that

$$E_{ii}(k; z) = z(u_*^{(i)})^2 F_{ii}(kz), \qquad E_{ij}(k; z) = z u_*^{(i)} u_*^{(j)} F_{ij}(kz) , \tag{4}$$

where $F_{ii}(kz)$ and $F_{ij}(kz)$ are universal functions of $kz = q$ for $\eta^{-1} \gg k \gg H_0^{-1}$ (here H_0 is the thickness of the neutrally stratified atmospheric surface layer). Therefore

$$\hat{R}_{ij}(\Delta x; z) = z \int_0^\infty \cos(k\Delta x) F_{ij}(kz) \, dk = \int_0^\infty \cos\left(q\frac{\Delta x}{z}\right) F_{ij}(q) \, dq , \tag{5}$$

for both $i = j$ and $i \neq j$ so that $\hat{R}_{ij}(\Delta x; z) = \hat{R}_{ij}(\Delta x/z) = R_{ij}(\xi)$ is the function of a single variable $\xi = \Delta x/z$.

In fact it has been already noted in the Introduction that numerous spectral measurements in logarithmic sublayers treated by Kader (1984, 1987) show that the following equations are valid with good accuracy:

$$F_{ii}(kz) = \begin{cases} C_{ii}(kz)^{-5/3} & \text{for } c_{ii} < kz < b_{ii} \\ G_{ii}(kz)^{-1} & \text{for } d_{ii} < kz < c_{ii} \end{cases}, \tag{6}$$

and

$$F_{ij}(kz) = G_{ij}(kz)^{-1} \qquad \text{for } d_{ij} < kz < c_{ij}. \tag{6a}$$

Here C_{ii}, G_{ii}, G_{ij}, c_{ii} and c_{ij} are universal constants but $b_{ii} \sim z/\eta$ depends on z, u_* and ν and d_{ii} and d_{ij} depend on z/H_0. Moreover, the functions $F_{ij}(kz)$, $i \neq j$ fall off with kz in the inertial subrange (where $F_{ii}(kz) \sim (kz)^{-5/3}$) much faster than $(kz)^{-5/3}$. (Strictly speaking, the Kolmogorov theory of locally isotropic turbulence implies that all the cospectra should vanish within the inertial subrange; see, e.g., Monin and Yaglom, 1975, Chapter 8. The data show, however, that this prediction is not quite correct.) Many observations agree well with the assumption that $E_{13}(k) \sim k^{-7/3}$ and $E_{34}(k) \sim k^{-7/3}$ in the inertial subrange; see, e.g., the papers by Kaimal et al. (1972) and Wyngaard and Coté (1972), the latter containing also some theoretical arguments which lead to the $-7/3$ power law for the Reynolds stress and heat flux cospectra $E_{13} = E_{uw}$ and $E_{34} = E_{w\theta}$. As for the cospectrum $E_{14} = E_{u\theta}$, it has been measured quite rarely and Kaimal et al. (1978) indicate that their measurements of E_{14} agree in the inertial subrange slightly better with the $-5/2$ power law than with the $-7/3$ power law. However, the exponents $-5/2$ and $-7/3$ are rather close to each other and the $-5/2$ power law for E_{14} is poorly established. Therefore for simplicity, we shall assume that the $-7/3$ power law is valid in the inertial subrange for all three cospectra E_{13}, E_{14}, and E_{34}. (Note that the change in the exponent of the power law implies a rather simple modification of the subsequent results which can be easily established.)

In the dissipation subrange beyond the small-scale limit of the inertial subrange, all the spectra and cospectra fall off with k very rapidly (apparently at the exponential rate). Note also that the main contribution to the variance $\sigma_i^2 = \langle u_i^2 \rangle = \int_0^\infty E_{ii}(k)\, dk$ and covariance $\langle u_i u_j \rangle = \int_0^\infty E_{ij}(k)\, dk$ is apparently due to the spectral subrange where the -1 power law is valid for E_{ii} and, respectively, for E_{ij}; see Kader (1984, 1987). There are no reliable data on the spectral shapes in the large-scale subrange where $kz < d_{ii}$ or $kz < d_{ij}$. Therefore in practical computations, the simplest assumption about the constancy of spectra and cospectra is often used in this subrange. For the correctness of the subsequent results, it is sufficient to assume that $E_{ii}(k) \sim k^{-\alpha_{ii}}$ for $k < d_{ii}/z$, $E_{ij}(k) \sim k^{-\alpha_{ij}}$ for $k < d_{ij}/z$, where $\alpha_{ii} < 1$, $\alpha_{ij} < 1$, and that the contribution of the considered large-scale subrange to the corresponding variance or covariance is rather small and can be neglected as a first approximation.

According to available data of numerous spectra measurements at the Tsimlyansk Field Station:

$$C_{11} \approx 0.95, \quad C_{22} \approx 1.0, \quad C_{33} \approx 1.2, \quad C_{44} \approx 1.1 ; \tag{7}$$

$$G_{11} \approx 0.95, \quad G_{22} \approx 0.6, \quad G_{33} \approx 0.35, \quad G_{44} \approx 0.9 ; \tag{8}$$

$$G_{13} \approx -0.2, \quad G_{34} \approx 0.2, \quad G_{14} \approx -0.6 . \tag{8a}$$

The estimates of C_{11}, C_{22}, C_{33}, and C_{44} agree quite satisfactorily with the results of all other measurements (both atmospheric and laboratory) of turbulent spectra in the inertial subrange of the logarithmic sublayer. Moreover, since

$$\epsilon = \frac{u_*^3}{\kappa z}, \qquad N = \frac{\mathrm{Pr}_t u_* \theta_*^2}{\kappa z}, \tag{9}$$

in the logarithmic sublayer (where N is the dissipation rate of $\theta^2/2$, $\kappa \approx 0.4$ is von Karman's constant and $\mathrm{Pr}_t \approx 0.85$ is the turbulent Prandtl number within the logarithmic sublayer; see, e.g., Yaglom, 1979, or Kader and Yaglom, 1980), the estimates (7) are also in excellent agreement with the available experimental information concerning Kolmogorov and Obukhov's constants C_K and C_O entering the universal $-5/3$ power laws for the velocity and temperature spectra. (These laws have the forms: $E_{11}(k) = 3E_{22}(k)/4 = 3E_{33}(k)/4 = C_K \epsilon^{-2/3} k^{-5/3}$, $E_{44}(k) = C_O N \epsilon^{-1/3} k^{-5/3}$ and numerous experimental estimates of C_K and C_O are collected by Yaglom, 1981.) Of course, according to the theory the equations $C_{22} = C_{33} = 4C_{11}/3$ should be valid. However only a few measurements of $E_{22}(k)$ were performed at the Tsimlyansk station and the estimate (7) of C_{22} is rather unreliable. Therefore there seems to be little reason for the conclusion that the estimates (7) contradict the Kolmogorov theory. The estimates (8) and (8a) of G_{11}, G_{22}, G_{33} and G_{13} agree satisfactorily with the rough estimates of the same coefficients in laboratory flows given by Kader (1984). The authors do not know of any laboratory estimates of the coefficients G_{14} and G_{34}, while the mean estimate of G_{44} obtained by the treatment of the laboratory data is closer to 0.3, i.e., it is three times smaller than the estimate G_{44} in (8). It is natural to assume that the striking discrepancy between the laboratory and atmospheric values of G_{44} is due to the influence of the thermal non-homogeneity of the Earth's surface which also leads, as we have already noted, to the considerable increase in the atmospheric value of $a_4 = \sigma_\theta / \theta_*$.

The Tsimlyansk spectral measurements at practically neutral thermal stratification imply the following estimates of the coefficients c_{ii}, d_{ii} and c_{ij}, d_{ij}:

$$c_{11} \approx 1, \quad d_{11} \approx 0.02; \quad c_{33} \approx 6.3, \quad d_{33} \approx 0.7; \quad c_{44} \approx 1.4, \quad d_{44} \approx 0.02 ; \tag{10}$$

$$c_{13} \approx 2, \quad d_{13} \approx 0.05; \quad c_{14} \approx 1, \quad d_{14} \approx 0.01; \quad c_{34} \approx 2, \quad d_{34} \approx 0.02 . \tag{10a}$$

Note also that by virtue of the continuity of the functions $E_{ii}(k)$, it follows from

(6) that

$$C_{ii} = G_{ii}(c_{ii})^{2/3}.$$ (11)

Let us now discuss the results of the longitudinal correlation functions which follow from the spectral data. We begin with the study of autocorrelation functions $B_{ii}(\Delta x; z)$ and $B_{ii}(\Delta x; z)/\sigma_i^2 = R_{ii}(\Delta x/z) = R_{ii}(\xi)$. If $\Delta x < \eta$, then $k\Delta x$ is greater than one only when $k > 1/\eta$, i.e., only for wave numbers k from the part of the dissipation subrange which is rather far beyond the large-scale limit of this subrange. Since this part of the turbulent spectrum gives only negligibly small contribution to the integral in the right-hand part of (3), we can assume, when $\Delta x < \eta$, that $k\Delta x < 1$ in the integrand and replace $\cos(k\Delta x)$ by Taylor's expansion of this function. Restricting ourselves to the two first terms of Taylor's series (i.e., replacing $\cos(k\Delta x)$ by $1 - (k\Delta x)^2/2)$) and using the well-known equations from the theory of locally isotropic turbulence

$$\int_0^\infty k^2 E_{ii}(k)\,\mathrm{d}k = \begin{cases} \epsilon/15\nu & \text{for } i = 1 \\ 2\epsilon/15\nu & \text{for } i = 2, 3, \\ N/3\chi & \text{for } i = 4, \end{cases}$$ (12)

we obtain

$$R_{ii}(\xi) = 1 - A_i\xi^2 \qquad \text{for } \xi < \eta/z,$$ (13)

where

$$A_1 = z^2\epsilon/30\nu\sigma_1^2, \quad A_2 = z^2\epsilon/15\nu\sigma_2^2, \quad A_3 = z^2\epsilon/15\nu\sigma_3^2,$$
$$A_4 = z^2N/6\chi\sigma_4^2.$$ (13a)

It is easy to see that Equations (13) and (13a) are equivalent to the equations describing the shape of the velocity and temperature structure functions for $\Delta x < \eta$ (see Kolmogorov, 1941; Obukhov, 1949; or Monin and Yaglom, 1975, Equations (21.16) and (21.86)).

Consider now the case when $z/b_{ii} < \Delta x < z/c_{ii}$, i.e., when Δx belongs to the inertial subrange. According to (6), Equation (5) with $i = j$ can be rewritten as

$$R_{ii}(\xi) = a_i^{-2} \int_0^\infty \cos q\xi\, F_{ii}(q)\,\mathrm{d}q$$

$$= a_i^{-2} \left\{ \int_0^\infty F_{ii}(q)\,\mathrm{d}q - \int_0^\infty (1 - \cos q\xi) F_{ii}(q)\,\mathrm{d}q \right\}$$

$$= 1 - a_i^{-2}\xi^{-1} \int_0^\infty (1 - \cos q_1) F_{ii}(q_1/\xi)\,\mathrm{d}q_1$$ (14)

$$= 1 - a_i^{-2} \xi^{-1} \int_0^{d_{ii}\xi} (1 - \cos q_1) F_{ii}(q_1/\xi) \, dq_1 \tag{14}$$

$$- a_i^{-2} G_{ii} \int_{d_{ii}\xi}^{c_{ii}\xi} \frac{1 - \cos q_1}{q_1} \, dq_1 - a_i^{-2} C_{ii} \xi^{2/3} \int_{c_{ii}\xi}^{b_{ii}\xi} \frac{1 - \cos q_1}{q_1^{5/3}} \, dq_1$$

$$- a_i^{-2} \xi^{-1} \int_{b_{ii}\xi}^{\infty} (1 - \cos q_1) F_{ii}\left(\frac{q_1}{\xi}\right) dq_1 = 1 - J_1 - J_2 - J_3 - J_4 \,,$$

where $q_1 = q\xi = k\Delta x$ and J_1, J_2, J_3 and J_4 are the contributions of spectral subranges $0 < k < d_{ii}/z$, $d_{ii}/z < k < c_{ii}/z$, $c_{ii}/z < k < b_{ii}/z$, and $b_{ii}/z < k < \infty$ to $1 - R_{ii}(\xi)$. It is easy to see that in the above case, when $\Delta x < \eta \sim z/b_{ii}$, the term J_4 plays the main part, i.e., the contribution of the dissipation subrange to the integral in the right-hand part of (5) is most important. If, however, Δx belongs to the inertial subrange, then $z > \Delta x > \eta$, $\xi = \Delta x/z < 1$ but $b_{ii}\xi \gg 1$ (since b_{ii} is usually considerably smaller than z/η) and at the same time $c_{ii}\xi \ll 1$, $d_{ii}\xi < c_{ii}\xi \ll 1$. Therefore in this case J_1, J_2 and J_4 are negligibly small and in the expression for J_3, the lower and upper limits of integration can be replaced by 0 and ∞ without appreciable error. Performing now the integration in the expression for J_3, we obtain

$$R_{ii}(\xi) = 1 - B_i \xi^{2/3} \qquad \text{for } 1 > \xi > \eta/z \,, \tag{15}$$

where $B_i = 3 C_{ii} \Gamma(1/3)/4 a_i^2$, $i = 1, 2, 3, 4$, and Γ is the Γ-function (cf. Monin and Yaglom, 1975, p. 355). This result is equivalent to the 2/3 power law for velocity and temperature structure functions; see Kolmogorov (1941), Obukhov (1949), or Monin and Yaglom (1975), Equations (21.17), (21.25) and (21.87), (21.90).

Let us now examine the case of even longer distances Δx satisfying the inequalities $\Delta x > z$ and $z/c_{ii} < \Delta x < z/d_{ii}$. Then $\xi > 1$, $c_{ii}\xi > 1$, $b_{ii}\xi \gg c_{ii}\xi \gg 1$, but $d_{ii}\xi < 1$; for simplicity, we assume that $c_{ii}\xi \gg 1$, $d_{ii}\xi \ll 1$ (it is well known that the asymptotic relations have usually a rather broad range of applicability). The terms J_1 and J_4 are clearly negligibly small here but some computations are needed to estimate J_2 and J_3. As for J_3, the asymptotic expansion of the higher function $C(x, -2/3)$ (see Erdélyi et al., 1953, Section 9.10) implies, as can be easily seen, that

$$\xi^{2/3} \int_{c_{ii}\xi}^{b_{ii}\xi} q_1^{-5/3} \cos q_1 \, dq_1 \,,$$

is very small here. By virtue of (6), it follows that

$$J_3 \approx 1.5 G_{ii} [1 - (c_{ii}/b_{ii})^{2/3}] \,. \tag{16}$$

Note that the extent of the inertial subrange is usually rather limited for laboratory boundary layers at moderately large Reynolds numbers. Therefore the ratio c_{ii}/b_{ii} is not too small for such layers (and depends on Re). This implies that the value of J_3 depends on Re in laboratory logarithmic sublayers, i.e., it is not universal. However, the inertial subrange is very broad in an atmospheric surface layer where Re is always very high; hence c_{ii}/b_{ii} is much smaller than unity here and $J_3 \approx 1.5 G_{ii}$. Moreover, Equation 9.8(9) from Erdélyi *et al.* (1953) can be used to estimate J_2 for $d_{ii}\xi \ll 1$, $c_{ii}\xi \gg 1$. This equation leads to the following result:

$$J_2 \approx G_{ii}(\gamma + \ln c_{ii}), \tag{16a}$$

where $\gamma \approx 0.58$ is Euler's constant. Equations (16), (16a) and the inequality $c_{ii}/b_{ii} \ll 1$ imply that

$$R_{ii}(\xi) \approx 1 - D_i - d_i \ln \xi \qquad \text{for } c_{ii}^{-1} < \xi < d_{ii}^{-1}, \tag{17}$$

where

$$D_i = G_{ii}(1.5 + \gamma + \ln c_{ii})/a_i^2, \qquad d_i = G_{ii}/a_i^2. \tag{17a}$$

The behaviour of the autocorrelation functions $R_{ii}(\xi)$ in the range of long distances $\Delta x > z/d_{ii}$ depends on that of spectra $E_{ii}(k)$ at $k < d_{ii}/z$. Since there are no reliable data on the shape of spectra at such large wave numbers, there is no reason for detailed discussion of the correlation function behaviour at $\Delta x > z/d_{ii}$. However, rough estimates of the integral length scales of turbulence in the neutrally stratified surface layer seem to be interesting. The integral length scale L_i of u_i fluctuations by definition is equal to the integral of $R_{ii}(\Delta x)$ from 0 to ∞; it differs from the spectrum $E_{ii}(k)$ at zero wave number by the constant factor $\pi/2$. Of course, the values of turbulent spectra at $k = 0$ are poorly known but we have already mentioned that the simplest assumption about the constancy of spectra in the large-scale range (beyond the large-scale limit of the range of validity of the -1 power law) is often used in practice and apparently does not lead to great errors. Therefore Equations (6), (8) and (10) combined with the values of a_i given in Section 2 permit one to obtain the following estimates of the integral length scale for the wind components and temperature:

$$L_1 = L_u = 10.3z, \quad L_2 = L_v = 7.5z, \quad L_3 = L_w = 0.5z,$$
$$L_4 = L_\theta = 8.4z. \tag{18}$$

We see that the longitudinal scales of horizontal wind components and temperature fluctuations are comparable with each other but the longitudinal length scale of the vertical velocity fluctuations is much shorter than the others.

The shape of the symmetrized longitudinal cross-correlation functions $R_{ij}(\xi)$ and $\hat{R}_{ij}(\xi) = a_i a_j R_{ij}(\xi)$ can be investigated quite similarly. In the case of very small distances $\Delta x < \eta$, the relation

$$\int_0^\infty k^2 E_{ij}(k; z)\, dk = \langle(\partial u_i/\partial x)(\partial u_j/\partial x)\rangle, \tag{19}$$

which follows from (3) and (1a) can be used. Similarly as in the derivation of (13), this relation leads to the following result:

$$\hat{R}_{ij}(\xi) = \hat{R}_{ij}(0) - A_{ij}\xi^2 \qquad \text{for } \xi < \eta/z, \tag{20}$$

where

$$A_{ij} = \langle(\partial u_i/\partial x)(\partial u_j/\partial x)\rangle z^2/2 u_i^* u_j^*. \tag{20a}$$

(The same result can also be derived from the identity

$$B_{ij}(0) - B_{ij}(\Delta x) = \tfrac{1}{2}\langle[u_i(x + \Delta x) - u_i(x)][u_j(x + \Delta x) - u_j(x)]\rangle; \tag{21}$$

cf. derivation of Equation (21.16) in Monin and Yaglom, 1975.) As for the inertial subrange of lengths Δx, the assumption of validity of the $-7/3$ power law for the cospectra $E_{ij}(k)$ in the inertial subrange (which at least for $E_{13}(k)$ and $E_{34}(k)$ agree well with the observations) implies a 4/3 power law for the cross-structure functions $D_{ij}(\Delta x) = 2[B_{ij}(0) - B_{ij}(\Delta x)]$. This 4/3 power law can be written as

$$\hat{R}_{ij}(\xi) = \hat{R}_{ij}(0) - \hat{B}_{ij}\xi^{4/3} \qquad \text{for } \eta/z \leqslant \xi \ll c_{ij}^{-1}, \tag{22}$$

where $\hat{B}_{ij} = [\Gamma(-4/3)/2]C_{ij}$, C_{ij} is the dimensionless coefficient in the equation $F_{ij}(kz) = C_{ij}(kz)^{-7/3}$ (cf. Equations (21.72), (21.76) and (21.77) in Monin and Yaglom, 1975). Finally the behaviour of cross-correlation functions $\hat{R}_{ij}(\xi)$ in the range $c_{ij}^{-1} < \xi < d_{ij}^{-1}$ (i.e., for $z/c_{ij} < \Delta x < z/d_{ij}$) can be investigated quite similarly as in the derivation of formulae (17) and (17a). Making again the assumption that the $-7/3$ power law is valid for the cospectra in the inertial subrange of wave numbers, we easily obtain

$$\hat{R}_{ij}(\xi) = \hat{R}_{ij}(0) - G_{ij}(0.75 + \gamma + \ln c_{ij}) - G_{ij} \ln \xi \quad \text{for } c_{ij}^{-1} < \xi < d_{ij}^{-1}. \tag{23}$$

Note that the $-7/3$ power law for the cospectrum E_{ij} is used here only in the computation of the first term in the parentheses on the right-hand side of (23). Possible deviations from this power law will lead only to a change of the constant 0.75 and this change can hardly be appreciable (e.g., the replacement of the $-7/3$ power law by the $-5/2$ power law implies the replacement of 0.75 by $2/3 \approx 0.67$). Let us also recall that

$$\hat{R}_{13}(0) = \frac{\langle uw\rangle}{u_*^2} = -1, \quad \hat{R}_{34}(0) = \frac{\langle w\theta\rangle}{Q} = 1, \quad \hat{R}_{14}(0) = \frac{\langle u\theta\rangle}{Q} \approx -3.5, \tag{24}$$

(the numerical value of $\hat{R}_{14}(0)$ is discussed, e.g., by Monin and Yaglom, 1971, Yaglom, 1979, Kader and Yaglom, 1980, and Kader, 1984, 1988).

Purely dimensional arguments which lead to the -1 power law for the turbulent spectra and cospectra in the neutrally stratified surface layer can also be applied to the correlation functions $R_{ii}(\xi)$ and $\hat{R}_{ij}(\xi)$ at $1 \ll \xi \ll H_0/z$ (where again H_0 is the height of the neutrally stratified surface layer). These arguments imply the very simple results

$$R_{ii}(\xi) = l_{ii} = \text{const}, \quad \hat{R}_{ij}(\xi) = \hat{l}_{ij} = \text{const} \qquad \text{for } 1 \ll \xi \ll H_0/z . \qquad (25)$$

The above Equations (17), (17a) and (23) permit one to express the constants l_{ii} and $\hat{l}_{ij}$ through the spectral constants (8), (8a), (10), (10a) and easily measured constants a_i and $\langle u\theta \rangle/Q$. Moreover, which is more important, these equations determine also the correction terms to formulae (25), which turn out to be logarithmic in ξ. The figures in the next section show that these corrections are quite essential: formulae (25) do not fit the observations satisfactorily but the corrected Equations (17) and (23) agree with the data rather well.

3. Comparison with Observations

The above equations for the correlation and cross-correlation functions have been compared with some measurements performed in the atmospheric surface layer above flat grassy ground at the Tsimlyansk Field Station during the summer of 1985. To measure the longitudinal correlation functions, a period was selected when the thermal stratification was practically neutral and the weather was so settled that the mean wind speed, wind direction and temperature did not change appreciably during three hours. Five sets of instruments were used, each of them containing a three-component sonic anemometer and a low-inertia resistance thermometer. The sets were placed in a line along the mean wind direction at the same height of 3.2 m. The distances between the first set and the others were selected to. be 3, 10, 30 and 100 m. The wind components and temperature fluctuations at five observation points were recorded on magnetic tape during a 150-minute run. Then the variances and the covariances for the fluctuations at each of five points and also the covariances between the fluctuations at each of ten pairs of points were calculated with the aid of an ES-1022 computer. A more detailed description of the instruments, their characteristics, and methods of observations and data treatment can be found in the papers by Zubkovskii and Fedorov (1986) and Zubkovskii and Sushko (1987). Here we only note that two methods of data processing were employed. At the same time as the simplest computations were made, when average values over the whole observation period (150 min) were subtracted from the observations to obtain turbulent fluctuations, a more sophisticated procedure which included the removal of possible trend was also used in many cases. For trend removal, the observed time series of horizontal wind components and temperature were approximated by the method of least squares by the second-order polynomial best fitting the data. Then differences between the observations and the polynomial values were used as the fluctuations. It turned out that such a method of trend removal did not change the computational results appreciably. Therefore we could conclude that the trend was practically lacking or, at least, it did not play an important part.

The results of the correlation function measurements are compared in Figures 1 and 2 with the equations given in Section 2. In Figure 1, the measured

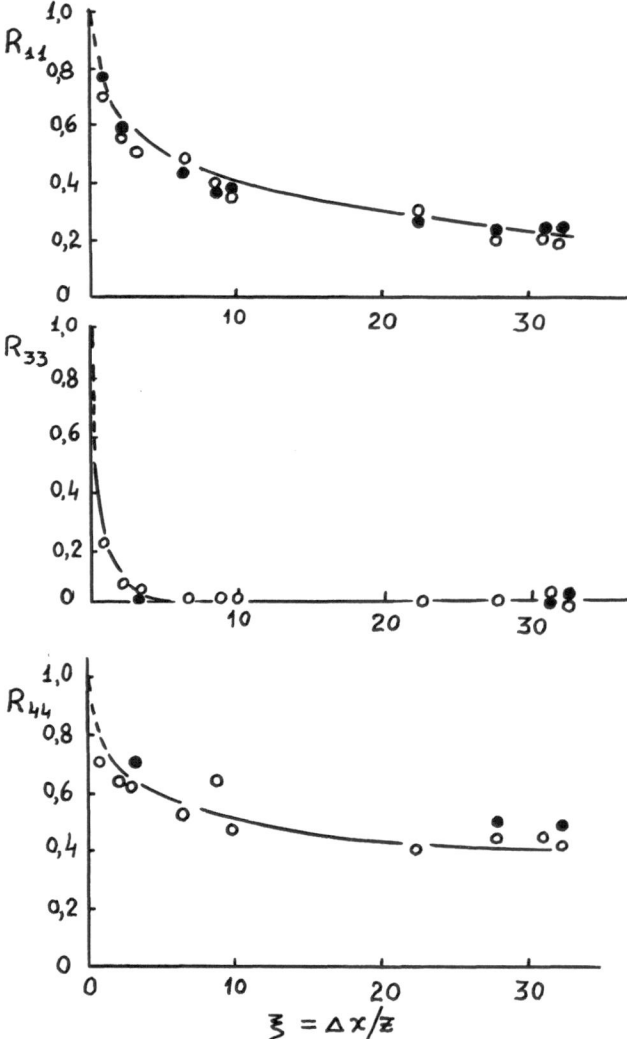

Fig. 1. Comparison between the measured and computed values of the longitudinal correlation functions R_{11}, R_{33}, and R_{44}. Open circles represent measured values obtained without trend removal and full circles are values obtained when the second-order-polynomial trend is eliminated. The dotted lines correspond to Equation (15) and solid lines to Equations (17) and (17a).

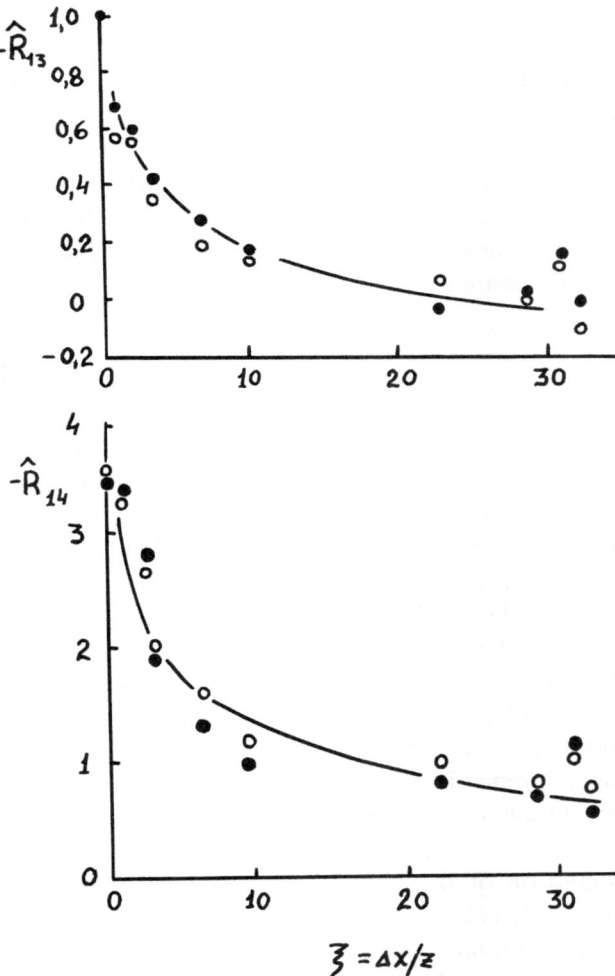

Fig. 2.   Comparison between measured and computed values of the symmetrized cross-correlation functions $\hat{R}_{13}$ and $\hat{R}_{14}$. The open and full circles have the same meaning as in Figure 1; the solid lines correspond to Equation (23).

autocorrelation functions $R_{11} = R_{uu}$, $R_{33} = R_{ww}$, $R_{44} = R_{\theta\theta}$ are shown. The open circles correspond to the computations without trend removal and full circles show the results obtained when the second-order-polynomial trend was subtracted from the raw data. It is seen that the method employed for trend removal does not affect the computation results significantly. The scatter of the points in Figure 1 is clearly due to the fact that the values of the correlation coefficients calculated from a single realization include considerable sampling errors. The dotted lines in Figure 1 correspond to the 2/3 power law (15) and the solid lines represent the curves (17) with the coefficients (17a). The constants C_{ii}, G_{ii} and c_{ii}

entering the expressions for B_i, D_i and d_i are taken from (7), (8) and (10), i.e., they are obtained by averaging of the results of many different experiments. Since, however, available experimental values of a_i^2 are very scattered, the values used for plotting the lines in Figure 1 are determined from the same 150 min run which is used for the correlation function computations. In order to do this, the computed local values of the variances $\sigma_1^2 = \sigma_u^2$, $\sigma_3^2 = \sigma_w^2$, and $\sigma_4^2 = \sigma_\theta^2$ and of the turbulent fluxes $\langle -uw \rangle$ and $\langle w\theta \rangle$ are averaged over five observation points and the obtained averaged values are used to compute $a_1^2 = \sigma_u^2/u_*^2$, $a_3^2 = \sigma_w^2/u_*^2$ and $a_4^2 = \sigma_\theta^2/\theta_*^2$. This procedure gives the values

$$a_1^2 \approx 7.0, \quad a_3^2 \approx 1.7, \quad a_4^2 \approx 8.4 , \tag{26}$$

whose square roots differ, but not strongly, from the mean atmospheric values of a_1, a_3 and a_4 given in Section 2.

Figure 1 shows that the experimental data covering the $3 \text{ m} \leqslant \Delta x \leqslant 100 \text{ m}$ range do not allow one to verify the 2/3 power law (15); measurements at shorter distances are needed. However, all the non-zero values of the measured correlation functions agree satisfactorily with the results (17) and (17a) implied by the spectral -1 power law. Large values of a_1^2 and a_4^2 are chiefly responsible for the slow decrease of the correlation function R_{11} and R_{44} in the range of applicability of (17) and (17a). Since the drop in these functions is slow, the correlation coefficients $R_{11} = R_{uu}$ and $R_{44} = R_{\theta\theta}$ prove to be considerable even at $\Delta x = 100 \text{ m}$, $\xi \approx 30$. At the same time, since $a_3^2 = a_w^2$ is much smaller than a_1^2 and a_4^2, the correlation between $w(x + \Delta x)$ and $w(x)$ is practically lacking for $\Delta x > 10 \text{ m}$, $\xi \geqslant 3$. This difference in the behaviour of the correlation functions agrees well with the striking difference in the corresponding integral length scales given in (18).

A similar comparison of the measured cross-correlation functions $\hat{R}_{13}(\xi) = \hat{R}_{uw}(\xi)$ and $\hat{R}_{14}(\xi) = \hat{R}_{u\theta}(\xi)$ with Equations (23) is shown in Figure 2. (The value of $\hat{R}_{14}(0) = \langle u\theta \rangle/Q$ in Equation (23) for $\hat{R}_{14}(\xi)$ is computed from the 150-min run of raw data and it proved to be close to -3.5.) It is seen that the measured cross-correlation functions $\hat{R}_{13}$ and $\hat{R}_{14}$ agree well with the theoretical equations for the whole range of distances Δx covered by the experiment.

Finally in Figure 3, values of the lateral autocorrelation functions $R_{11}(\Delta y/z) = \langle u_1(x, y + \Delta y, z, t)u_1(x, y, z, t)\rangle/\sigma_1^2$ and $R_{44}(\Delta y/z) = \langle \theta(x, y + \Delta y, z, t)\theta(x, y, z, t)\rangle/\sigma_4^2$ measured from two other (60- and 120-min) runs of observations are shown. All the measurements were again performed at a height of 3.2 m in neutral stratification and under stable weather conditions, but the instruments were placed here in a line perpendicular to the mean wind direction. The number of observation points was smaller in these experiments than in the course of the longitudinal correlation measurements: during the 60-min run, four sonic anemometers and three resistance thermometers were used while during the 120-min run only three anemometers and two thermometers were employed. The distances between instruments were chosen to be shorter than the longest distance

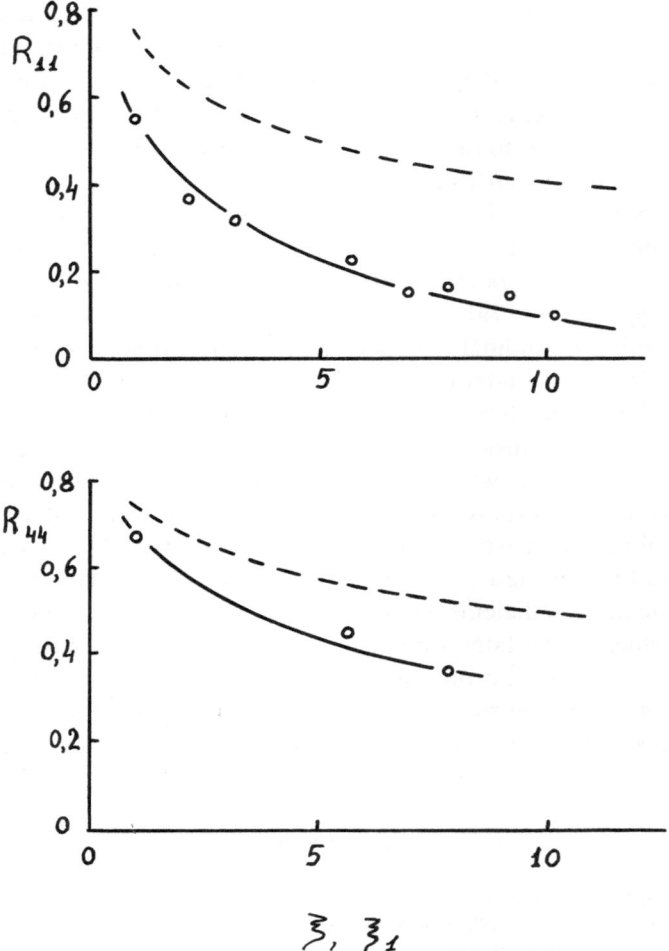

Fig. 3. Measured values of the lateral correlation functions $R_{11}(\xi_1)$ and $R_{44}(\xi_1)$ (open circles) and the curves of form (17) approximating these values (solid lines). The dotted lines represent the longitudinal correlation functions $R_{11}(\xi)$ and $R_{44}(\xi)$.

used in longitudinal correlation measurements since it is known that the lateral length scales of atmospheric turbulence are considerably shorter than the longitudinal scales. The collection of distances between instruments of the same type covered a range of Δy values from 3 to 32 m for the anemometers and from 3 to 25 m for thermometers. (There were eight different distances between the pairs of anemometers used in two runs and four such distances between the pairs of thermometers.) The dotted lines in Figure 3 show the shape of the corresponding longitudinal correlation function $R_{11}(\xi)$ and $R_{44}(\xi)$ and the points represent the measured values of lateral correlation functions $R_{11}(\xi_1)$ and $R_{44}(\xi_1)$ computed by the simplest method (without trend removal). We see that the drop in the

lateral correlation functions with distance is much faster than that of the
corresponding longitudinal correlation functions; this confirms that the lateral
length scales are shorter than the longitudinal ones. Note now that the values of
a_1^2 and a_4^2 must not depend on the arrangement of the sensors, though there can
be a small difference in these values for different runs due to sampling errors (in
the case considered, mean values of a_1^2 and a_4^2 for two runs of lateral correlation
measurements were $a_1^2 \approx 6.7$, $a_4^2 \approx 9$). There is also no reason to think that the
values of the constants c_{ii} in Equations (6) can differ significantly for the
longitudinal and lateral spectra (and, moreover, the term $\ln c_{ii}$ plays only a minor
part in the expression (17a) for D_i). Hence, if Equations (17) and (17a) can be
applied to both longitudinal and lateral correlation functions R_{11} and R_{44}, the
much faster drop of the lateral correlations can be explained only by considerably
larger values of the coefficients G_{ii} for the lateral spectra than the values given in
(8). To verify this conclusion, we approximate the points in Figure 3 by functions
of the form (17)–(17a), where the values of a_1^2 and a_4^2 are determined from
observations, c_{11} and c_{44} are taken from (10), and the values of G_{11} and G_{44} are
selected to obtain the best fit. Such a fit is obtained for $G_{11} = 1.4$ and $G_{44} = 1.4$
and the solid lines in Figure 3, which are described by Equations (17) and (17a)
with the indicated coefficients, show that these equations also agree well with the
measured values of the lateral correlation functions. The resulting values $G_{11} \approx$
1.4 and $G_{44} \approx 1.4$ are the only existing estimates of the coefficients in the -1
power law for the lateral space spectra. Unfortunately, no data on lateral spectra
of surface-layer turbulence are available for direct verification of these estimates.

References

Erdélyi, A., Magnus, W., Oberhettinger, F., and Tricomi, F. G.: 1953, 'Higher Transcendental
    Functions', Vol. 2, McGraw-Hill, New York, 396 pp.
Kader, B. A.: 1984, 'Structure of Anisotropic Velocity and Temperature Fluctuations in a Developed
    Turbulent Boundary Layer', *Izv. Akad. Nauk S.S.S.R., Ser. Mech. Liquid and Gas*, No. 4, pp.
    47–56.
Kader, B. A.: 1987, 'Anisotropic Fluctuations of Wind Velocity and Temperature in a Neutrally
    Stratified Atmospheric Surface Layer', in L. R. Tsvang and B. A. Kader (eds.), *Meteorological
    Researches*, No. 28, Acad. Sci. U.S.S.R., Soviet Geophysical Committee, Moscow, pp. 26–35 (in
    Russian).
Kader, B. A.: 1988, 'Three-Layer Structure of Unstable Atmospheric Surface Layer', *Izv. Akad.
    Nauk S.S.S.R., Ser. Atmos. and Oceanic Phys.* **24**, 1235–1250.
Kader, B. A. and Yaglom, A. M.: 1980, 'Similarity Laws for Turbulent Wall Flows', in *Developments
    in Science and Technology, Ser. Mech. Liquid and Gas*, Vol. 15, Soviet Inst. of Scient. and Eng.
    Inform., Moscow, pp. 81–155 (in Russian).
Kader, B.A. and Yaglom, A. M.: 1984, 'Influence of the Roughness and Longitudinal Pressure
    Gradient on Turbulent Boundary Layers', in *Developments in Science and Technology, Ser. Mech.
    Liquid and Gas*, Vol. 18, Soviet Inst. of Scient. and Eng. Inform., Moscow, pp. 3–111 (in Russian).
Kaimal, J. C., Wyngaard, J. C., Izumi, Y., and Coté, O. R.: 1972, 'Spectral Characteristics of Surface
    Layer Turbulence', *Quart. J. Roy. Meteorol. Soc.* **98**, 563–589.
Kolmogorov, A. N.: 1941, 'Local Structure of Turbulence in an Incompressible Fluid at Very High
    Reynolds Numbers', *Dokl. Akad. Nauk S.S.S.R.* **30**, 299–303. (English translation in S. K.

Friedlander and L. Topper (eds.), *Turbulence. Classical Papers on Statistical Theory*, Interscience, New York, 1961, pp. 151–155).

Koprov, B. M. and Sokolov, D. Yu.: 1973, 'Spatial Correlation Functions of Velocity Components and Temperature in the Atmospheric Surface Layer', *Izv. Akad. Nauk S.S.S.R., Ser. Atmos. and Oceanic Phys.*, **9**, 178–182.

Korotkov, B. N.: 1976, 'Some Types of Local Self-Similarity of the Velocity Field of Wall Turbulent Flows', *Izv. Akad. Nauk S.S.S.R., Ser. Mech. Liquid and Gas*, No. 6, 35–42.

Monin, A. S. and Yaglom, A. M.: 1971, *Statistical Fluid Mechanics*, Vol. 1, MIT-Press, Cambridge, Mass., 769 pp.

Monin, A. S. and Yaglom, A. M.: 1975, *Statistical Fluid Mechanics*, Vol. 2, MIT-Press, Cambridge, Mass., 874 pp.

Obukhov, A. M.: 1949, 'Structure of the Temperature Field in a Turbulent Flow', *Izv. Akad. Nauk S.S.S.R., Ser. Geogr. i Geophys.* **13**, 56–69 (reprinted in Obukhov, A. M., *Turbulence and Dynamics of Atmosphere*, Gidrometeoizdat, Leningrad, 1988, pp. 117–130).

Panofsky, H. A.: 1962, 'Scale Analysis of Atmospheric Turbulence at 2 m', *Quart. J. Roy. Meteorol. Soc.* **88**, 57–69.

Panofsky, H. A. and Dutton, J. A.: 1984, 'Atmospheric Turbulence', Wiley-Interscience, New York, 397 pp.

Perry, A. E. and Abell, S. J.: 1975, 'Scaling Laws for Pipe-Flow Turbulence', *J. Fluid Mech.* **67**, 257–271.

Volkov, Yu. A., Koprov, B. M., and Kravchenko, T. K.: 1975, 'Vertical Correlation Functions of the Turbulent Fields in the Atmospheric Surface Layer', *Izv. Akad. Nauk S.S.S.R., Ser. Atmos. and Oceanic Phys.* **11**, 794–801.

Wyngaard, J. C. and Coté, O. R.: 1972, 'Cospectral Similarity in the Atmospheric Surface Layer', *Quart. J. Roy. Meteorol. Soc.* **98**, 590–603.

Yaglom, A. M.: 1979, 'Similarity Laws for Constant-Pressure and Pressure-Gradient Turbulent Wall Flows', *Ann. Rev. Fluid Mech.* **11**, 505–540.

Yaglom, A. M.: 1981, 'Laws of Small-Scale Turbulence in Atmosphere and Ocean', *Izv. Akad. Nauk S.S.S.R., Ser. Atmos. and Oceanic Phys.* **17**, 1235–1257 (pp. 919–935 of the English edition).

Zubkovskii, S. L. and Fedorov, M. M.: 1986, 'On Large-Scale Structure of Wind Velocity Field in the Atmospheric Surface Layer', *Izv. Akad. Nauk S.S.S.R., Ser. Atmos. and Oceanic Phys.* **22**, 909–916.

Zubkovskii, S. L. and Sushko, A. A.: 1987, 'Experimental Investigation of the Spatial Structure of Temperature in the Atmospheric Surface Layer', in L. R. Tsvang and B. A. Kader (eds.), *Meteorological Researches*, No. 28, Acad. Sci. of the U.S.S.R., Soviet Geophysical Committee, Moscow, pp. 36–41 (in Russian).

INTERNAL BOUNDARY-LAYER HEIGHT FORMULAE
– A COMPARISON WITH ATMOSPHERIC DATA

JOHN L. WALMSLEY

Atmospheric Environment Service, Downsview, Ontario, Canada

(Received in final form 6 October, 1988)

Abstract. The height of the internal boundary layer (IBL) downwind of a step change in surface roughness is computed using formulae of Elliott (1958), Jackson (1976) and Panofsky and Dutton (1984). The results are compared with neutral-stratification atmospheric data extracted from the set of wind-tunnel and atmospheric data summarized by Jackson (1976) as well as neutral-stratification data presented by Peterson *et al.* (1979) and new data measured at Cherrywood, Ontario. It is found that the Panofsky-Dutton formulation gives the least root-mean-square (RMS) absolute errors for atmospheric applications.

1. Introduction

The empirical formula of Elliott (1958) is often used to compute IBL heights downwind of a change in surface roughness in neutrally-stratified flow. More recent formulae are based on a theoretical derivation of Miyake (1965). These are perhaps not as well known as Elliott's and possibly have not been extensively used because they require an iterative solution for the IBL height.

Jackson (1976) made a comparison between his formula and published neutral-stratification data from various sources. The data, however, were dominated by short-range atmospheric data and even shorter-range wind tunnel data and his comparison was essentially qualitative. More recently, Panofsky and Dutton (1984) introduced a similar formula without evaluating its accuracy.

The objective of the present study is to evaluate the above-mentioned formulae by comparison with a neutral-stratification atmospheric data set extracted from Jackson (1976) and supplemented by the longer-range data of Peterson *et al.* (1979) and from Cherrywood, Ontario.

2. Field Data

Data were extracted from Figure 1 of Jackson (1976), using his Table 1 to compute values of IBL height, δ, as a function of distance downwind, x, from a step-change in surface roughness (z_{01} to z_{02}). To these were added results given in Peterson *et al.* (1979) for Bognaes, Denmark and some values obtained recently (P. A. Taylor, personal communication) at Cherrywood, Ontario. All of these data, including the wind tunnel results, are plotted in Figure 1a. The symbols used are tabulated in Table I. The x-values range from about 10^{-2} m to

Boundary-Layer Meteorology **47**: 251–262, 1989.
© 1989 *Kluwer Academic Publishers.*

160 m. Corresponding δ-values range from 10^{-4} m to about 14 m. Roughness ratios, $m = z_{02}/z_{01}$, range from 8.6 to 167 for the smooth-to-rough cases and from 2.3×10^{-3} to 2.9×10^{-1} for the rough-to-smooth cases. There are also some wind tunnel data for $m = 1$. Note that the present definition of m is the inverse of the values tabulated by Jackson.

There are three points about Figure 1a which should be noted. First, the data are dominated by wind-tunnel results ($x < 1$–2 m). Second, there are no data at large distances ($x > 160$ m). Third, the logarithmic display tends to mask some fairly significant differences in observed values of δ at fixed x. We are interested in evaluating various formulae for δ as a function of x and any two of z_{01}, z_{02} and m in the real atmosphere. We shall therefore present data and formulae plotted in dimensional coordinates, with the wind-tunnel data eliminated from consideration.

Figure 1b shows the 29 atmospheric data points plotted in dimensional coordinates with a vertical exaggeration of $5:1$. The short-range data ($x < 10$ m) are still dominant, but the long-range data (say $x > 40$ m) now have relatively more influence. It is apparent, however, that there is a need for more measurements at distances $x > 40$ m. The linear scale reveals more clearly than in Figure 1a that there is significant scatter in the results. It remains to be seen whether some of this scatter can be explained by differing values of the roughness

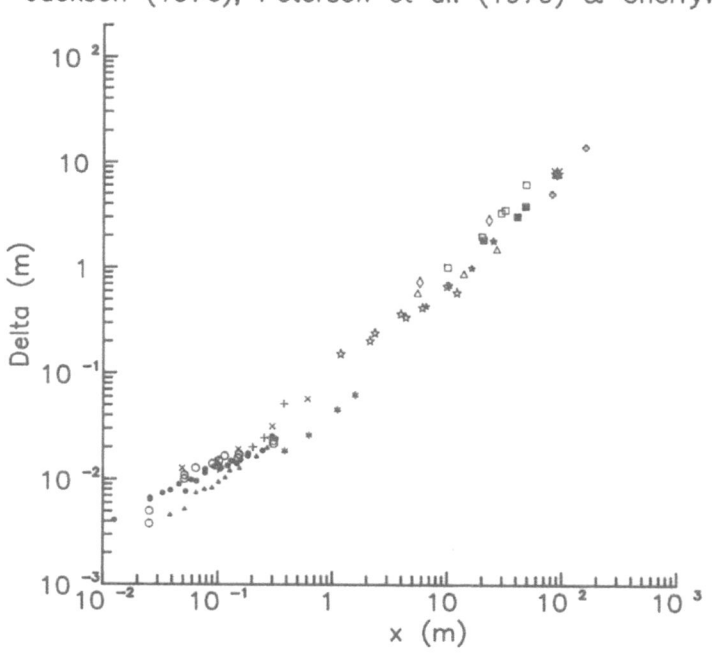

Fig. 1a.

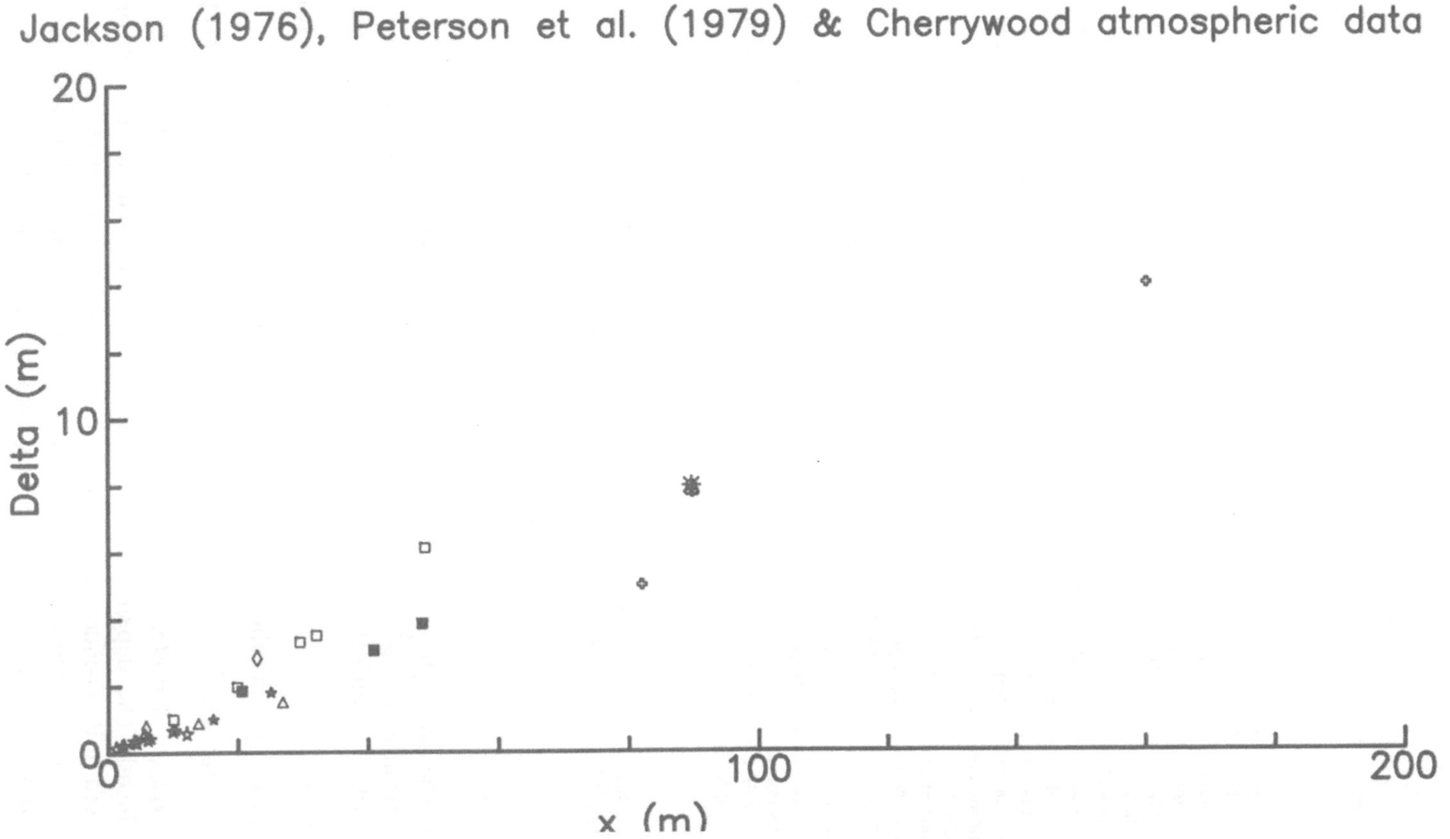

Jackson (1976), Peterson et al. (1979) & Cherrywood atmospheric data

Fig. 1. Jackson's atmospheric and wind tunnel data replotted as IBL height vs. distance downwind of a step-change in surface roughness. The data have been supplemented by those of Peterson et al. (1979) and some recent measurements at Cherrywood, Ontario. Symbols are defined in Table I. (a) Atmospheric and wind-tunnel data. Logarithmic scale. (b) Atmospheric data only. Linear scale, vertical exaggeration = 5:1.

TABLE I

Summary of experimental data

Symbol used in figures	Source	Type	z_{01} (mm)	z_{02} (mm)	m	d_2 (mm)
■	Kutzbach (1961)	A	0.1	12.2	122	100
◊	Blackadar et al. (1967)	A	35	10	0.29	15
*	Plate and Hidy (1967)	W	0.061	0.061	1.0	0
★	Bradley (1968) tarmac to spikes	A	0.02	2.5	125	2.5
☆	Bradley (1968) spikes to tarmac	A	2.5	0.02	0.008	0
△	Bradley (1968) grass to tarmac	A	5.0	0.02	0.004	0
×	Yeh and Nickerson (1970) raised crests	W	0.006	1.0	167	−2.0
+	Yeh and Nickerson (1970) flush crests	W	0.006	1.0	167	−2.0
●	Antonia and Luxton (1971a)	W	0.009	0.35	38	−1.86
▲	Antonia and Luxton (1971b)	W	0.009	0.35	38	−1.86
○	Antonia and Luxton (1972)	W	0.35	0.009	0.026	0
□	Angle (1973)	A	120	10	0.083	0
⋄	Peterson et al. (1979)	A	0.7	6	8.6	0
⚘	Cherrywood, trees to grass	A	219	4.4	0.020	0
☼	Cherrywood, trees to snow	A	293	0.68	0.0023	0

Type: A = atmospheric, W = wind tunnel.
Information on all but the last three experiments was obtained from Jackson (1976) which contains a more detailed summary.
z_{01} = upstream roughness length, z_{02} = downstream roughness length, $m = z_{02}/z_{01}$, d_2 = downstream displacement height.

parameters, z_{01}, z_{02} and/or m and which of the formulae best duplicates the observations. These questions will be examined in the following sections.

3. Formulae

In the formulae which follow, subscripts 1 and 2 will be used to denote the regions upstream and downstream, respectively, of a step-change in roughness.

Elliott's (1958) empirical formula may be written in its most familiar form as:

$$\delta' = A(x')^{0.8} \tag{1a}$$

but for comparison with other formulae, it may be recast as:

$$x' = (\delta'/A)^{1.25}. \tag{1b}$$

Here $x' = x/z_{02}$, $\delta' = \delta/z_{02}$ and, as originally given by Elliott, $A = 0.75 - 0.03 \ln(m)$. The dependence on m is rather weak and the formula is frequently used in its m-independent ($m = 1$) form, i.e., with $A = 0.75$.

In order to make Elliott's formula behave similarly to other formulae as x approaches zero, (1a) and (1b), respectively, may be modified as follows:

$$\delta' = A(x')^{0.8} + 1 \tag{1c}$$

$$x' = [(\delta' - 1)/A]^{1.25}. \tag{1d}$$

This change will have only a small effect ($< 4\%$ in δ') for $x' > 100$ and a negligible effect ($< 0.6\%$ in δ') for $x' > 1000$.

The following formulae can only be written as expressions for the downwind distance, x. They have to be solved iteratively for the IBL height, δ.

Jackson (1976) followed the derivation originally given by Miyake (1965), adding displacement height effects to arrive at the formula:

$$x'' = \{\delta''[\ln(\delta'') - 1] - \delta_0''[\ln(\delta_0'') - 1]\}/(\lambda\kappa) \tag{2a}$$

where $x'' = x/z_0'$, $\delta'' = (\delta - d_2)/z_0'$, $\delta'' = \delta_0''$ at $x'' = 0$, $\lambda = V'/u_*$ and κ is Von Karman's constant. Here $z_0' = [(z_{01}^2 + z_{02}^2)/2]^{1/2}$ is Jackson's somewhat arbitrary length scale, d_2 is the displacement height, V' is the characteristic vertical velocity of propagation of a smoke puff and u_* is the friction velocity. Jackson assumed $\lambda = 0.75$, whereas Miyake used $\lambda = 1.73$ with x' and δ' in place of x'' and δ'', respectively. If $\delta_0'' = 1$, then (2a) becomes:

$$x'' = \{\delta''[\ln(\delta'') - 1] + 1\}/(\lambda\kappa). \tag{2b}$$

Jackson's formula is dependent on m through the definition of z_0'. In order to compare with the other formulae, it will be assumed that $d_2 = 0$.

Panofsky and Dutton's (1984) formula is similar to (2b), the differences being in the definition of the length scale (z_{02}, instead of Jackson's z_0') and a different value for the constant.

$$x' = \{\delta'[\ln(\delta') - 1] + 1\}/(B\kappa). \tag{3}$$

The constant is defined as $B = \sigma_{w2}/u_{*2}$, where σ_w is the standard deviation of vertical fluctuations. It can be shown to be equivalent to Jackson's λ but, according to Panofsky and Dutton, is of order $B = 1.3$. Businger (1988) used $B = 1.25$, in accord with Panofsky and Dutton's Table 7.1, and that is the value adopted here. The Panofsky–Dutton formula, in the form given above, is independent of m, but note from Figure 2 that there is a dependence on m when plotted in dimensional units or as δ/z_{01} vs. x/z_{01}.

4. Comparison

Figure 2 shows a comparison between the formulae of Elliott (Equation 1b, $m = 1$) and Panofsky and Dutton (Equation 3). The greatest differences occur in Figure 2a where z_{02} and m are large. Elliott's δ-results are significantly higher than those of Panofsky and Dutton. Similar results, with smaller differences, occur in Figure 2b where z_{02} is moderate and $m = 1$. The smallest differences occur in Figure 2c where z_{02} and m are small. For this case, Elliott's formula

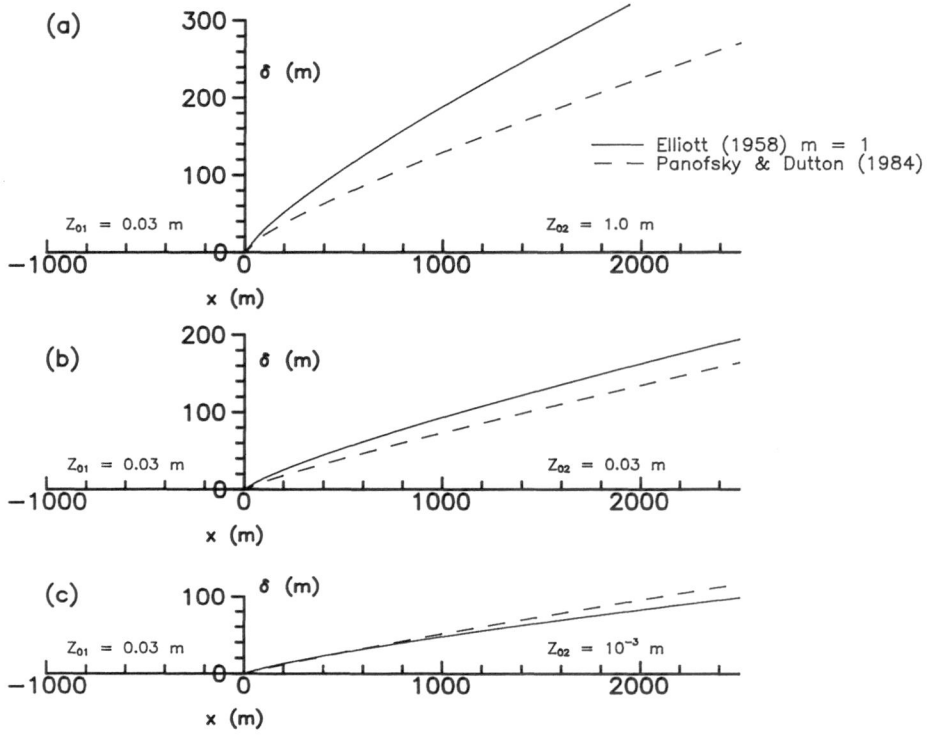

Fig. 2. IBL height (m) *vs.* downwind distance (m): comparison of formulae of Elliott (1958), Equation (1b), $m = 1$, and Panofsky and Dutton (1984), Equation (3). $z_{01} = 0.03$ m: (a) $z_{02} = 1.0$ m, $m = 33.3$, (b) $z_{02} = 0.03$ m, $m = 1$, (c) $z_{02} = 0.001$ m, $m = 0.033$.

gives smaller values for distances larger than 250 m, and about the same result as Panofsky and Dutton's formula for shorter distances.

Figures 3a and 3b indicate that Elliott's formula, (1b), tends to overestimate observed values of δ, especially at distances sufficiently large that $\delta > 1$ m. A visual comparison suggests that the m-dependent form (Figure 3a) gives slightly better agreement than the m-independent form (Figure 3b), but both give results that, in general, are too high.

Figures 3c and 3d show that both the Jackson and Panofsky–Dutton formulae, (2b) and (3), respectively, yield improved comparisons with the measured values of δ. A visual inspection suggests that the Panofsky–Dutton form is the better of the two.

Table II confirms the qualitative assessment of Figure 3. First, it can be seen that the differences between (1b) and (1d) are insignificant for this atmospheric data set. Second, the Elliott m-dependent form is slightly superior to the m-independent form. Third, incorporating the d_2 term in the Jackson formula has very little impact for this data set. Fourth, although the Jackson formula gives the

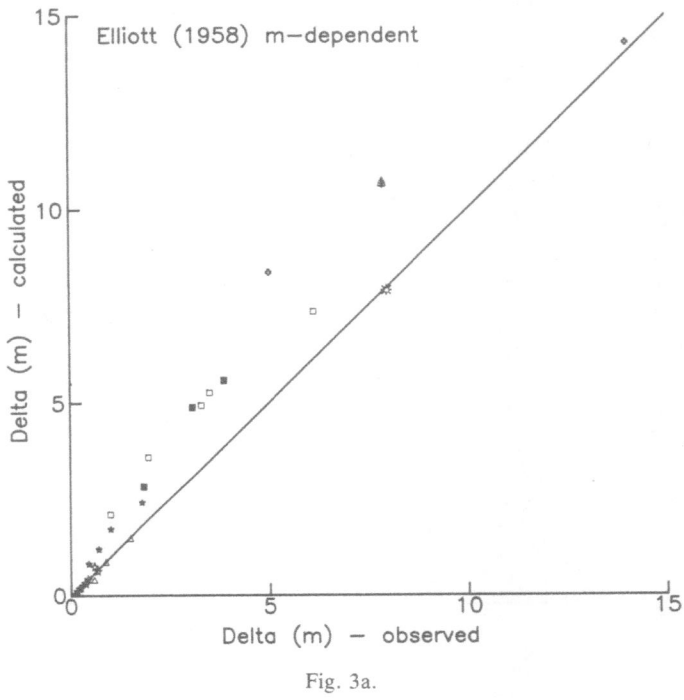

Fig. 3a.

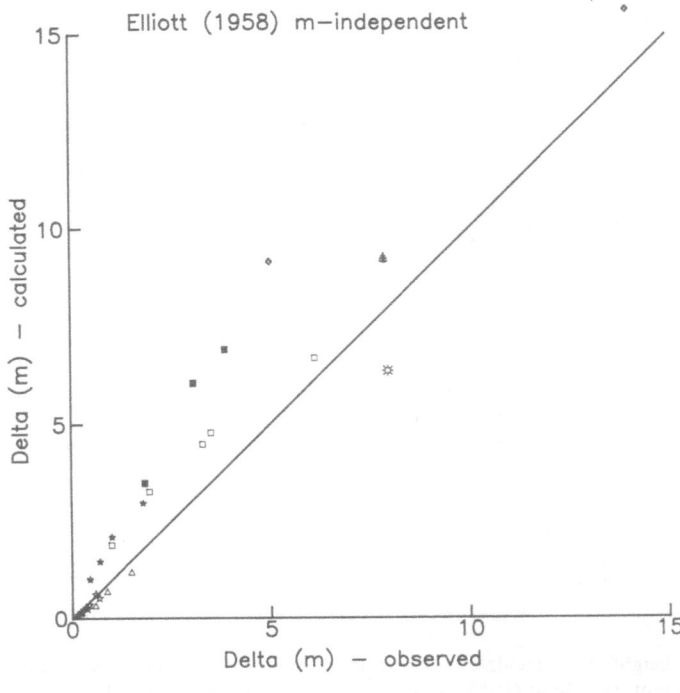

Fig. 3b.

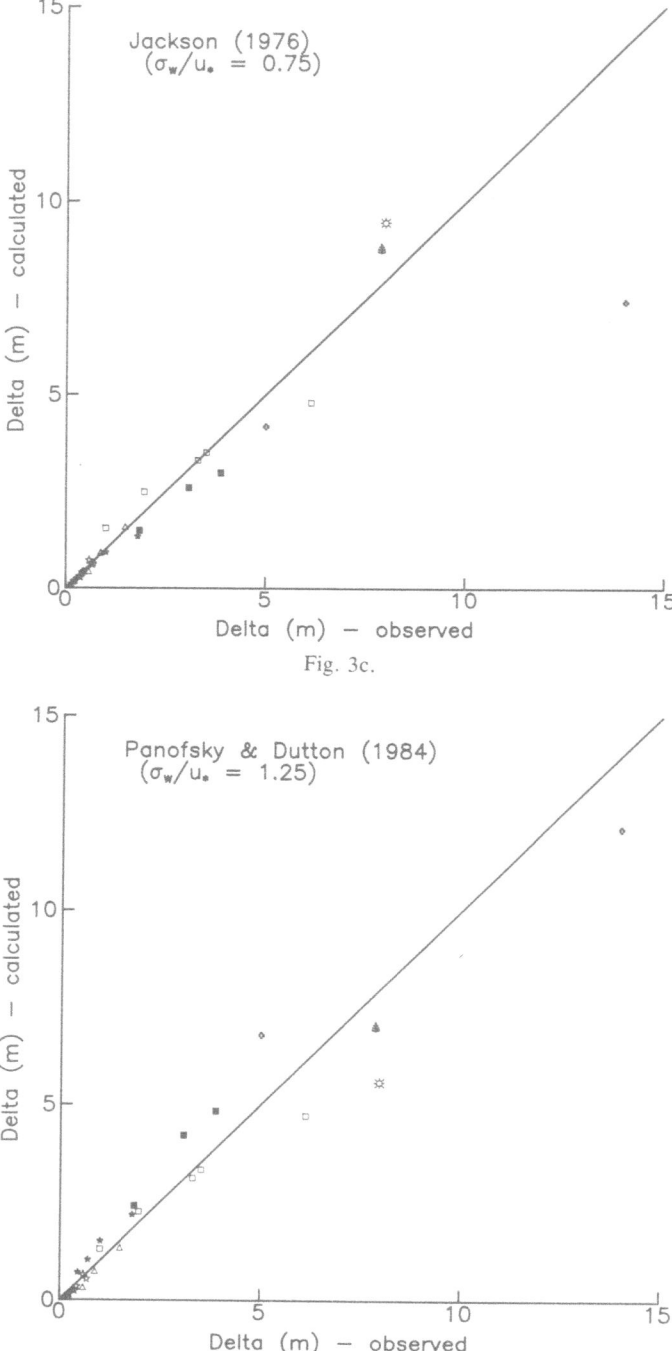

Fig. 3. IBL height (m): calculated *vs.* observed, atmospheric data only. Straight line represents perfect agreement: (a) Elliott (1958), Equation (1b), *m*-dependent, (b) Elliott (1958), Equation (1b), *m* = 1, (c) Jackson (1976), Equation (2b), $d_2 = 0$, (d) Panofsky and Dutton (1984), Equation 3.

TABLE II

Accuracy of IBL formulae: atmospheric data

	RMS Error (m)	RMS Error (%)	Correlation Coefficient
Elliott, Equation (1b), m-dependent	1.2	48	0.97
Elliott, Equation (1d), m-dependent	1.2	48	0.97
Elliott, Equation (1b), $m = 1$	1.4	61	0.95
Elliott, Equation (1d), $m = 1$	1.4	61	0.95
Jackson, Equation (2b), $d_2 = 0$	1.3	21	0.91
Jackson, Equation (2b), d_2 incorporated	1.3	21	0.91
Panofsky and Dutton, Equation (3)	0.8	30	0.97

lowest root-mean-square (RMS) percentage error, it also yields the poorest correlation coefficient and an RMS absolute error second in magnitude only to that of Elliott's m-independent form. Finally, the Panofsky–Dutton formula produces an RMS absolute error much lower than the others, a correlation coefficient that is not significantly different from the highest and an RMS percentage error much better than Elliott's and second only to Jackson's.

It should be noted that the relatively low RMS percentage error for Jackson's formula is undoubtedly due to the good agreement for $\delta < 1$ m (Figure 3c). The slightly poorer performance of the Panofsky–Dutton formula in that range (Figure 3d) is magnified both by the sensitivity of the RMS percentage error to small values and further by the relatively large number of small values. The RMS absolute error is therefore a more reliable measure of performance of the formulae for these data.

Figure 4 presents a further confirmation that the Panofsky–Dutton formulation gives the best overall agreement with the atmospheric data set. These best-fit lines derived from the data of Figure 3 all suggest a slight overestimation of the observations for $\delta < 2$ m, despite the fact that the agreement is generally quite good in that range, especially for the Jackson formula. The best-fit lines are apparently influenced by a few slightly high calculated values in the interval $\delta = 1$–2 m. The exact location of the lines at the low end of the scale is probably not important. For $\delta > 2$ m, Figure 4 shows clearly the tendency of the Elliott formula to overestimate, while the other two formulae have a tendency to underestimate. The poor performance of the Jackson formula (see Figure 3c and Table I) for the Bognaes, $x = 160$ m, $\delta = 14$ m data point of Peterson *et al.* (1979) is mainly responsible for the low slope of its best-fit curve in Figure 4 as well as its relatively high RMS absolute error and low correlation coefficient in Table II.

A comparison of (2b) and (3) shows that the only differences between the formulae of Jackson and Panofsky–Dutton are in the definition of the length scale (z_0' and z_{02}, respectively) and in the coefficient (λ and B, respectively). It is

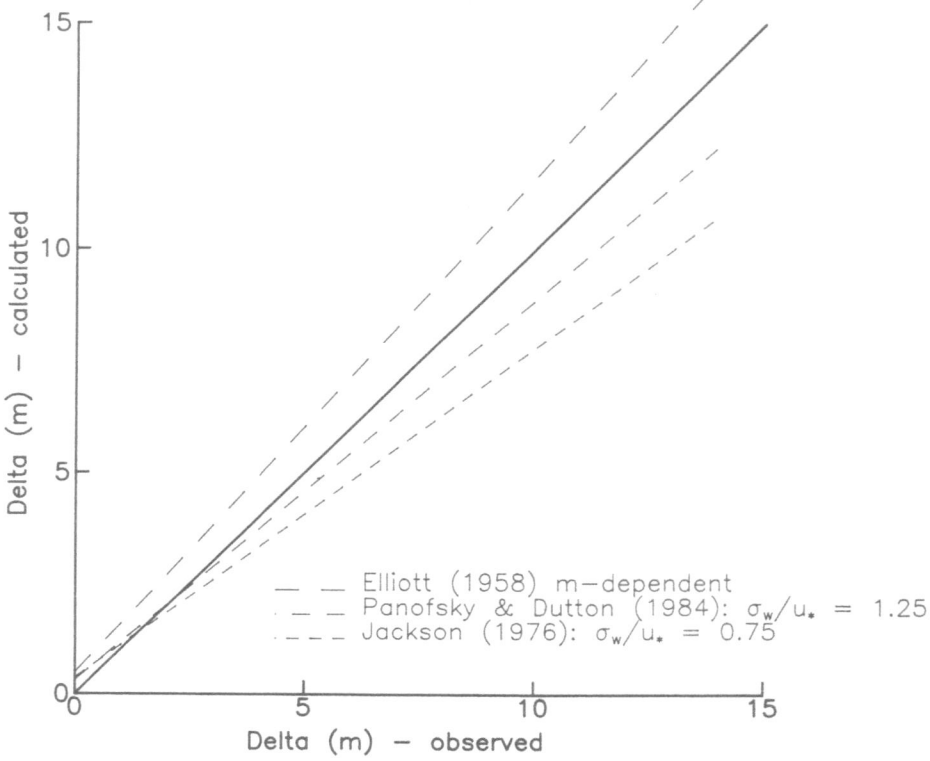

Fig. 4. IBL height (m): calculated *vs.* observed, atmospheric data only. Dashed lines represent best fit through the data plotted in Figures 3a, 3c and 3d. Solid line represents perfect agreement.

instructive, therefore, to test the sensitivity of the δ results to these two parameters. Since the two formulae in question give large differences in δ for the above-mentioned Bognaes data point, sensitivity to the two parameters will be investigated for $x = 160$ m with the roughness lengths listed for Peterson *et al.* (1979) in Table I. Note that this is a smooth-to-rough case.

It can be seen from Table III, that the results are not very sensitive to the admittedly small difference in length scale, but *are* quite sensitive to the coefficient. It is the small value of λ which causes the Jackson formula to seriously underestimate the height of this measurement.

Greater sensitivity to the length scale may be expected for a rough-to-smooth case, as z_0' is more strongly dependent on the larger than on the smaller roughness length. The difference between z_0' and z_{02} will thus be more marked for a rough-to-smooth than for a smooth-to-rough case. The value of the coefficient, nevertheless, will continue to have significant influence on the results.

TABLE III

Sensitivity of δ to parameters in Equation (2b) and (3).

Length Scale (mm)	Coefficient	δ^1 (m)
6.0	1.25	12.1[2]
4.27	1.25	11.6
6.0	0.75	7.78
4.27	0.75	7.43[3]

[1] Measured $\delta = 14$ m (Peterson *et al.*, 1979).
[2] Plotted in Figure 3d.
[3] Plotted in Figure 3c.

5. Summary and Conclusion

The IBL height formulae of Elliott (1958), Jackson (1976) and Panofsky and Dutton (1984) have been evaluated by comparison with an atmospheric data set consisting of 29 measurements. The data set is not ideal, being dominated by measurements at short range and being somewhat limited in the range of m-values, particularly for smooth-to-rough cases. Evaluation of the accuracy of the formulae was accomplished by plotting calculated *vs.* observed values, by drawing best-fit lines through the resulting plots and by tabulating RMS errors and correlation coefficients.

The relatively low *percentage* errors achieved by the Jackson formula are attributed to good agreement (though only slightly better than the other formulae) with the fairly large subset of short-range data. On the other hand, the Jackson formula gave significant errors for the longest-range data point. This poor performance is attributed mainly to the value of Jackson's coefficient.

The Panofsky–Dutton formula gave the best overall agreement with the data set. It yielded an RMS percentage error second only to that of the Jackson formula, a correlation coefficient essentially equal to the highest, a best-fit curve closest to perfect agreement and the lowest RMS absolute error.

Acknowledgements

My thanks to Peter Taylor for reviewing this manuscript and for providing the Cherrywood data. Jim Salmon, Jim Arnold and Paul Stalker were also involved in the field measurement program at Cherrywood.

References

Angle, R. P.: 1973, M.Sc. thesis, Dept. of Geography, Univ. of Alberta, Edmonton.

Antonia, R. A. and Luxton, R. E.: 1971a, 'The Response of a Turbulent Boundary Layer to an Upstanding Change in Surface Roughness', *Trans. ASME J. Basic Engineering* **93**, 22–34.

Antonia, R. A. and Luxton, R. E.: 1971b, 'The Response of a Turbulent Boundary Layer to a Step Change in Surface Roughness. Part 1. Smooth to Rough', *J. Fluid Mech.* **48**, 721–761.

Antonia, R. A. and Luxton, R. E.: 1972, 'The Response of a Turbulent Boundary Layer to a Step Change in Surface Roughness. Part 2. Rough to Smooth', *J. Fluid Mech.* **53**, 737–757.

Blackadar, A. K., Panofsky, H. A., Glass, P. E., and Boogaard, J. F.: 1967, 'Determination of the Effect of Roughness Change on the Wind Profile', *Phys. Fluids Suppl. Boundary Layers and Turbulence*, pp. S209–S211.

Bradley, E. F.: 1968, 'A Micrometeorological Study of Velocity Profiles and Surface Drag in the Region Modified by a Change in Surface Roughness', *Quart. J. Roy. Meteorol. Soc.* **94**, 361–379.

Businger, J. A.: 1988, 'Some Effects of Change of Terrain, Mesoscale Divergence, and Entrainment on the Structure of the Surface Layer and Its Consequences for Dry Deposition', in H. van Dop (ed.), *Air Pollution Modeling and Its Application*, VI, Plenum, New York, pp. 3–13.

Elliott, W. P.: 1958, 'The Growth of the Atmospheric Internal Boundary Layer', *Trans. Amer. Geophys. Union* **39**, 1048–1054.

Jackson, N. A.: 1976, 'The Propagation of Modified Flow Downstream of a Change in Roughness', *Quart. J. Roy. Meteorol. Soc.* **102**, 924–933.

Kutzbach, J. E.: 1961, 'Investigations of the Modification of Wind Profiles by Artificially Controlled Surface Roughness', Univ. Wisconsin Dept. of Meteorol. Annual Rep., pp. 71–113.

Miyake, M.: 1965, 'Transformation of the Atmospheric Boundary Layer Over Inhomogeneous Surfaces', Sci. Rep. 5R-6, Univ. of Washington, Seattle.

Panofsky, H. A. and Dutton, J. A.: 1984, *Atmospheric Turbulence: Models and Methods for Engineering Applications*, John Wiley & Sons, New York, 397 pp.

Peterson, E. W., Jensen, N. O. and Hojstrup, J.: 1979, 'Observations of Downwind Development of Wind Speed and Variance Profiles at Bognaes and Comparison with Theory', *Quart. J. Roy. Meteorol. Soc.* **105**, 521–529.

Plate, E. J. and Hidy, G. M.: 1967, 'Laboratory Study of Air Flowing Over a Smooth Surface onto Small Water Waves', *J. Geophys. Res.* **72**, 4627–4641.

Yeh, F. F. and Nickerson, E. C.: 1970, 'Air Flow Over Roughness Discontinuity', Project Themis Tech. Rep. No. 8, Colorado State University, Fort Collins.

RECURRENCE OF HIGH CONCENTRATION VALUES
IN A DIFFUSING, FLUCTUATING SCALAR FIELD

L. KRISTENSEN,[1] J. C. WEIL[2] and J. C. WYNGAARD

National Center for Atmospheric Research,[3] Boulder, CO 80307, U.S.A.

(Received in final form 10 October, 1988)

Abstract. We propose a simple model for esimating the average number of occurrences per unit time $\eta(c_0)$ that a threshold concentration c_0 is exceeded. It is based on the joint probability density of the observed concentration $c(t)$ and its time derivative $\dot{c}(t)$ under the assumption that $c(t)$ is a stationary time series; this assumption leads to the hypothesis that $c(t)$ and $\dot{c}(t)$ are statistically independent. Adopting plausible forms of the frequency distributions of c and $\dot{c}$, we apply the model to diffusion from an infinite area source and from an elevated point source, both in the neutral boundary layer, and obtain simple results for $\eta(c_0)$ and the average duration of one excursion above c_0 as functions of c_0, the mean and the standard deviation of the concentration, and surface-layer variables.

1. Introduction

In recent years it has become increasingly clear that air pollution models should be more informative; i.e., they should predict more than simply mean concentrations. Consequently, there has been a mounting effort to supplement predictions of the mean concentration C with the standard deviation σ_c and even the concentration frequency distribution (e.g., see Deardorff and Willis, 1988; Sykes, 1988). While this is extremely important and certainly a step in the right direction, we suggest a slight extension. Instead of asking for the probability that the concentration exceeds a certain level at any given instant, we should ask: "How long must one wait in exposure to a pollutant before there is an even chance that a fluctuation in the concentration exceeding a particular value will actually occur?" (Deardorff and Willis, 1988) or, in short, "How often will a threshold c_0 be exceeded on average per unit time?" (See Figure 1.) If we can answer this question, we can also estimate the average duration of a high-concentration event as well as the probability for such an event occurring during a specified duration of time.

In the following, we shall formulate our suggestion in more rigorous terms and then use standard information about the atmospheric surface layer for its implementation under simplifying assumptions.

[1] Permanent affiliation: Risø National Laboratory, 4000 Roskilde, Denmark.
[2] Permanent affiliation: Cooperative Institute for Research in Environmental Sciences, University of Colorado, Boulder, CO 80309.
[3] The National Center for Atmospheric Research is sponsored by the National Science Foundation.

Boundary-Layer Meteorology **47**: 263–276, 1989.
© 1989 *Kluwer Academic Publishers.*

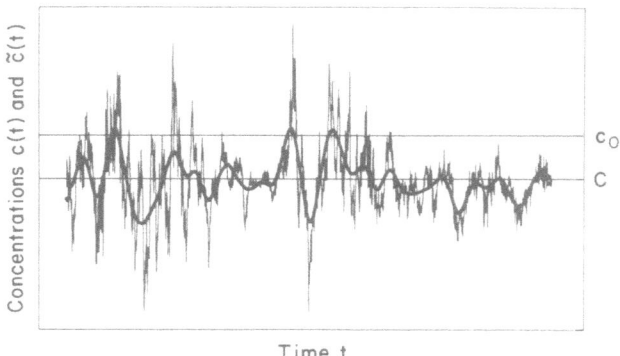

Fig. 1. Sketch of a concentration record, unfiltered ($c(t)$, thin line) and filtered ($\bar{c}(t)$, thick line). Note the dramatic effect of the filtering; $c(t)$ shows 17 excursions above c_0 whereas $\bar{c}(t)$ shows only 3. Since Equation (15) is valid for the unfiltered as well as filtered concentration time series, and since we assume that the filtering has little effect on the frequency distribution $Q_0(c)$, we expect that the average "pulse" duration will be larger for the filtered than for the unfiltered concentration signal. The figure indicates that this seems in fact to be the case.

2. Rice's Theory

We define one excursion beyond the concentration c_0 as one up-crossing followed by one down-crossing of the level c_0. According to Rice (1939, 1944–45), it is possible to determine the average number of such excursions per unit time $\eta(c_0)$, provided that the joint probability density $P(c, \dot{c})$ of the concentration c and its time derivative $\dot{c}$ is known. A useful summary of Rice's theory is given by Panofsky and Dutton (1984). They derive, under the assumption that $c(t)$ is ergodic, the general relation

$$\eta_0 \equiv \eta(c_0) = \int\limits_0^\infty \dot{c} P(c_0, \dot{c}) \, d\dot{c} \ . \tag{1}$$

For a fixed value of c_0, $\eta(c_0)$ is proportional to the average time derivative. This is shown in Figure 1, where a prototype (artificial) high-frequency signal $c(t)$ is given (thin line) together with the same signal subjected to low-pass filtering $\bar{c}(t)$ (thick line). The low-pass filtered signal has smaller time derivatives and fewer excursions beyond c_0 than $c(t)$.

If c_0 is sufficiently large, the individual excursions will be so infrequent that they can be considered statistically independent, and consequently Poisson distributed. That is, the probability $P_T[n]$ that the concentration c_0 is exceeded n times during the period T is

$$P_T[n] = \frac{e^{-\eta_0 T}}{n!} (\eta_0 T)^n \ . \tag{2}$$

Equation (1) will allow us to specify the so-called recurrence of high-concentration events as follows: determine the concentration $c_1(T)$ which on average is exceeded once in the period T. We find $c_1 = c_1(T)$ as the solution to the equation

$$1 \equiv T\eta(c_1) = T \int_0^\infty \dot{c} P(c_1, \dot{c}) \, d\dot{c} . \tag{3}$$

If T is large compared to the integral time scale of $c(t)$, the individual excursions beyond c_1 are statistically independent and we may use (2) with $\eta_0 T = \eta(c_1) T = 1$ to evaluate the probabilities for zero, one, two or several incidents of high-concentration events. ($P_T[0] \geqslant e^{-1}$, $P_T[1] = e^{-1}$, $P_T[2] = e^{-1}/2$, $P_T[n] = e^{-1}/n!$).

It is also possible to estimate the average duration $\theta(c_0)$ of one excursion, i.e., the average lifetime of a "pulse" greater than c_0. During the period T, the total average time $\Theta(c_0)$ that the concentration c is greater than c_0 is equal to T times the probability of c exceeding c_0, i.e.,

$$\Theta_0 \equiv \Theta(c_0) = T \int_{c_0}^\infty dc \int_{-\infty}^\infty P(c, \dot{c}) \, d\dot{c} . \tag{4}$$

Since the average number of excursions in the same period is $\eta_0 T$, we have

$$\theta_0 \equiv \theta(c_0) \approx \frac{\Theta_0}{\eta_0 T} = \frac{\int_{c_0}^\infty dc \int_{-\infty}^\infty P(c, \dot{c}) \, d\dot{c}}{\int_0^\infty \dot{c} P(c_0, \dot{c}) \, d\dot{c}} . \tag{5}$$

We note that θ_0 is independent of the observation period.

Thus, once it is determined, the joint probability density $P(c, \dot{c})$ provides much more information than does the probability density of c alone. In the following, we shall construct a simple model for $P(c, \dot{c})$.

3. A Simple Model

We first note that if $c(t)$ is a stationary time series,[4] then $c(t)$ and $\dot{c}(t)$ are uncorrelated. Denoting ensemble or long-term averaging by angle brackets, we have

$$\langle c(t) \rangle = C = \text{constant} \tag{6}$$

and

$$\langle \dot{c}(t) \rangle = \frac{d}{dt} \langle c(t) \rangle = \frac{dC}{dt} = 0 . \tag{7}$$

Defining

$$c'(t) = c(t) - C , \tag{8}$$

---

[4] In fact, for the theory outlined in section (2) to be valid, we must assume ergodicity which implies stationarity.

the covariance becomes

$$\langle c'(t)\dot{c}(t)\rangle = \langle c(t)\dot{c}(t)\rangle = \frac{1}{2}\frac{d}{dt}\langle c^2(t)\rangle = 0, \tag{9}$$

which proves that $c(t)$ and $\dot{c}(t)$ are uncorrelated.

If $P(c, \dot{c})$ were a joint Gaussian distribution, i.e., if

$$P(c, \dot{c}) = \frac{1}{2\pi\sigma_c\sigma_{\dot{c}}\sqrt{1-\rho^2}}\exp\left(-\frac{1}{2(1-\rho^2)}\left\{\frac{c'^2}{\sigma_c^2}-2\rho\frac{c'\dot{c}}{\sigma_c\sigma_{\dot{c}}}+\frac{\dot{c}^2}{\sigma_{\dot{c}}^2}\right\}\right), \tag{10}$$

where $\sigma_c^2 = \langle c'^2\rangle$ and $\sigma_{\dot{c}}^2 = \langle \dot{c}^2\rangle$ are the variances of c and $\dot{c}$, respectively, and where $\rho = \langle c'\dot{c}\rangle/(\sigma_c\sigma_{\dot{c}})$ is their correlation coefficient, then (9) would imply that c and $\dot{c}$ are statistically independent. This is the case because then $\rho = 0$, so that $P(c, \dot{c})$ can be decomposed according to

$$P(c, \dot{c}) = Q_0(c)Q_1(\dot{c}), \tag{11}$$

where $Q_0(c)$ and $Q_1(\dot{c})$ are the frequency distributions of c and $\dot{c}$, respectively. Unfortunately, there are no observations to guide a choice of $P(c, \dot{c})$. However, laboratory measurements of concentrations from an elevated point source in simulated convective and neutral boundary layers show that a gamma distribution is appropriate for $Q_0(c)$ (Deardorff and Willis, 1988). In addition, when $\sigma_c/C = 1$, the gamma distribution reduces to an exponential form which has been observed for near-surface releases in the atmospheric boundary layer (Hanna, 1984). Therefore, we believe that this distribution has a useful range of applicability.

A gamma distribution for $Q_0(c)$ means that $P(c, \dot{c})$ is not joint Gaussian and therefore (9) represents only circumstantial evidence that $c(t)$ and $\dot{c}(t)$ are statistically independent. Nevertheless, we shall assume this independence and, hence, (11) in the following.

In addition, we shall assume for simplicity that $Q_1(\dot{c})$ is Gaussian, i.e.,

$$Q_1(\dot{c}) = \frac{1}{\sqrt{2\pi}\sigma_{\dot{c}}}\exp\left(-\frac{\dot{c}^2}{2\sigma_{\dot{c}}^2}\right). \tag{12}$$

Substituting (12) into (1) and (5) yields

$$\eta_0 = \frac{\sigma_{\dot{c}}Q_0(c_0)}{\sqrt{2\pi}} \tag{13}$$

and

$$\theta_0 = \frac{\sqrt{2\pi}}{\sigma_{\dot{c}}Q_0(c_0)}\int\limits_{c_0}^{\infty} Q_0(c)\,dc. \tag{14}$$

It follows quite generally from (1) and (5) that if we consider (5) as the definition of η_0, then

$$\eta_0 \theta_0 = \int_{c_0}^{\infty} Q_0(c) \, dc \,, \tag{15}$$

which says that the product of the average number of occurrences per unit time that a threshold concentration c_0 is exceeded and the average duration of those excursions is simply the probability that c exceeds c_0. The assumption (11) that c and $\dot{c}$ are statistically independent is not necessary for the validity of (15).

We also derived the expression for η_0 under the assumptions that $Q_1(\dot{c})$ is symmetric and that $\dot{c}^2/\sigma_{\dot{c}}^2$ has a log-normal frequency distribution (Gurvich and Yaglom, 1967). In this case the denominator in (13) changes from $\sqrt{2\pi}$ to $2K^{-1/8}$, where $K = \langle \dot{c}^4 \rangle / \langle \dot{c}^2 \rangle^2$ is the flatness factor of $\dot{c}$. The arguments of Gurvich and Yaglom (1967) and Wyngaard and Tennekes (1970) indicate that K should increase monotonically with Reynolds number. The data of Gibson et al. (1970) for the temperature derivative in the atmospheric surface layer are consistent with this, showing K increasing from about 25 to 40 as the Reynolds number increases. Even for $K = 40$, however, the denominators differ by only about 25%. This is not important for our purposes, so we shall use the Gaussian distribution.

4. Applications

To demonstrate how the model might work, we apply it to concentration fields near the surface and assume that the concentration fluctuations have a gamma distribution (Deardorff and Willis, 1988):

$$Q_0(c) = \frac{\alpha}{\Gamma(\alpha)} \frac{1}{C} \left(\alpha \frac{c}{C} \right)^{\alpha-1} \exp\left(-\alpha \frac{c}{C} \right), \tag{16}$$

where

$$\alpha = \frac{C^2}{\sigma_c^2}. \tag{17}$$

One other quantity that we need to determine is the variance $\sigma_{\dot{c}}^2$ of $\dot{c}(t)$, which we can relate to micrometeorological parameters through the budget equation for the scalar variance. Assuming the governing equation

$$\frac{\partial c}{\partial t} + u_i \frac{\partial c}{\partial x_i} = \gamma_c \frac{\partial^2 c}{\partial x_i \partial x_i}, \tag{18}$$

where u_i is the velocity field and γ_c the molecular diffusivity, and where repeated indices imply summation, this budget equation is:

$$\frac{\partial}{\partial t} \langle c'^2 \rangle = -2 \langle u_i' c' \rangle \frac{\partial C}{\partial x_i} - U_i \frac{\partial}{\partial x_i} \langle c'^2 \rangle - \chi_c$$

$$- \frac{\partial}{\partial x_i} \langle u_i' c'^2 \rangle + \gamma_c \frac{\partial^2}{\partial x_i \partial x_i} \langle c'^2 \rangle. \tag{19}$$

The terms on the right side of (19) represent gradient production, mean advection, molecular destruction, turbulent transport, and molecular diffusion, respectively (Panofsky and Dutton, 1984). We have decomposed u_i into the mean U_i and the fluctuating component u_i' and introduced the molecular destruction rate

$$\chi_c = 2\gamma_c \left\langle \frac{\partial c'}{\partial x_i} \frac{\partial c'}{\partial x_i} \right\rangle . \tag{20}$$

Assuming steady state and a horizontally homogeneous velocity field with the mean velocity $U = U_1$ in the direction of the x-axis (x_1-axis), (19) becomes

$$\chi_c = -2\langle u_i' c' \rangle \frac{\partial C}{\partial x_i} - U \frac{\partial}{\partial x} \langle c'^2 \rangle - \frac{\partial}{\partial x_i} \langle u_i' c'^2 \rangle , \tag{21}$$

where we have dropped the molecular diffusion term because it is negligible in the atmospheric boundary layer. With the aid of Taylor's hypothesis, we have, due to local isotropy

$$\chi_c = 6\gamma_c \left\langle \left(\frac{\partial c'}{\partial x} \right)^2 \right\rangle = 6\gamma_c U^{-2} \sigma_{\dot{c}}^2 , \tag{22}$$

so that

$$\sigma_{\dot{c}}^2 = \frac{\chi_c}{6\gamma_c} U^2 . \tag{23}$$

It follows that direct measurements of $\sigma_{\dot{c}}$ can be related to other surface-layer properties through (21) and (23). Direct determination of $\sigma_{\dot{c}}$, however, requires that $c(t)$ be measured by an instrument capable of resolving the Kolmogorov microscale-size eddies (Wyngaard, 1971). Most instruments average over much larger space and time scales. We shall illustrate the effect of time averaging by assuming that the sensor is sufficiently small that spatial filtering can be neglected and that it acts as a first-order temporal filter. The autocovariance function $\tilde{R}_c(\tau)$ of the filtered concentration signal $\tilde{c}(t)$ is expressed in terms of the filtered spectrum $\tilde{\Phi}_c(\omega)$ as

$$\tilde{R}_c(\tau) \equiv \langle \tilde{c}(t)\tilde{c}(t+\tau) \rangle = \int_{-\infty}^{\infty} \tilde{\Phi}_c(\omega) e^{i\omega\tau} d\omega . \tag{24}$$

It is easily seen that

$$\langle \dot{\tilde{c}}(t) \dot{\tilde{c}}(t+\tau) \rangle = -\ddot{\tilde{R}}(\tau) \tag{25}$$

so that

$$\tilde{\sigma}_{\dot{c}}^2 \equiv -\ddot{\tilde{R}}(0) = 2 \int_0^{\infty} \omega^2 \tilde{\Phi}_c(\omega) d\omega . \tag{26}$$

If the spectral transfer function of the filter is $H(\omega)$, we can write (26) as

$$\tilde{\sigma}_{\dot{c}}^2 = 2 \int_0^{\infty} \omega^2 H(\omega) \Phi_c(\omega) \, d\omega .$$ (27)

Panofsky and Dutton (1984) discuss the spectrum of a conservative scalar. Their expression for its inertial subrange is all we need since $\Phi_c(\omega)$ is weighted by ω^2 in (27), and therefore the exact spectral behavior at low frequencies is unimportant. Transforming their expression from the wave-number domain to the frequency domain, using Taylor's hypothesis, and multiplying the result by the first-order, low-pass filter transfer function $H(\omega) = 1/(1 + \omega^2 \tau_c^2)$, we obtain the following expression for the filtered frequency spectrum $\tilde{\Phi}_c(\omega)$:

$$\tilde{\Phi}_c(\omega) = \frac{\Phi_c(\omega)}{1 + \omega^2 \tau_c^2} = b \chi_c \epsilon^{-1/3} U^{2/3} \frac{\omega^{-5/3}}{1 + \omega^2 \tau_c^2},$$ (28)

where $b \approx 0.2$ is a dimensionless constant,[5] ϵ is the rate of dissipation of turbulent kinetic energy, and τ_c is the instrument time constant. We assume that the production and dissipation rates of turbulent kinetic energy are in local balance, which means that in the neutral surface layer, ϵ can be expressed in terms of the height $x_3 = z$ and the surface friction velocity

$$u_* = \lim_{z \to 0} \sqrt{-\langle u_3' u_1' \rangle} .$$ (29)

Substituting

$$\epsilon = \frac{u_*^3}{\kappa z},$$ (30)

where κ is the von Kármán constant, into (28) and carrying out the integration (26), we find the low-pass filtered counterpart of (23):

$$\tilde{\sigma}_{\dot{c}}^2 = \frac{2\pi}{\sqrt{3}} b \chi_c \left(\frac{\kappa z}{U}\right)^{1/3} \frac{U}{u_*} \tau_c^{-4/3} .$$ (31)

Evaluation of the exceedence rate η_0 or $\bar{\eta}_0$ and the average excursion duration θ_0 is, under the given assumptions, straightforward with $Q_0(c)$ and $\sigma_{\dot{c}}^2$ or $\tilde{\sigma}_{\dot{c}}^2$ given; the last two quantities will require knowledge of χ_c which we shall obtain for two experimental situations from the variance budget (21).

4.1. INFINITE AREA SOURCE

If we consider concentration fluctuations inside a large area with a homogeneous surface-source distribution, then the scalar field is horizontally homogeneous.

---

[5] The value of b we use is 1/4 of that used by Kaimal *et al.* (1972) because (1) their molecular destruction rate pertains to the half variance; and (2) they define the spectrum as the integral from 0 to ∞, whereas ours is the integral from $-\infty$ to ∞, as implied by (24).

This means that the second term on the right-hand side of (21), mean advection, is identically zero. We shall assume that the third term, turbulent transport, is small compared to the first term, gradient production. This is substantiated to some extent by the analysis of Wyngaard *et al.* (1978) that shows turbulent transport of squared temperature fluctuations in the unstable surface layer to be less than 20% of the gradient production. Thus, we have:

$$\chi_c \approx -2\langle u_3' c' \rangle \frac{\partial C}{\partial z} \approx 2 u_* c_* \frac{dC}{dz}. \tag{32}$$

We have introduced the surface-concentration scale

$$c_* = -\frac{\lim_{z \to 0} \langle u_3' c' \rangle}{u_*}. \tag{33}$$

In the neutral surface layer we assume

$$\frac{dC}{dz} = \frac{c_*}{\kappa_c z}, \tag{34}$$

where κ_c is a dimensionless constant, which can be related to the von Kármán constant κ and the turbulent Schmidt number Sc_T by

$$\kappa_c = \kappa / Sc_T. \tag{35}$$

Based on the data from the 1976 International Turbulence Comparison Experiment (ITCE) in New South Wales, Australia, Dyer and Bradley (1982) found that the corresponding constants κ_H and κ_W for heat and water vapor, respectively, within an accuracy of 5% are equal to κ, implying $Sc_T = 1$. If we follow the suggestion by Panofsky and Dutton (1984) (p. 148) that scalars in general follow the same flux-gradient relation as heat, then it follows from Businger *et al.* (1971) that $Sc_T = 0.74$ for neutral stratification so that $\kappa_c \approx 0.5$ if $\kappa = 0.4$.

Substituting (34) into (32) and subsequently (32) into (23) yields

$$\sigma_{\dot{c}}^2 = \frac{Pe_z}{3\kappa_c} \left(\frac{c_* U}{z} \right)^2. \tag{36}$$

We have introduced a Péclet number based on u_*, z and γ_c:

$$Pe_z = \frac{u_* z}{\gamma_c} \tag{37}$$

and will assume that this is approximately equal to the Reynolds number, which is based on the kinematic viscosity ν. Inserting (16) and (36) in (13), we obtain

$$\eta_0 = \frac{Pe_z^{1/2}}{\sqrt{6\pi\kappa_c}} \left(\frac{c_* U}{Cz} \right) \frac{\alpha}{\Gamma(\alpha)} \left(\alpha \frac{c_0}{C} \right)^{\alpha-1} e^{-\alpha(c_0/C)}. \tag{38}$$

Alternatively, using (31) instead of (23) we get, in the case where $c(t)$ is low-pass filtered

$$\tilde{\sigma}_{\dot{c}}^2 = \frac{4\pi b\kappa^{1/3}}{\sqrt{3}\kappa_c}\left(\frac{c_* U}{z}\right)^2\left(\frac{z}{U\tau_c}\right)^{4/3}. \tag{39}$$

The low-pass filtering will in principle also affect $Q_0(c)$, the probability density of c; the simplest effect is a narrowing of Q_0 through reduction of the variance σ_c^2. However, if the filter time constant τ_c is not too large, the variance will be negligibly reduced. We take this to be the case, and assume that Q_0 is unchanged by the filtering. Thus, under our assumptions the influence of filtering on the exceedance parameters η_0 and θ_0, Equations (13) and (14), is felt only through $\sigma_{\dot{c}}$. The expression for the filtered η_0, for example, becomes

$$\tilde{\eta}_0 = A\left(\frac{c_* U}{Cz}\right)\left(\frac{z}{U\tau_c}\right)^{2/3}\frac{\alpha}{\Gamma(\alpha)}\left(\alpha\frac{c_0}{C}\right)^{\alpha-1}e^{-\alpha(c_0/C)}, \tag{40}$$

where

$$A = \kappa^{-1/3}\left(\frac{4}{3}\right)^{1/4}\left(\frac{b\kappa}{\kappa_c}\right)^{1/2} \approx 0.6. \tag{41}$$

If we use τ_c and $U\tau_c$ as time and length scales, respectively, we can, without making an explicit assumption about the form of the frequency distribution $Q_0(c)$, write (40) as

$$\tilde{\eta}_0\tau_c = Ac_* Q_0(c_0)\left(\frac{z}{U\tau_c}\right)^{-1/3}. \tag{42}$$

This form has a simple interpretation: for given turbulent surface fluxes of momentum and scalar c, and for a given threshold c_0, the number of excursions beyond c_0 during a period of time equal to the instrument time constant is inversely proportional to the height, normalized by the length $U\tau_c$ of the column of air which has gone through the instrument in the period τ_c, to the power one third.

Let us now examine more closely the effect of filtering on $\sigma_{\dot{c}}$. From Equations (23) and (31), we have in general

$$\frac{\sigma_{\dot{c}}^2}{\tilde{\sigma}_{\dot{c}}^2} = \frac{Pe_z}{4\pi b\sqrt{3}\kappa^{1/3}}\left(\frac{U\tau_c}{z}\right)^{4/3}. \tag{43}$$

It follows from (43) that the filtered and unfiltered derivative variances are equal if the time constant has the value

$$\tau_{c0} = (4\sqrt{3}\pi b)^{3/4}\kappa^{1/4}Pe_z^{-3/4}\frac{z}{U} \approx 2.4Pe_z^{-3/4}\frac{z}{U}. \tag{44}$$

If $u_* = 0.2\text{ m s}^{-1}$, $U = 3\text{ m s}^{-1}$ and $z = 2\text{ m}$, then $\tau_{c0} = 7.7 \times 10^{-4}\text{ s}$, which is much less than the time constant of any available scalar sensor with the possible exception of the fastest fine-wire thermometers. If, by contrast, we assume that τ_c is chosen such that $U\tau_c/z = O(1)$, so that we filter out those concentration "eddies" smaller than distance to the surface, we see from (43) that the filtering reduces the derivative variance by more than a factor of 8000. In our model, where we assume that $Q_0(c)$ is unaffected by the filtering, we have $\theta_0/\bar{\theta}_0 = \tilde{\eta}_0/\eta_0 = \bar{\sigma}_{\dot{c}}/\sigma_{\dot{c}}$. In this case, we therefore have $\tilde{\eta}_0 \approx 0.01\,\eta_0$. In cases of practical importance $\tilde{\eta}_0 \ll \eta_0$ and $\bar{\theta}_0 \gg \theta_0$. Figure 1 illustrates this qualitatively.

This extreme sensitivity of the scalar derivative variance to low-pass filtering is due, of course, to the $\omega^{-5/3}$ inertial subrange behavior of the frequency spectrum of scalar turbulence. The scalar derivative spectrum increases as $\omega^{1/3}$ in that range [as shown by (26)], which means that it peaks near the highest frequencies present in a turbulent flow, those corresponding to the diffusive cutoff at Kolmogorov scales. Thus, Equation (43) tells us that measuring all the derivative variance requires extremely good temporal resolution.

We conclude that the unfiltered scalar derivative signal $\dot{c}(t)$ is strongly influenced by very short duration "bursts" of concentration microstructure, as the steep flanks of $c(t)$, Figure 1, depict graphically. As Equation (36) shows, the variance of $\dot{c}(t)$ is proportional to Pe_z, the Péclet number based on height z, which is quite large in the surface layer; for $z = 2$ m and $u_* = 0.2$ m s^{-1}, for example, $Pe_z \sim 10^4$. Since the Péclet numbers attainable in laboratory flows are very much smaller, the intensity of the scalar microstructure intermittency in the surface layer cannot be replicated in laboratory simulations. This is not a great loss, however, since the exceedance parameters η_0 and θ_0 for the unfiltered time series $c(t)$ are likely to be of little practical interest. As Figure 1 suggests, those for the filtered series are apt to be more meaningful, and Equation (39) indicates that the filtered exceedance parameters can be simulated in laboratory diffusion experiments by matching the scaled time constant $\tau_c U/z$ with the atmospheric value.

The average "pulse duration" becomes

$$\theta_0 = \eta_0^{-1}\,\frac{\Gamma\!\left(\alpha,\,\alpha\,\dfrac{c_0}{C}\right)}{\Gamma(\alpha)}, \tag{45}$$

where η_0 represents both η_0 and $\tilde{\eta}_0$ given by (38) and (40), respectively, and where

$$\Gamma(\alpha, x) = \int_x^\infty t^{\alpha-1}\,e^{-t}\,dt \tag{46}$$

is the complementary incomplete Gamma function.

In the limit $\alpha c_0/C \to \infty$, (45) becomes

$$\theta_0 \approx \frac{\sqrt{2\pi}}{\sigma_{\dot{c}}}\left(\frac{\sigma_c}{C}\right)^2, \tag{47}$$

where $\sigma_{\dot{c}}$ is given by either (23) or (39). It may seem surprising that θ_0 ceases to be a function of c_0. This behavior is intimately tied to the choice of the concentration frequency distribution. If $Q_0(c)$ is a normal rather than gamma distribution, θ_0 is inversely proportional to the difference $c_0 - C$ for $c_0 \to \infty$, whereas for a log-normal distribution $\theta_0 \to \infty$ in this limit.

4.2. ELEVATED POINT SOURCE

In this example, we consider concentration fluctuations of effluents from an elevated point source in a neutral boundary layer. Based on wind-tunnel measurements, Fackrell and Robins (1982) showed that the steady-state variance budget Equation (21) can be simplified because the production term can be neglected entirely. In their experiments, where the source height was about one fifth of the boundary-layer thickness h, they also found that if the downstream distance x was greater than $3h$, the turbulent diffusion term could also be neglected. In fact, their measurements showed that most of the fluctuations are produced very near the source and that the level of fluctuation then decays downstream with an approximate balance between advection and dissipation. If we limit ourselves to source heights less than $0.2h$, the budget equation consequently reduces to

$$\chi_c = -U \frac{\partial}{\partial x} \langle c'^2 \rangle . \tag{48}$$

Csanady (1967) has shown that if $\langle c'^2 \rangle$ develops downstream in a self-similar way such that

$$\langle c'^2 \rangle = f(x) g(y, z) , \tag{49}$$

then the right-hand side of (48) is proportional to $\langle c'^2 \rangle$ divided by the flight time x/U, where U is the wind speed at source height. Netterville (1979) found from wind-tunnel investigations that, to a good approximation, the factor of proportionality can be set equal to one. This means that (48) can be written

$$\chi_c = U \frac{\langle c'^2 \rangle}{x} . \tag{50}$$

We assume that (50) holds in the atmosphere, even when the plume has grown so that it touches the ground.

Substituting (50) into (23) and interpreting, for simplicity, U in (23) as the wind speed at source height, we get in the case when $c(t)$ is unfiltered

$$\sigma_{\dot{c}}^2 = \frac{U^3}{6\gamma_c x} \langle c'^2 \rangle . \tag{51}$$

If $c(t)$ is low-pass filtered with a time constant τ_c, we substitute (50) into (31) to obtain

$$\bar{\sigma}_{\dot{c}}^2 = \frac{2\pi b \kappa^{1/3}}{\sqrt{3}} \left(\frac{\sigma_c U}{x}\right)^2 \left(\frac{z^{1/3} x}{U^{1/3} u_* \tau_c^{4/3}}\right).$$ (52)

The results in this case are

$$\eta_0 = \frac{1}{\sqrt{12\pi}} \left(\frac{U^3}{\gamma_c x}\right)^{1/2} \frac{\alpha^{1/2}}{\Gamma(\alpha)} \left(\alpha \frac{c_0}{C}\right)^{\alpha-1} e^{-\alpha(c_0/C)},$$ (53)

for the unfiltered and

$$\bar{\eta}_0 = B \frac{U}{x} \left(\frac{z^{1/4} x^{3/4}}{U^{1/4} u_*^{3/4} \tau_c}\right)^{2/3} \frac{\alpha^{1/2}}{\Gamma(\alpha)} \left(\alpha - \frac{c_0}{C}\right)^{\alpha-1} e^{-\alpha(c_0/C)},$$ (54)

with

$$B = \left(\frac{b\kappa^{1/3}}{\sqrt{3}}\right)^{1/2} \approx 0.3$$ (55)

for the filtered time series $c(t)$. Equations (53) and (54) apply only to concentrations in the plume, i.e., when there is no intermittency due to plume meandering.

As in the case of a homogeneous surface source, we can rewrite (54) in dimensionless form:

$$\bar{\eta}_0 \tau_c = B \left(\frac{u_*}{U}\right)^{-1/2} \sigma_c Q_0(c_0) \left(\frac{x}{U\tau_c}\right)^{1/2} \left(\frac{z}{U\tau_c}\right)^{1/6}.$$ (56)

We see that in this case the average number of excursions is an increasing function of the distance x from the source as well as the height z.

5. Summary and Recommendations

The purpose of this paper is to extend the study of concentration fluctuations to include recurrence statistics, i.e., to estimate the average number of occurrences per unit time $\eta(c_0)$ that a threshold c_0 is exceeded. By assuming $c(t)$ and $\dot{c}(t)$ to be statistically independent and choosing plausible forms of their frequency distributions, we derived simple expressions for $\eta(c_0)$ and the average duration θ_0 for one excursion above c_0. Applications of the model to an infinite area source and an elevated point source showed how $\eta(c_0)$ depends on C/σ_c, c_0/C, atmospheric variables, and the time constant τ_c of the instrument used to determine $c(t)$. The time averaging was found to have a profound effect on the variance of $\dot{c}$ and hence on $\eta(c_0)$. These functional dependencies entered through the assumed forms of the frequency distributions of $c(t)$ and $\dot{c}(t)$. We believe that the model includes the key parameters that control $\eta(c_0)$, but we found that $\theta(c_0)$ is sensitive to the assumed frequency distribution $Q_0(c)$. As a result of this and the lack of observations of $P(c, \dot{c})$ and $Q_1(\dot{c})$, we suggest that future experimental investigations focus on the forms of these frequency distributions as well as recurrence statistics in general.

There are several ways in which the present study could be extended. One is to account for the effects of diabatic stratification in the boundary layer. A second is to include the probability distribution of the pulse duration above c_0 instead of presenting only the average duration. However, these extensions are beyond the intent of the present work which is to encourage future investigators to include recurrence statistics in the study of concentration fluctuations.

Acknowledgements

The senior authors wish to express our gratitude to NCAR for accepting us as visitors and for the privilege of pursuing this line of research. Also, we would like to thank our colleague Don Lenschow of NCAR for his encouragement, useful discussions and suggestions; P. C. Gallacher, visiting scientist at NCAR, for a very thorough review of a draft manuscript, leading to several modifications; and to Art Isbell for preparing Figure 1. J.C. Weil is grateful for support from the U.S. Army Research Office under Contract No. DAAL03-88-K-0036 and from the Electric Power Research Institute under Contract No. RP1616-32 during the period of this work.

References

Businger, J. A., Wyngaard, J. C., Izumi, Y., and Bradley, E.F.: 1971, 'Flux-Profile Relationships in the Atmospheric Surface Layer', *J. Atmos. Sci.* **28**, 181–189.

Csanady, G. T.: 1967, 'Concentration Fluctuations in Turbulent Diffusion', *J. Atmos. Sci.* **24**, 21–28.

Deardorff, J. W. and Willis, G. E.: 1988, 'Concentration Fluctuations Within a Laboratory Convectively Mixed Layer', in A. Venkatram and J. C. Wyngaard (eds.), *Lectures on Air Pollution Modeling*, American Meteorological Society, Boston, MA, 357–384.

Dyer, A. J. and Bradley, E. F.: 1982, 'An Alternative Analysis of Flux-Gradient Relationships at the 1976 ITCE', *Boundary-Layer Meteorol.* **22**, 3–19.

Fackrell, J. E. and Robins, A. G.: 1982, 'Concentration Fluctuations and Fluxes in Plumes from Point Sources in a Turbulent Boundary Layer', *J. Fluid Mech.* **117**, 1–26.

Gibson, C. H., Stegun, G. R., and Williams, R. B.: 1970, 'Statistics of the Fine Structure of Turbulent Velocity and Temperature Fields Measured at High Reynolds Number', *J. Fluid Mech.* **41**, 153–167.

Gurvich, A. S. and Yaglom, A. M.: 1967, 'Breakdown of Eddies and Probability Distributions for Small-Scale Turbulence', *Phys. Fluids* **10**, Supplement (Proc. Kyoto Sympos. on Boundary Layers and Turbulence), S59–S65.

Hanna, S. R.: 1984, 'The Exponential Probability Distribution Function and Concentration Fluctuations in Smoke Plumes', *Boundary-Layer Meteorol.* **29**, 361–375.

Kaimal, J. C., Wyngaard, J. C., Izumi, Y., and Coté, O. R.: 1972, 'Spectral Characteristics of Surface Layer Turbulence', *Quart. J. Roy. Meteorol. Soc.* **98**, 563–589.

Netterville, D. D. J.: 1979, 'Concentration Fluctuations in Plumes', Syncrude Environmental Research Monograph 1979-4. Syncrude Canada Ltd., Edmonton, Alberta, 288 pp.

Panofsky, H. A. and Dutton, J. A.: 1984, *Atmospheric Turbulence*, pp. 144–148, 182, 319–321. John Wiley & Sons, New York.

Rice, S. O.: 1939, 1944–45, 'Mathematical Analysis of Random Noise', *Bell Syst. Tech. J.* **23**, 282–332, **24**, 46–156.

Sykes, R. I.: 1988, 'Concentration Fluctuations in Dispersing Plumes', in A. Venkatram and J. C. Wyngaard (eds.), *Lectures on Air Pollution Modeling*, American Meteorological Society, Boston, MA, 325–356.

Wyngaard, J. C. and Tennekes, H.: 1970, 'Measurements of the Small-Scale Structure of Turbulence at Moderate Reynolds Numbers', *Phys. Fluids* **13**, 1962–1969.

Wyngaard, J. C.: 1971, 'Spatial Resolution of a Resistance Wire Temperature Sensor', *Phys. Fluids* **14**, 2052–2054.

Wyngaard, J. C., Pennell, W. T., Lenschow, D. H., and LeMone, M. A.: 1978, 'The Temperature-Humidity Covariance Budget in the Convective Boundary Layer', *J. Atmos. Sci.* **35**, 47–58.

WATER VAPOR FLUX AT THE SEA SURFACE

(Review paper)

STUART D. SMITH*

Department of Fisheries and Oceans, Bedford Institute of Oceanography, P.O. Box 1006, Dartmouth, N.S., Canada B2Y 4A2

(Received 18 October, 1988)

Abstract. Methods and instrumentation for determining the rate of evaporation at the sea surface are reviewed. At experimental sites free of local influences, there is a consensus that the evaporation coefficient in neutral conditions $C_{EN} = 1.2 \times 10^{-3}$ at low and moderate wind speeds. Combining this with Businger–Dyer flux-gradient formulas, a parameterization scheme is proposed. Evaporation of spray droplets from breaking waves is expected to cause C_{EN} to increase at high wind speeds, but no direct observations of this are found. Recently it has become possible to estimate water vapor flux in tropical regions from satellite data, opening the possiblity of studying large-scale evaporative events as a function of both time and space.

1. Introduction

Fluxes of water, heat and momentum at the sea surface together with radiative fluxes, control the formation and mixing of water and air masses and are responsible for the dynamic and thermodynamic forcing of the oceans and much of the atmosphere. Knowledge of these fluxes is essential in applications ranging from local studies to global climate change. Remote sensing techniques, which rely on linking radiative properties of the sea surface, such as radar backscatter and infared emissions to desired parameters, place additional demands on the understanding of wind stress, waves and water vapor and heat fluxes at the sea surface.

2. Methods of Measuring Evaporation over the Sea Surface

A direct measure of vertical water vapor flux in the atmospheric boundary layer

$$E = \overline{\rho q u_3} = \overline{\rho_v u_3},$$ (1)

is obtained by *eddy correlation*, where u_3 is the fluctuating vertical component of the wind, $\rho_v = \rho q$ is the fluctuation of the partial density of water vapor in air of density ρ; q is the fluctuating part of the specific humidity and the overbar

---

* Visiting scholar, Department of Atmospheric Sciences AK-40, University of Washington, Seattle, Washington 98195, U.S.A. (until June, 1989).

Boundary-Layer Meteorology **47**: 277–293, 1989.

denotes a time average. If there are negligible sources of vapor between the measuring height and the surface, then (1) also gives the evaporation at the surface. Averaging times of 20 minutes to an hour are usually chosen for measurements at heights of 4 to 20 m above sea level (Haugen, 1973).

We shall assume that the water vapor flux, sensible heat flux and wind stress are nearly invariant with height in a *surface layer* of height $h \ll H$ (say $h \leq 0.1H$), where H is the boundary-layer or inversion height (Kraus, 1972). Unfortunately, surface observations often do not give H, and marine radiosonde data are sparse. Nicholls (1985) suggested $H \simeq 0.2 \, u_*/f$ for mid-latitudes, where $f = 0.145 \, s^{-1} \times 10^{-3} \sin \phi$ is the Coriolis parameter for latitude ϕ. For example, if $U_{10} = 10$ m/s at 45°, we can estimate $H = 654$ m (Smith, 1988) and the surface-layer height can be taken to be 65 m.

Time series of vertical wind and water vapor density (or humidity) fluctuations are required to measure evaporation by the *eddy correlation method*. Fluctuations of horizontal wind and temperature are usually measured concurrently so that the wind stress and the sensible heat flux $H = \rho c_p \overline{tu_3}$, where c_p is the specific heat of air and t is the temperature fluctuation, can also be determined. Instruments measuring the vertical wind require a rigid, slender supporting structure so that u_3 will not be contaminated by motion of the support, nor by distortion of the wind flow as it passes around the structure. Ships and buoys are therefore generally unsuitable for eddy correlation measurements. Specially designed ships such as FLIP (Pond *et al.*, 1971) or stabilized buoys (e.g., Kruspe, 1977) are an integral part of marine eddy correlation measuring systems, and these have limited the sites and the range of conditions in which measurements are possible. Aircraft can also be adapted to carry eddy correlation systems (e.g., Bean *et al.*, 1972; Nicholls, 1978; Nicholls and Readings, 1979).

Water vapor, heat and momentum fluxes are estimated from measurements of vertical gradients of mean water vapor density $\overline{\rho_v}$, temperature T and wind speed U, by the profile method. If the vertical density gradient is neutral (adiabatic), then the profiles in the surface layer are logarithmic with height z above mean sea level,

$$U = (u_*/K) \ln(z/z_0) \tag{2}$$

$$\overline{\rho_v} = \rho_s + (\rho_*/K) \ln(z/z_{oq}) , \tag{3}$$

where $u_* = (\tau/\rho)^{1/2}$ is the friction velocity, τ is the wind stress on the surface, z_o and z_{oq} are laminar sublayer heights for velocity and vapor, ρ_s is the saturation vapor density at the surface and $K = 0.4$ is von Karman's constant. The scaling water vapor density ρ_* and the scaling humidity q_*, defined by

$$E = -\rho_* u_* = -\rho q_* u_* \tag{4}$$

can be obtained by fitting (2) and (3) to measured profiles. If the vertical density gradient is not adiabatic, then the wind and vapor profiles deviate from the logarithmic forms (2) and (3), and iterative solutions using empirical flux-gradient functions are required. The profile method was the standard technique until a generation ago, when the advent of modern computers facilitated the analysis of the long time series of data needed to compute eddy fluxes.

Because turbulent fluctuations of humidity and wind are not necessarily in phase at all levels of the profiles, the data must be averaged over long enough periods to remove most of the turbulent variability. Averaging periods of 10 to 30 minutes are typically chosen (Haugen, 1973).

The *spectral dissipation* or *inertial-dissipation* method is becoming more widely used. It is based on an assumption that turbulent fluxes of momentum and water vapor are linked to the spectra of velocity $\phi_u(f)$ and humidity $\phi_q(f)$ through the dissipation of turbulent kinetic energy and humidity variance (e.g., Fairall and Larsen, 1986). The spectra of wind and vapor density follow a power law ($\phi_u(f)$, $\phi_q(f) \propto f^{-5/3}$) over an inertial subrange of frequencies. Fluxes can be estimated from measurements of spectra within this subrange, which typically is found to apply at normalized frequencies

$$f_n = fz/U \geq 1 , \tag{5}$$

e.g., at frequencies $f \geq 1$ Hz for a height $z = 10$ m and a wind speed $U = 10$ m/s (Phelps and Pond, 1971; Smith and Anderson, 1984). The upper limit of the inertial subrange, typically several hundred Hz, is determined by the onset of viscous or diffusive dissipation and is beyond the response range of most sensors.

In neutral stability, the dissipation formula for evaporation (e.g., Large and Pond, 1982) simplifies to

$$E = \rho(2\pi Kz/U)^{2/3} f^{5/3} [\phi_u(f)\phi_q(f)/(K'\beta_q')]^{1/2} , \tag{6}$$

where the one-dimensional Kolmogoroff constants are $K' = 0.55$ for velocity and $\beta_q' = 0.80$ for humidity (e.g., Paquin and Pond, 1971; Large and Pond, 1982).

The advantage of the inertial-dissipation method over the eddy correlation method is that the vertical component of the wind is not required, and so constraints on the rigidity and flow-distortion of the supporting structure can be relaxed. The inertial subrange occurs at higher frequencies than the dominant waves and associated ship or buoy motions. Suitably placed masts on ships or reasonably stable buoys can be used, greatly expanding the range of potential experimental sites and conditions. The computing power needed for spectral dissipation analysis is not substantially less than for eddy correlation analysis, since long time series (typically 10 to 60 min) must still be processed.

Since the required data are a subset of the data needed for eddy correlation fluxes, the dissipation method can be tested or "calibrated" with existing data. Even if some of the assumptions made in the derivation of (6) are questionable, the method can be justified as an empirical extension of the eddy correlation method.

The *bulk method* estimates sea surface fluxes from standard meteorological observations using empirical coefficients derived from more direct measurements. The bulk formula for evaporation is

$$E = C_E U_{10}(\rho_s - \bar{\rho}_v) = \rho C_E U_{10}(Q_s - Q) , \tag{7}$$

where C_E is the evaporation coefficient or Dalton number, $\bar{\rho}_v = \rho Q$ is the mean vapor density, U_{10} is the mean wind speed at a 10 m reference height and the saturation vapor density $\rho_s = \rho Q_s$ at the surface temperature T_s is about 2% less for sea water than for fresh water. Since measurement of E requires specialized skill and effort, most applications rely on the bulk method.

Much of this review will focus on experimental determinations of E and C_E by various methods, and on parameterization schemes which estimate C_E from standard observations of U_{10}, Q, T_s, and the air temperature T.

The evaporation coefficient is expected to depend on wind speed and the stability,

$$\zeta = z/L , \tag{8a}$$

where the Monin–Obukhov length is

$$L = \frac{-u_*^3(T + 273°)}{gK\overline{t_v u_3}} , \tag{8b}$$

where g is the acceleration of gravity and the fluctuation of virtual temperature is

$$t_v = t + 0.61q(T + 273°) . \tag{8c}$$

Dry air at the virtual temperature T_v has the same density as ambient air at temperature T and humidity Q. The logarithmic profiles (2), (3) are modified to

$$U = (u_*/K)[\ln(z/z_0 - \Phi(\zeta))] \tag{9}$$

$$\rho_v = \rho_s + (\rho_*/K)[\ln(z/z_{oq}) - \Phi_q(\zeta)] , \tag{10}$$

where $\Phi(\zeta)$ and $\Phi_q(\zeta)$ are dimensionless empirical functions of stability (Lumley and Panofsky, 1964), which have been determined mainly over land surfaces. Among the many forms proposed, the Businger-Dyer functions (e.g., Dyer, 1974) are most widely accepted. To compare results of measurements in various stability conditions, a neutral evaporation coefficient

$$C_{EN} = E/[U_{10N}(\rho_s - \rho_N)] , \tag{11}$$

is used, where U_{10N} and ρ_N are the equivalent neutral wind speed and vapor density at 10 m, obtained by determining u_*, z_o, z_{oq} and ρ_* from (9) and (10) and then substituting these in the neutral profile formulas (2) and (3).

For a more comprehensive introduction to the atmospheric boundary layer, see Hasse and Dobson (1986), Haugen (1973), Kraus (1972, 1977) and Panofsky and Dutton (1984).

3. Instrumentation

Standard meteorological observations of humidity are made with wet- and dry-bulb *psychrometers*, often using a sling or a fan to ventilate the wet bulb. Psychrometers are adapted for continuous electrical readout by replacing mercury thermometers with thermocouples, platinum resistance wires or thermistors. Fine wire psychrometers can have high enough frequency response (up to several Hz) to be used in eddy correlation or spectral dissipation measurements of vapor flux.

Although it is transparent to light, water vapor has absorption bands at both longer and shorter wavelengths which can be used to detect its concentration. Microwave *refractometers* tuned to a water vapor absorption band have a frequency response which is limited mainly by ventilation of the resonant cavity. Their output can be calibrated in water vapor density; close attention is needed to maintain this calibration while operating in the marine atmosphere. Microwave refractometers are often used in airborne flux systems (e.g., Nicholls, 1978).

The Lyman α ultraviolet line of hydrogen is strongly absorbed by water vapor. A *Lyman α humidiometer* typically consists of a source tube aimed at a detector tube, separated by a sampling volume which may be open to the airflow or may be aspirated (Buck, 1976). The required pathlength is typically about 1 cm, and the compact sampling volume allows rapid ventilation and makes the Lyman α humidiometer a good choice for near-surface studies (e.g., Mestayer *et al.*, 1986, 1988). The source and detector windows, in order to transmit ultraviolet, are made of salt crystals such as MgF_2, which is slightly soluble in water and is vulnerable to calibration drift and damage caused by accumulation of salty water droplets and organic material from the marine atmosphere. Even a window which appears clear on visual examination can be contaminated for ultraviolet transmission. Experience has shown that the windows can often be restored by promptly polishing with a paste of fine alumina powder in alcohol on a cotton swab.

Many types of *anemometers* have been used to sense horizontal and vertical winds for eddy flux, profile and spectral dissipation measurements, including sonic, propeller, cup, hot film and thrust anemometers. Instruments for flux studies are reviewed by Dobson *et al.* (1980).

Even with careful calibration, it has not been possible to maintain the sensitivity of any fast-response humidiometer in the marine atmosphere to better than about ±5%, and often there have been larger calibration uncertainties. Similarly allowing about ±3% sensitivity variation in vertical or horizontal turbulent wind sensors, we can anticipate systematic calibration errors of ±8% in eddy flux or dissipation estimates of evaporation, in addition to scatter of ±10 to 20% among runs due to using finite samples (run durations) of ongoing time series processes. Errors in determining the bulk parameters are comparable: 1% systematic errors in determining the mean relative humidity and the surface saturation humidity gives 10% systematic error in $(Q_s - Q)$ at 80% relative humidity. Wind measurements from ships can have >10% systematic error due to flow distortion by the ship's structure.

Profile studies make severe demands on the precision of mean humidity measurement, since small differences of only a few percent in relative humidity between measuring levels must be resolved. A large part of the mean humidity difference occurs very close to the surface, where waves and spray threaten the sensors. If a surface-following buoy is used for support, the sensor traverses a part of the boundary layer where the humidity difference varies logarithmically with height so that the effective level of the lower sensors differs from their mean height above sea level (e.g., Dunckel et al., 1974). Systematic errors of ±10% can be anticipated in any given series of profile determinations of evaporation, in addition to scatter introduced by averaging individual profiles over finite intervals of time.

4. Eddy Flux Studies

The classic data of Pond et al. (1971) were taken with a Lyman α humidiometer on the special-purpose R.V. FLIP (FLoating Instrumentation Platform) in the 1969 BOMEX (Barbados Oceanographic and Meteorological Experiment) and in a preliminary experiment off San Diego. They found $10^3 C_E = 1.2 \pm 0.2$ for wind speeds from 4 to 7 m/s in generally unstable stratification, corresponding to a value in neutral conditions of $C_{EN} = 1.1 \times 10^{-3}$ (Figure 1). Paulson et al. (1972) reported 25% higher evaporation from BOMEX profile measurements, which was considered to be "fair agreement". Holland (1972) reported that the airborne eddy correlation determinations of evaporation (Bean et al., 1972) in BOMEX at altitudes of 18, 45 and 152 m and the Pond et al. near-surface eddy fluxes were in good agreement with estimates from both atmospheric and oceanic regional budgets, each giving a daily evaporation of 5–6 mm of water (Holland, 1972). Friehe and Schmitt (1976) combined the Pond et al. data with results from two other eddy flux experiments from FLIP to obtain $C_E = 1.32 \times 10^{-3}$, corresponding to $C_{EN} = 1.2 \times 10^{-3}$ (Figure 1).

Kruspe (1977) added eddy fluxes in the German Bight and the Baltic Sea with a microwave refractometer and sonic anemometer to the data of Pond et al. (1971), finding $10^3 C_E = 1.36 \pm 0.25$ but with a negative intercept suggesting zero evaporation when $U_{10}(Q_s - Q) = 5$ (m/s)(g/kg). He suggested that a cool surface microlayer may have existed, so that Q_s was overestimated from the bulk water temperature.

Garratt and Hyson (1975) used an infrared hygrometer, a vertical Gill propeller anemometer and "fluxatron" analog eddy correlation analysis in AMTEX 74 (Air Mass Transformation Experiment) at towers on the shorelines of islands

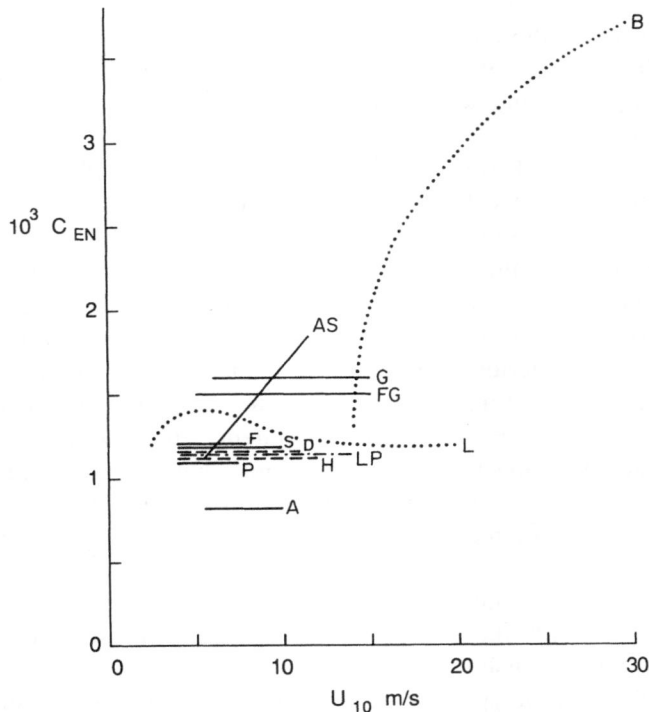

Fig. 1. Evaporation coefficients in neutral conditions. Eddy correlation method (solid lines): A, Antonia *et al.* (1978); F, Friehe and Schmitt, (1976); P, Pond *et al.* (1971); S, Smith (1974). Sites with surf zones: AS, Anderson and Smith (1981); FG, Francey and Garratt (1978); G, Garratt and Hyson (1975). Profile method (dashed lines): D, Dunckel *et al.* (1974); H, Hasse *et al.* (1978). Dissipation method, (dashed/dotted line): LP, Large and Pond (1982). Theoretical projections (dotted lines): Vapor flux only, L, Liu *et al.* (1979); spray and vapor flux, B, Bortkovskii (1987).

in the North China Sea. Their value of $10^3 C_{EN} = 1.6 \pm 0.3$ is higher than the above-mentioned open sea values, and may have been enhanced by spray from breaking waves on coral reefs upwind of the towers. Francey and Garratt (1978) found $C_{EN} = 1.5 \times 10^{-3}$ in AMTEX 75. Fujitani (1981) reported $C_E = 1.05 \times 10^{-3}$ from a shipboard eddy flux system in AMTEX 74 and 75. He compensated the measured wind vector for rotational motion of the ship and for the translation of the sonic anemometer relative to the center of gravity of the ship. In spite of the possibility of errors due to wind deflection by the ship, varying with ship motion, this does not differ greatly from the open-sea BOMEX results. The AMTEX experiments cover a relatively high wind speed range of 5–15 m/s.

Antonia *et al.* (1978) used a Lyman α humidiometer and a Gill propeller array at a boom on an oil platform 80 km offshore and 70 m in depth in Bass Strait, Australia. They reported $10^3 C_E = 0.82 \pm 0.15$ for a 5 m reference height. This result is the lowest of those reviewed here (Figure 1). Limited frequency response of the propeller anemometer may possibly have caused the evaporation to be underestimated, and wind flow distortion by the platform is another possible source of error.

In the International Field Year for the Great Lakes (IFYGL), 1971–1972, Smith (1974) reported 9 values of C_E averaging 1.2×10^{-3} at a tower in Lake Ontario. Anderson and Smith (1981) found $10^3 C_{EN} = 1.27 \pm 0.26$ using a Lyman α humidiometer and sonic and thrust anemometers during periods of onshore winds at a 10 m tower on the beach of Sable Island, a low sand island 160 km off the coast of Nova Scotia in the North Atlantic. They reported both upward and downward fluxes, the latter corresponding to negative values of $\rho_s - \rho_v$. Wind speeds ranged from 5–11 m/s, and C_{EN} was found to increase with rising wind speed (Figure 1). As in the AMTEX experiments, there were shoals upwind of the tower where breaking waves and spray may have contributed more to the evaporation than in otherwise similar conditions over the open sea. The size and intensity of the surf zone increased with wind speed, possibly accounting for the variation of C_{EN}.

Preliminary results of the 1986 HEXOS (Humidity Exchange over the Sea) Main Experiment with the Lyman α and a sonic anemometer at Meetpost Noordwijk, a Dutch offshore research platform in the North Sea, indicate $C_{EN} = 1.2 \times 10^{-3}$ at wind speeds from 5 to 18 m/s (Smith and Anderson, 1988). Much of the extensive data set from the HEXOS program (Katsaros et al., 1987) remains to be analyzed at the time of writing. The data include eddy correlation and spectral dissipation measurements of evaporation from a platform, a ship and an aircraft at substantially higher wind speeds than previously attained and concurrent observations of breaking waves, aerosol droplet distributions and regional meteorology and oceanography. Some of the data (not yet published) extend to wind speeds over 20 m/s.

Thorpe et al. (1973) reported from eddy correlation studies with a Lyman α humidiometer that sublimation from the frozen surface of the Beaufort Sea followed a diurnal cycle with $C_E = 0.55 \times 10^{-3}$. At the low ambient temperatures $(-16$ to $-28°C)$ the surface vapor pressure and the rate of sublimation were much smaller than over water surfaces.

5. Profile Studies

Numerous humidity, temperature and wind profiles were obtained in the 1969 Atlantic Tradewind Experiment (ATEX); aspirated platinum-resistance psychrometers were placed at heights between 1 and 8 m on a surface-following buoy (Dunckel et al., 1974). It was recognized that following the waves through the nonlinear (logarithmic) profile significantly increased the apparent profile gradient at the lowest sensors (see also Krügermeyer et al., 1978). This effect may help explain discrepancies in certain earlier profile-derived fluxes, such as the BOMEX evaporation rates (Paulson *et al.*, 1972) which were some 25% higher than the rates obtained by the eddy correlation method (Pond *et al.*, 1971). Although the wind speeds (4–11 m/s) were not appreciably stronger, large tropical humidity gradients in ATEX resulted in an approximate doubling of the range of U_{10} $(Q_s - Q)$ from the BOMEX experiment. The overall ATEX result, $C_E = 1.28 \times 10^{-3}$, agreed very well with the above mentioned BOMEX eddy fluxes. Hasse *et al.* (1978) similarly obtained 10^3 $C_{EN} = 1.15$ from profiles in the 1974 GATE (GARP Atlantic Tropical Experiment) of the Global Atmospheric Research Project (Figure 1).

6. Spectral Dissipation Studies

Large and Pond (1982) reported 117 dissipation estimates of evaporation from Lyman α humidiometers and Gill propeller anemometers. These were installed on temporary masts on the bow of R.V. *Meteor* in the JASIN experiment in the North Atlantic off Great Britain, and of CSS *Parizeau* in the North Pacific off the Canadian coast. Wind speeds were 4–9 m/s in the Atlantic and 4–14 m/s in the Pacific. The mean value was $10^3 C_{EN} = 1.15 \pm 0.22$ (Figure 1). The few (15) values at wind speeds above 10 m/s suggest that the evaporation coefficient may increase with wind speed about half as fast as the drag coefficient, but this possible increase is not clearly resolved. The dissipation formula was verified by comparison with eddy correlation heat (but not vapor) fluxes at an offshore tower.

7. Bulk Parameterization

Several parameterization schemes have been proposed to specify the variation of C_E with the local, short-term bulk parameters U, T, T_s and Q, usually at a reference height of 10 m. Because the stability ζ depends on the wind stress and the sensible heat flux in addition to the evaporation, these schemes must explicitly or implicitly estimate all three fluxes. Iterative solutions or further empirical parameterizations are required. These solutions are sometimes specified for individual sets of observations, or they may be tabulated or graphed for a range of conditions.

Bunker (1976) tabulated evaporation coefficients for ranges of wind speed and air–sea temperature differences, based on the then available data and on the premise that $10^3 C_{EN}$ varies about half as fast with wind speed as the drag coefficient, i.e.; from 1.32 at low wind speeds to 1.86 for $U > 20$ m/s. Noting that these are higher than recent experimental values, Isemer and Hasse (1987) reduced C_E by 13% in their re-creation and expanded presentation of Bunker's surface flux climatology of the North Atlantic ocean, but made two other changes which partially compensate: they used slightly higher wind speeds and they allowed for the unstable stratification due to vertical gradients of humidity by using $(T_v - T_s)$ instead of $(T - T_s)$.

Bunker *et al.* (1982), on the other hand, found that C_E should be higher than

Bunker's original values by 0.4×10^{-3} to balance the surface heat budget of the Mediterranean and Red seas with oceanic net heat transport through straits. A partial explanation of this paradox lies in their not using virtual temperature to estimate the influence of stability on C_E. The notably rapid evaporation in these areas is accompanied by contributions of water vapor gradients to unstable stratification and increased C_E.

Kondo (1975) postulated a nonlinear variation of C_E and other flux coefficients which remains within the range of experimental results. He has graphed the coefficients as functions of wind speed and sea–air temperature difference.

Blanc (1985, 1987) has compared ten parameterizations including Bunker's and Kondo's, a similar scheme of Masagutov (1981), and coefficients derived from several of the above-mentioned experimental studies. Using data from Ocean Weather Station C in 1973, Blanc found typical variations of about 25% in E among the schemes, and expressed concern at this state of affairs. As we have just seen, selection of experiments at sites free from local influences can greatly reduce this scatter. Launiainen (1983) has reviewed vapor fluxes over water surfaces and suggests that the neutral evaporation coefficient varies with the neutral drag coefficient, $C_{EN} = 0.63 C_{10N} + 0.32 \times 10^{-3}$.

We have reviewed the results of several experiments (Figure 1) that find on average that $10^3 C_{EN} = 1.2 \pm 0.1$ for winds of 4–14 m/s, with good agreement among eddy correlation, profile and spectral dissipation methods. Although there is considerable scatter among data runs within each experiment, the scatter among different experiments is so small that our previous estimates of possible

TABLE I

Evaporation coefficients $10^3 C_E$ as recommended by Smith (1981, 1988)

U_{10} (m/s) $T_s - T_v$ (°C)	2	3	4	5	7	10	15
−20						0.58	0.98
−15						0.73	1.04
−10					0.50	0.90	1.10
−5				0.50	0.84	1.06	1.15
−4			0.37	0.61	0.91	1.08	1.16
−3			0.53	0.74	0.98	1.12	1.17
−2		0.46	0.72	0.89	1.06	1.14	1.18
−1	0.41	0.72	0.95	1.04	1.13	1.17	1.19
0	1.20	1.20	1.20	1.20	1.20	1.20	1.20
1	1.56	1.44	1.37	1.32	1.26	1.23	1.21
2	1.70	1.55	1.45	1.38	1.31	1.25	1.22
3	1.80	1.62	1.51	1.43	1.33	1.27	1.23
4	1.88	1.68	1.56	1.48	1.37	1.29	1.24
5	1.96	1.73	1.60	1.50	1.39	1.31	1.25
10	2.18	1.91	1.74	1.62	1.48	1.33	1.27
15	2.35	2.04	1.84	1.70	1.54	1.40	1.30
20	2.48	2.14	1.92	1.78	1.58	1.44	1.32

errors appear pessimistic. Higher values were obtained at sites with a surf zone upwind of the sensors, and lower values in one experiment were not fully explained. C_{EN} either increases modestly with wind speed or remains constant. Preliminary results from the 1984 and 1986 Humidity Exchange over the Sea (HEXOS) experiments (Katsaros *et al.*, 1987; Smith and Anderson, 1988) suggest no increase with wind speed up to 18 m/s; Smith (1988) sets $C_{EN} = 1.2 \times 10^{-3}$. From an iterative solution (Smith, 1981) of the Businger-Dyer flux-gradient formulas (Dyer, 1974), drag and heat flux coefficients are given as functions of U_{10} and $T_s - T_v$. The corresponding values of C_E are listed in Table I and shown in Figure 2. In very stable conditions, the solution breaks down (blank area in Table I), corresponding to conditions in which insufficient turbulence is generated by shear flow to maintain a constant-flux turbulent boundary layer extending up to the 10 m reference height. In applying this parameterization to study the surface heat budget at Ocean Weather Station Bravo, Smith and Dobson (1984) set the missing coefficients to zero. Since this parameterization does not include an

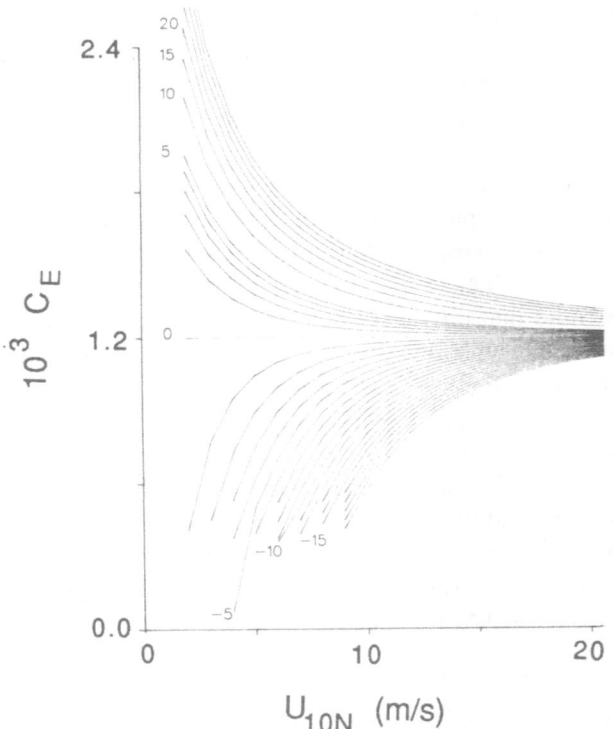

Fig. 2. Stability dependence of the evaporation coefficient according to Dyer (1974) with $C_{EN} = 1.2 \times 10^{-3}$. Lines from $(T_s - T_v) = +40\,°C$ (unstable) to $-20\,°C$ (stable). Influence of spray not included.

increase in C_E due to the influence of spray from breaking waves, application to wind speeds above 14 m/s is uncertain.

8. Modelling the Influence of Spray

We have so far considered water vapor flux only. At low wind speeds (except near shoals), there is negligible spray and virtually all of the vertical water transport is in the vapor phase. At higher wind speeds, with substantial volumes of water entering the atmosphere as droplets from the bursting of air bubbles entrained by breaking waves, the picture becomes more complicated. The precise, fast-response humidity sensors needed for flux measurements are notoriously vulnerable to damage from accumulation of salty droplets, and it is no coincidence that few data sets include vapor fluxes at wind speeds above 15 m/s.

In a spray layer, evaporating droplets increase the humidity and decrease the temperature. The vapor flux increases with height, while to supply latent heat, the sensible heat flux must decrease with height. Above the spray layer, we again expect constant eddy fluxes with height, and the net effect of evaporation of droplets can be expressed by an enhanced value of C_E. Even within the spray layer, it would be theoretically possible to measure eddy vapor fluxes if droplets could somehow be excluded. Profiles of humidity and temperature extending into the spray layer could not be used to derive fluxes in the usual way by (2), (3) and (4).

Although we are not aware of direct measurements of the vertical transport of liquid water by spray droplets, several models have predicted this transport by accounting for sources, transport, thermodynamics and evaporation of droplets as a function of height. The Ling and Kao (1976) model follows the above principles. They note that only a rough estimate of the droplet source function is possible. Their sample calculations are for a wind speed of 15 m/s, and they conclude that droplets play a dominant role. The bulk of experimental work cited above does not, however, suggest a dominance of droplet water flux at this speed.

The model was refined by Ling et al. (1980) to include the size distribution of droplets in five categories and their free-fall velocities. The influences of wind speed, humidity and stability are not formulated in the same way as in other models, but numerical examples given correspond to $10^3 C_E = -0.13$ to 4.7 for moderate winds and a wide range of humidity. Subsequently Ling and Kao (1981) examined data from the JASIN experiment and proposed a simplified formula

$$E = C_1(e_{ws} - e_{10}) U_{10} + C_2(e_{10s} - e_{10}) U_{10}^2, \tag{12}$$

where E is in g/cm^2 s, U_{10} is in m/s and e_{10}, e_{10s} and e_{ws} are the water vapor pressures in millibars at 10 m, for saturation at the temperature at 10 m and at the surface, respectively. The values of the constants are given as $C_1 = 5.58 \times 10^{-8}$ g/(mb cm^2 m) and $C_2 = 6.54 \times 10^{-8}$ g s/(mb cm^2 m^2). The first term in (12) is the vapor flux at the surface and the second term allows for the influence of evaporating droplets. This series of papers represents a valuable pioneering effort, but this reviewer shares the widely-held view that the influence of spray is introduced at too low a wind speed and that unconventional treatment of the influence of stability and humidity may limit the range of applicable conditions.

Bortkovski (1987) has also analyzed the contribution of spray from bursting bubbles of whitecaps. He begins with a systematic review of sea spray and its generation by breaking waves and whitecaps. Extensive references to the Soviet literature are cited. He constructs a numerical model to simulate trajectories, thermodynamics and evaporation of individual droplets of various sizes, and then integrates over droplet source distributions to estimate the droplet evaporation at wind speeds up to 30 m/s. He adds this to the vapor transport to obtain a net evaporation coefficient

$$C_E = 1.3 \times 10^{-3} + C_{Ed} , \tag{13}$$

where the droplet contribution C_{Ed} varies nonlinearly from 0 at 14 m/s to 2.3×10^{-3} at 30 m/s (Figure 1). Finally, Bortkovskii explores the contributions of C_{Ed} to the surface flux climate of the North Atlantic Ocean and along storm tracks. In order to reach a solution, several assumptions and simplifications were made, particularly regarding the dynamics of droplets in boundary-layer turbulence and the source and distribution of relatively large droplets close to the surface. Although reasonable estimates have gone into the model, there is a rather wide latitude to adjust the outcome. In particular, a better knowledge of the source and distribution of larger droplets is needed (Stramska, 1987).

Bortkovskii disputes Ling and Kao's (1976) model in which even the larger droplets are diffused by atmospheric turbulence in much the same way as vapor. Instead, he assumes that the larger droplets have an initial velocity imparted by bubble bursting, which causes them to hop through the lower part of the temperature and humidity profiles and fall back to the surface without losing a great part of their mass. Bortkovskii's analysis also predicts that the heat transfer coefficient should increase rapidly at wind speeds above 14 m/s, which is not in keeping with experimental results (e.g., Large and Pond, 1982). In applying Bortkovskii's parameterization, one must be aware that more refined theories can be expected as new data on droplet sources and dynamics become available.

9. Water Vapor Flux Derived from Satellite Data

The availability of surface data effectively limits surface flux studies to either long-term climatology from compilations of ship weather reports (e.g., Bunker, 1976; Isemer and Hasse, 1987) or time series studies at the relatively few sites of Ocean Weather Stations (e.g., Smith and Dobson, 1984) or data buoys. Recently, satellites have made it possible to obtain continuous global data provided that algorithms can be developed to deduce the relevant variables from passive or active radiation measurements.

Liu and Niiler (1984) demonstrated that data from the Seasat scanning multi-channel microwave radiometer (SMMR) can be used to estimate evaporation by the bulk method, with near-surface humidity Q related to precipitable water W, at least on a monthly time scale. Liu (1986) derived a relation between W and Q using 17 years of data from 46 globally distributed ocean radiosonde stations. Liu (1988) demonstrates that the standard deviation in monthly mean Q derived from SMMR and from the lowest level of radiosondes is about 0.4 g/kg at tropical stations, and 1 g/kg at mid-latitude stations where the seasonal variability is also greater. He also finds that, at least for one tropical data buoy site with persistent weather patterns, the bulk parameterizations of Liu et al. (1979) can be applied to monthly means instead of hourly observations of U, T_s, and Q with a negligible change in the monthly mean evaporation. (Isemer and Hasse (1987) find that this simplification cannot be applied in general.) He compares winds, surface temperatures and the resultant monthly latent heat fluxes derived from Nimbus 7 SMMR and from ship reports in the tropical and sub-tropical Pacific Ocean from 1980–1983 for $2° \times 2°$ bins having at least 10 ship reports in a given month. After allowing for an expected $+0.5\,°C$ bias in ship T_s, he reports a standard deviation of about $30\,W/m^2$ between coincident monthly latent heat fluxes from satellite and ship data. Finally, he examines the spatial and seasonal variations of satellite-derived latent heat flux and of the bulk parameters for non-ENSO (El Niño/Southern Oscillation) years 1980–1981 and the differences for ENSO years 1982–1983. The relatively large variations of the ENSO offer an ideal case study to demonstrate this important new capability of simultaneously studying temporal and spatial variability of the fluxes. Further testing, development and extension of this method will be an important new direction in the study of sea surface fluxes.

10. Conclusion

Measurements of evaporation at the sea surface by eddy correlation, profile and spectral dissipation methods agree and give a neutral evaporation coefficient $C_{EN} = 1.2 \times 10^{-3}$ for wind speeds at least up to 14 m/s. Higher coefficients are obtained at sites with surf zones, and are also anticipated at higher wind speeds due to spray from breaking waves. At the present time, we do not have reliable direct flux measurements at wind speeds above 14 m/s. The observational deficiency is due in large measure to the onset of sea spray. There is a need for more information on the source, size distribution and height distribution of spray droplets from breaking waves in order to refine estimates of their contribution to the water flux.

The data from the Humidity Exchange over the Sea (HEXOS) program (Katsaros et al., 1987) are expected to extend the range of wind speeds with vapor flux measurements and to provide further information on the contribution of spray droplets. Further investigation is still needed, particularly to obtain data at even higher wind speeds, to investigate the influence of atmospheric stratification and to elucidate the contribution of spray droplets from breaking waves to the transfer process.

The advent of SMMR data from satellites opens the possibility of studying evaporation at monthly and longer periods on regional to global scales. Particular attention is needed to exploring a relationship between surface humidity and precipitable water.

Acknowledgements

This review was prepared at the request of the Working Group on Boundary-layer Dynamics and Air–Sea Interaction of the International Commission of Dynamic Meteorology. This work was supported by the Department of Fisheries and Oceans, Canada, and by the United States Office of Naval Research, Marine Meteorology, grant N00014-85-K-0123. This is Contribution 14 of the Humidity Exchange over the Sea (HEXOS) Program.

References

Anderson, R. J. and Smith, S. D.: 1981, 'Evaporation Coefficient for the Sea Surface from Eddy Flux Measurements', *J. Geophys. Res.* **86**, 449–456.

Antonia, R. A., Chambers, A. J., Rajagopalan, S., and Sreenivasan, K. R.: 1978, 'Measurements of Turbulent Fluxes in the Bass Strait', *J. Phys. Oceanogr.* **8**, 28–37.

Bean, B. R., Gilmer, R., Grossman, R. L., McGavin, R. and Travis, C.: 1972, 'An Analysis of Airborne Measurements of Vertical Water Vapor Flux during BOMEX', *J. Atmos. Sci.* **29**, 860–869.

Blanc, T. V.: 1985, 'Variation of Bulk-derived Surface Flux, Stability and Roughness Results due to Different Transfer Coefficient Schemes', *J. Phys. Oceanogr.* **15**, 650–669.

Blanc, T. V.: 1987, 'Accuracy of Bulk-Method-Determined Flux, Stability and Sea Surface Roughness', *J. Geophys. Res.* **92**, 3867–3876.

Bortkovskii, R. S.: 1987, *Air-Sea Exchange of Heat and Moisture during Storms*, D. Reidel Publ. Co., Dordrecht, Holland, xiii + 194 pp.

Buck, A. L.: 1976, 'The Variable-path Lyman-alpha Humidiometer and its Operating Characteristics', *Bull. Amer. Meteorol. Soc.* **57**, 1113–1118.

Bunker, A. F.: 1976, 'Computations of Surface Energy Flux and Annual Air–Sea Interaction Cycles of the North Atlantic Ocean', *Mon. Wea. Rev.* **104**, 1122–1140.

Bunker, A. F., Charnock, H. and Goldsmith, R. A.: 1982, 'A Note on the Heat Balance of the Mediterranean and Red Seas', *J. Marine Res.* **40**, supplement, 73–84.

Dobson, F. W., Hasse, L. and Davis, R.: 1980, *Air-sea Interaction Instruments and Methods*, Plenum Press, New York, 801 pp.

Dunckel, M., Hasse, L., Krügermeyer, L., Schreiver, D., and Wucknitz, J.: 1974, 'Turbulent Fluxes of Momentum, Heat and Water Vapor in the Atmospheric Surface Layer at Sea during ATEX', *Boundary-Layer Meteorol.* **6**, 81–106.

Dyer, A. J.: 1974, 'A Review of Flux-profile Relationships', *Boundary-Layer Meteorol.* **7**, 363–372.

Fairall, C. W. and Larsen, S. E.: 1986, 'Inertial-Dissipation Methods and Turbulent Fluxes at the Air-Ocean Interface', *Boundary-Layer Meteorol.* **34**, 287–301.

Francey, R. J. and Garratt, J. R.: 1978, 'Eddy Flux Measurements over the Ocean and Related Transfer Coefficients', *Boundary-Layer Meteorol.* **14**, 153–166.

Friehe, C. A. and Schmitt, K. F.: 1976, 'Parameterization of Air-Sea Interface Fluxes of Sensible Heat and Moisture by the Bulk Aerodynamic Formulas', *J. Phys. Oceanogr.* **6**, 801–809.

Fujitani, T.: 1981, 'Direct Measurement of Turbulent Fluxes over the Sea during AMTEX', *Papers in Meteorology and Geophysics (Japan)* **32**, 119–134.

Garratt, J. R. and Hyson, P.: 1975, 'Vertical Fluxes of Momentum, Sensible Heat and Water Vapor during the Air Mass Transformation Experiment (AMTEX) 1974', *J. Meteorol. Soc. Japan* **53**, 149–160.

Hasse L. and Dobson, F.: 1986, *Introductory Physics of the Atmosphere and Ocean*, D. Reidel Publ. Co., Dordrecht, Holland, viii + 126 pp.

Hasse, L., Grünewald, M., Wucknitz, J., Dunckel, M., and Schreiver, D.: 1978, 'Profile Derived Turbulent Fluxes in the Surface Layer under Disturbed and Undisturbed Conditions during GATE', *Meteor-Forschungsergebnisse B* **13**, 24–40.

Haugen, D. A. (ed.): 1973, *Workship on Micrometeorology*, American Meteorological Society, Boston, xi + 392 pp.

Holland, J. Z.: 1972, 'Comparative Evaluation of some BOMEX Measurements of Sea Surface Evaporation, Energy Flux and Stress', *J. Phys. Oceanogr.* **2**, 476–486.

Isemer, H.-J. and Hasse, L.: 1987, *The Bunker Climate Atlas of the North Atlantic Ocean, Vol. 2: Air-Sea Interactions*, Springer-Verlag, Berlin, vii + 252 pp.

Katsaros, K. B., Smith, S. D. and Oost, W. A.: 1987, 'HEXOS – Humidity Exchange over the Sea, a Program for Research on Water – Vapor and Droplet Fluxes from Sea to Air at Moderate to High Wind Speeds', *Bull. Amer. Meteorol. Soc.* **68**, 466–476.

Kondo, J.: 1975, 'Air-Sea Bulk Transfer Coefficients in Diabatic Conditions', *Boundary-Layer Meteorol.* **9**, 91–112.

Kraus, E. B.: 1972, *Atmosphere-Ocean Interaction*, Clarendon Press, Oxford, 275 pp.

Kraus, E. B. (ed.): 1977, *Modelling and Prediction of the Upper Layers of the Ocean*, Pergamon Press, Oxford, 325 pp.

Krügermeyer, L., Grunewald, M., and Dunckel, M.: 1978, 'The Influence of Sea Waves on the Wind Profile', *Boundary-Layer Meteorol.* **14**, 403–414.

Kruspe, G.: 1977, 'On Moisture-Flux Parameterization', *Boundary-Layer Meteorol.* **11**, 55–63.

Large, W. G. and Pond, S.: 1982, 'Sensible and Latent Heat Flux Measurements over the Ocean', *J. Phys. Oceanogr.* **12**, 464–482.

Launiainen, J.: 1983, 'Parameterization of the Water Vapor Flux over a Water Surface by the Bulk Aerodynamic Method', *Annales Geophysicae* **1**, 481–492.

Ling, S. C. and Kao, T. W.: 1976, 'Parameterization of the Moisture and Heat Transfer Process over the Ocean under Whitecap Sea States', *J. Phys. Oceanogr.* **6**, 306–315.

Ling, S. C. and Kao, T. W.: 1981, 'Multiphase Fluxes over the North Atlantic Ocean – JASIN 1978 Experiment', Unpublished Report, Catholic Univ. of America, Washington D.C.

Ling, S. C., Kao, T. W. and Saad, A. I.: 1980, 'Microdroplets and Transfer of Moisture from the Ocean', *J. Engineering Mechanics Div. Amer. Soc. of Civil Engineers* **106**, 1327–1339.

Liu, W. T.: 1988, 'Moisture and Latent Heat Flux Variabilities in the Tropical Pacific Derived from Satellite Data', *J. Geophys. Res.*, **93**, 6749–6760 and 6965–6968.

Liu, W. T.: 1986, 'Statistical Relation between Monthly Precipitable Water and Surface-Level Humidity over Global Oceans', *Mon. Wea. Rev.* **114**, 1591–1602.

Liu, W. T., Katsaros, K. B. and Businger, J. A.: 1979, 'Bulk Parameterization of Air-Sea Exchanges of Heat and Water Vapor including the Molecular Constraints at the Interface', *J. Atmos. Sci.* **36**, 1722–1735.

Liu, W. T. and Niiler, P. P.: 1984, 'Determination of Monthly Mean Humidity in the Atmosphere Surface Layer over Oceans from Satellite Data', *J. Phys. Oceanogr.* **14**, 1451–1457.

Lumley, J. L. and Panofsky, H. A.: 1964, 'The Structure of Atmospheric Turbulence', *Interscience* (John Wiley and Sons), New York, 239 pp.

Masagutov, T. F.: 1981, 'Calculation of Vertical Turbulent Fluxes in the Near-Water Layer of the Ocean in Tropical Latitudes', *Meteorologiya i Gidrologiya (Meteorology and Hydrology, Moscow)* **12**, 61–68; Russian.

Mestayer, P. G., Rabatett, C. and Goutail, F.: 1986, 'Improved Lyman-α Humidiometer for Small-Scale Atmospheric Turbulence Measurements, Part I: Miniaturizing the Sampling Volume', *Rev. Sci. Instr.* **57**, 20–25.

Mestayer, P. G., Goutail, F. and Larsen, S. E.: 1989, 'Improved Lyman-α Humidiometer. Part III: Performance of a Field Version for Small Scale Atmospheric Turbulence Measurements over the Sea', *Rev. Sci. Instr.* **60**, in press.

Nicholls, S.: 1978, 'Measurements of Turbulence by an Instrumented Aircraft in a Convective Atmospheric Boundary Boundary Layer over the Sea', *Quart. J. Roy. Meteorol. Soc.* **104**, 653–676.

Nicholls, S. and Reading, C. J.: 1979, 'Aircraft Observations of the Structure of the Lower Boundary Layer over the Sea', *Quart. J. Roy. Meteorol. Soc.* **105**, 785–802.

Nicholls, S.: 1985, 'Aircraft Observations of the Ekman Layer during the Joint Air-Sea Interaction Experiment', *Quart. J. Roy. Meteorol. Soc.* **111**, 391–426.

Panofsky, H. A. and Dutton, J. A.: 1984, *Atmospheric Turbulence Models and Methods for Engineering Application*, John Wiley and Sons, New York, xix + 397 pp.

Paquin, J. E. and Pond, S.: 1971, 'The Determination of the Kolmogoroff Constants for Velocity, Temperature and Humidity Fluctuations from Second and Third-order Structure Functions', *J. Fluid Mech.* **50**, part 2, 257–269.

Paulson, C. A., Leavitt, E. and Fleagle, R. G.: 1972, 'Air-Sea Transfer of Momentum, Heat and Water Determined from Profile Measurements during BOMEX', *J. Phys. Oceanogr.* **2**, 487–497.

Phelps, G. T. and Pond, S.: 1971, 'Spectra of the Temperature and Humidity Fluctuations and of the Fluxes of Moisture and Sensible Heat in the Marine Boundary Layer', *J. Atmos. Sci.* **28**, 918–928.

Pond, S., Phelps, G. T., Paquin, J. E., McBean, G., and Stewart, R. W.: 1971, 'Measurements of the Turbulent Fluxes of Momentum, Moisture and Sensible Heat over the Ocean', *J. Atmos. Sci.* **28**, 901–917.

Smith, S. D.: 1974, 'Eddy Flux Measurements over Lake Ontario', *Boundary-Layer Meteorol.* **6**, 235–255.

Smith, S. D.: 1981, 'Coefficients for Sea-Surface Wind Stress and Heat Exchange', Rep BI-R-81-19, Bedford Institute of Oceanography, Dartmouth, N.S. 31 pp.

Smith, S. D.: 1989, 'Coefficients for Sea Surface Wind Stress, Heat Flux and Wind Profiles as a Function of Wind Speed and Temperature', *J. Geophys. Res.* **93**, (in press).

Smith, S. D. and Anderson, R. J.: 1984, 'Spectra of Humidity, Temperature and Wind over the Sea at Sable Island, Nova Scotia', *J. Geophys. Res.* **89**, 2029–2040.

Smith, S. D. and Anderson, R. J.: 1988, 'Bedford Institute of Oceanography Eddy Flux Measurements during HEXMAX', in Proceedings of the NATO Advanced Workshop, Humidity Exchange over the Sea Main Experiment (HEXMAX) Analysis and Interpretation, April 25–29, 1988. Tech. Rep., Dept. of Atmospheric Sciences, Univ. of Washington, Seattle.

Smith, S. D. and Dobson, F. W.: 1984, 'The Heat Budget at Ocean Weather Station *Bravo*', *Atmosphere-Ocean* **22**, 1–22.

Stramska, M.: 1987, 'Vertical Profiles of Sea Salt Aerosol in the Atmospheric Surface Layer: A Numerical Model', *Acta Geophysica Polonica* **35**, 87–99.

Thorpe, M. R., Banke, E. G. and Smith, S. D.: 1973, 'Eddy Correlation Measurements of Evaporation and Sensible Heat flux over Arctic Sea Ice', *J. Geophys. Res.* **78**, 3573–3584.

AN ANALYSIS SCHEME FOR DETERMINATION OF TRUE SURFACE WINDS AT SEA FROM SHIP SYNOPTIC WIND AND PRESSURE OBSERVATIONS

K. BUMKE and L. HASSE

Institut für Meereskunde, Düsternbrooker Weg 20, D 2300 Kiel, F.R.G.

(Received in final form 24 October, 1988)

Abstract. For oceanographic studies, a high resolution description of the wind field at the sea surface is required. In order to estimate the air-sea interaction fluxes correctly, it is important that the derived wind field has the characteristics of actual surface winds. Ship synoptic observations are at irregular positions, while applications require boundary conditions at a regular grid. The problem is solved by locally fitting a second-order pressure surface to both wind and pressure observations, with the aid of a boundary-layer formulation. The results show that fitting wind and pressure data together provides better spatial resolution than using wind or pressure data alone, because more information is available. Wind data provide at least the same accuracy and less bias than pressure data, tested against independent wind observations. Also a stability and wind-dependent boundary-layer formulation results in less bias than a constant one. For the ship synoptic data available via GTS for the North Atlantic Ocean, spatial resolution, as expressed by an equivalent filter half-width, is 220 km, considerably improved compared to typical data assimilation schemes used in numerical forecast models.

1. Introduction

Even with satellites and numerical weather prediction, there is still a considerable demand for a good description of the surface wind at sea. From the practical point of view, weather routing is an important aspect of economical shipping. On the scientific side, air-sea interaction and hence most of oceanography and climate prediction depends on the wind and turbulence field at sea. Also, algorithms for estimation of wind speeds and directions from satellite-borne microwave scatterometers need ground-truth data for the verification. We report on an analysis scheme that endeavours to make optimal use of ship synoptic surface observations to obtain a high resolution unbiased surface wind field at sea.

Modern numerical analysis and forecast models provide wind fields at the surface. Through data assimilation techniques including prognosis fields, data-sparse regions are interpolated in space and time. Hence, it may seem that there is no need for a separate surface wind field analysis scheme. This is not entirely true. The assimilation schemes of operational forecast models are usually fairly coarse. Often a linear, weighted average over a large area of influence is used (for a review see, e.g., Gustafsson, 1981). More severe, the data-assimilation schemes adapt observed winds and/or pressure fields to conform with the physics and the numerical schemes of the model, e.g., the balance equation expressed in terms of grid-point differences. The outcome of these procedures for the lowest computational level may then still be called the surface wind by the modeller –

but such wind may be severely biased compared to real surface winds. This bias may not be important for short- and medium-range forecasting, but it is a matter of concern when calculating air-sea interactions.

The problem is as follows: experimental knowledge of the turbulent exchanges between sea and air is usually obtained with the aid of fairly sophisticated instrumentation. Hence, such measurements are obtained as point measurements, using short-term averages of, say, surface wind, temperature, humidity, and sea surface temperature as parameters – referred to a 10 m reference height. These are local parameters. Modellers may choose the same type of parameterisation, but since they are using a larger scale grid net, their parameterisation coefficients have to include the average effects of subgrid-scale processes (and of model physics, too), while in experimental determinations, these processes are explicitly represented by the local parameters. Hence, though the variables carry the same name and parameterisation formulae look alike, in fact they include some differences in physics.

It is the opinion of the present authors that some of these inconsistencies can be alleviated if a detailed analysis of the surface wind field is provided. There are, of course, practical limits to this. The data coverage over the world oceans is rather sparse. Also, it is difficult to obtain good wind measurements from ships: measured shipboard winds have to be corrected for course and speed of the ships. Hence it is often believed that surface air pressure is more reliably measured at sea than winds. That may be true, but it has to be noted that for determination of the surface wind field, the pressure gradient matters, not the pressure itself. With few pressure observations and some knowledge of the synoptic situation, it might always be possible to interpolate a reasonable pressure pattern, but it is much less certain that the gradients from interpolated isobars provide a reasonably accurate surface wind field.

In the present paper, a method is presented to analyse surface air pressures, surface wind speeds and directions simultaneously. The method is designed for studies of air-sea interactions and oceanic circulations. Since the quasigeostrophic eddies of the ocean have smaller scales than the corresponding systems in the atmosphere, a high spatial resolution is desired. This, and the demand to secure the character of a surface wind, was the primary reason to choose a local polynomial fit together with a non-geostrophic wind-pressure relationship.

In some sense, our work is an extension of a method used 40 years ago by H. A. Panofsky (1949) for the objective analysis of synoptic weather maps. Panofsky used the assumption of geostrophy in order to analyse wind and pressure observations consistently. He fitted wind components and dynamic heights simultaneously for the 700 hPa level. Aside from differences in technical details, scale of resolution, and primary objectives, our work is an extension of H. A. Panofsky's in the sense that we use a polynomial fit to analyse pressure and wind observations simultaneously; instead of the free atmosphere, we deal with the wind field in the surface layer. Hence we use some boundary-layer formulations in order to determine geostrophic winds from the observed surface winds, before the polynomial fit of geostrophic wind components and pressure observations is made.

2. Data Base

In the following, we test our analysis scheme with synoptic data from the North Atlantic Ocean. International voluntary observing ships transmit their observations via national shore stations into the GTS (Global Telecommunication System) maintained by the member states of WMO. The observations available at the German Weather Service of Offenbach at the main observing hours (0000, 0600, 1200, 1800 GMT) were transferred to magnetic tape and analysed thereafter. These data were available to us without prior editing or quality control. Part of our analysis scheme therefore is an error control scheme. Errors in this sense are from quite different sources. There are operational errors induced by estimation of wind speed and direction from a travelling ship (Godshall *et al.*, 1976), coding and transmitting errors, including erroneous coding of positions, and there is also the meso-scale and local variability which may appear as error though perfectly correctly observed, coded and transmitted. A typical data coverage over the North Atlantic Ocean is shown in Figure 1. Most of the ships provide pressure measurements and wind observations. In order to avoid spurious influences, pressure observations are taken only from land and island stations. Usually daytime (1200 and 1800 GMT) observations are more plentiful than nighttime ones (0000 and 0600 GMT). It is obvious from Figure 1 that resolution of mesoscale and local-scale wind fields is not possible with such data coverage. Hence, if there are strong local effects (say at a front or in the vicinity of a cumulonimbus), these might appear as inconsistent with neighbouring information and be treated as an error.

The first run of the analysis was used to detect erroneous observations. For this purpose the analysis scheme was simplified by applying a second-order fit, without weighting, to all available information within 600 km for wind and 800 km for pressure from a grid point. Then those observations are taken as erroneous where the deviations from the analysed field exceed either 3 hPa for pressure, or 45 deg for wind direction, or 5 m/s for wind speed. This results in a smoother analysed field by decreasing the influence of one single (might be erroneous) observation on the calculated grid point value.

We may mention one peculiarity of ship synoptic data. Due to known difficulties to obtain reliable measurements of true winds from ships, most marine weather services require that surface winds are obtained by estimation of Beaufort strength and wind direction. This estimation is fairly independent of observing height and, of course, of speed of the ship. On the other hand, the conversion of Beaufort strength to wind speed has been a source of continuing debate. The conversion of synoptic data is governed by WMO code 1100. It is

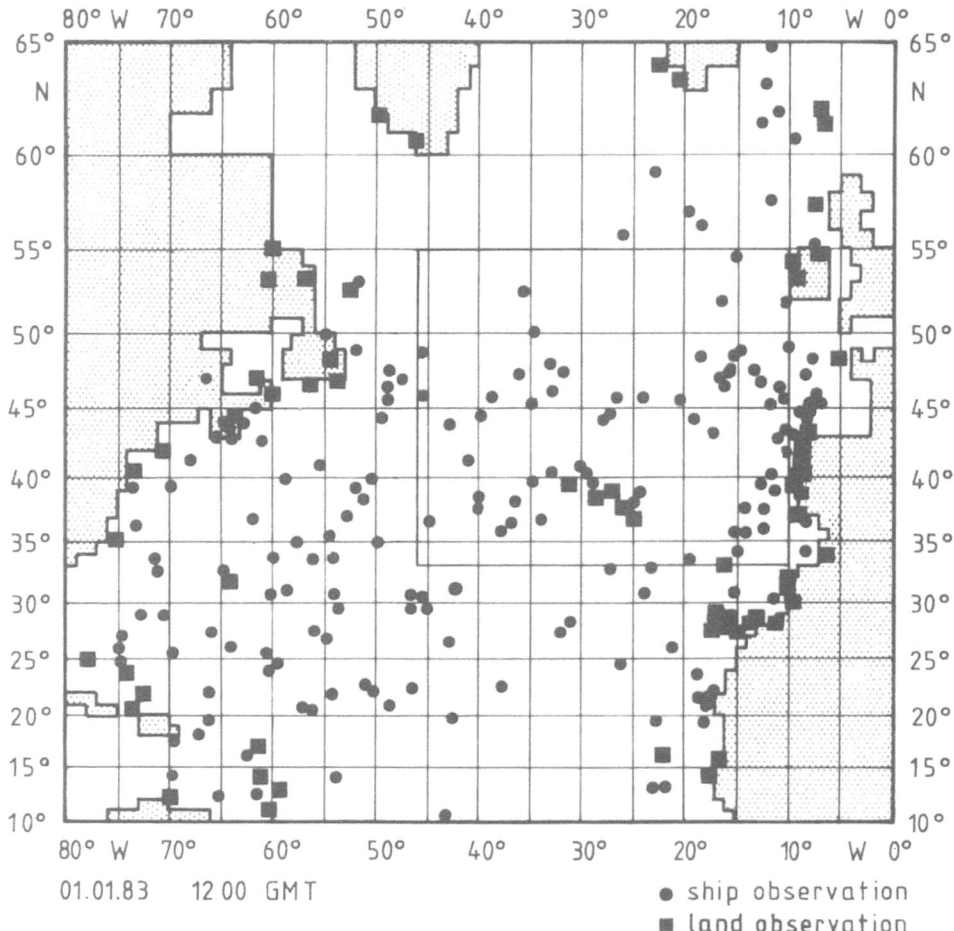

Fig. 1. Typical data density at 1200 GMT over the North Atlantic Ocean as received via GTS (1st January, 1983). Dots, ships; squares, land and island stations (for pressure data only). Test area indicated by full line.

known (Kaufeld, 1981; WMO, 1970) that this scale is not optimal (speeds too low below Beaufort 8, and speeds too high above). The resulting bias has to be considered when calculating air-sea fluxes. In our analysis scheme, we convert winds marked as estimated by a conversion scale derived from Kaufeld (1981).

3. Theory

Our analysis scheme is based on the polynomial method (Panofsky, 1949; Gilchrist and Cressman, 1954) and fits locally a second-order pressure surface to both wind and pressure observations:

$$p' = a_{00} + a_{10}x + a_{20}x^2 + a_{11}xy + a_{01}y + a_{02}y^2,$$ (1)

where x, y are distances between points of observations and the grid point in the east and north directions, and a prime denotes an estimated variable.

In geostrophic equilibrium then

$$u'_g = -(a_{01} + 2a_{02}y + a_{11}x)/f\rho, \tag{2a}$$

$$v'_g = +(a_{10} + 2a_{20}x + a_{11}y)/f\rho. \tag{2b}$$

We cannot assume that wind and pressure are in geostrophic balance at the surface. Hence, surface wind observations are related to geostrophic winds with help of simplified boundary-layer theory; see below.

Since the observations may include errors, we use redundant information and solve the polynomial (1) by minimizing the sum S

$$S = (1 - W) \sum_{k=1}^{n} C^2(p - p')^2 + W \sum_{i=1}^{m} C^2[(u - u')^2 + (v - v')^2], \tag{3}$$

using the wind and pressure observations simultaneously to estimate the coefficients a_{ik} of (1). Here, u, v are the x, y surface wind components obtained from ship wind speeds and directions, while u', v' are the estimated wind components. The weights C in (3) have been introduced to provide decreasing weight for observations at increasing distance to the grid point (Figure 2). For this purpose a Cressman function (Cressman, 1959) was selected:

$$C(r) = (R^2 - r^2)/(R^2 + r^2), \tag{4}$$

where R is the radius of influence and r is the distance of observation to the grid point. The radius of influence R is held at 600 km for wind and 800 km for

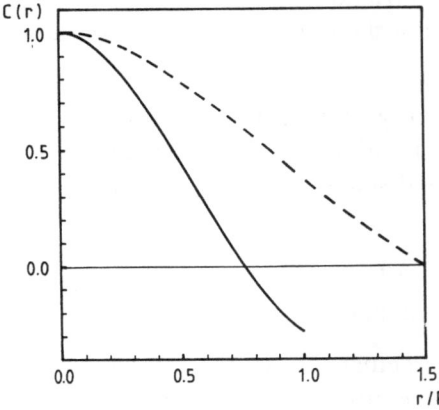

Fig. 2.   Weight factors as a function of distance from grid point. Full line shows the combined effect of Cressman weight and second-order least-square fitting. The implicit weights of second-order fit depend on the distribution of data points. The line shown is calculated assuming an homogeneous distribution of observations over the area within R. The Cressman weights (dashed line) are shown for comparison.

pressure observations. The weight W serves to optimize the influence of wind and pressure information ($W = 1$, wind only; $W = 0$, pressure only). In order to obtain reliable wind estimates at a grid point, a minimum amount of information from well distributed observations should be available. In the routine version of the scheme, we require at least 12 items of information (counting pressure, wind speed, and wind direction as one item each).

In order to achieve high spatial resolution, at each grid point we used only those observations nearest the gridpoint under consideration that together yield the minimum number of 12 items of information. The radius enclosing these observations is taken as characterizing the resolution.

Polynomial fits are known to yield spurious estimates if used outside the range covered by data. Hence, a distribution criterion was introduced. Test calculations showed, that for this purpose, a distribution of observations is sufficient if the grid point is enclosed by a triangle of observations. If the minimum number of observations is not available or the distribution criterion is not fulfilled, analysis is deferred for an additional run. In this run, analysed values from neighbouring grid points are introduced as pseudo-observations with half weights. Rarely are there areas with observations missing where even use of pseudo-observations does not help to satisfy the distribution criterion. For these grid points, the radius for inclusion of pressure observations is extended to a maximal 1200 km. When pseudo-observations are included, only those within the radius of resolution are considered. If the distribution criterion cannot be fulfilled even including pseudo-observations, we fit a first-order pressure surface locally. That is the case for less than 5% of the grid points.

In the following, all calculations are made for a test area as indicated in Figure 1.

The analysis includes frictional influences. For the relation of surface wind to geostrophic wind, we use the following parameterisations, called model I, II and III in this study.

(I) a constant ratio γ of actual to geostrophic wind speed at the surface $\gamma = 0.7$ together with an ageostrophic angle $\alpha = 17°$;

(II) a wind speed and stability-dependent empirical relationship (Luthard and Hasse, 1981):

$$\alpha = 14.5° + 4.4 \cdot \Delta T$$
$$U = G \cdot (0.59 - 0.03\Delta T) + (2.13 - 0.05\Delta T) , \tag{5}$$

where U is the actual, G the geostrophic wind speed in m/sec, and ΔT the air-sea temperature difference in deg C. For $G \leq 4$ m/s, $G = 4$ was used instead for determination of γ. For $|\Delta T| \geq 3$ K, ΔT was taken to be constant as $+$ or -3 K, respectively.

(III) a simplified form of the boundary-layer formulation of Brown et al. (1986) as shown in Figure 3.

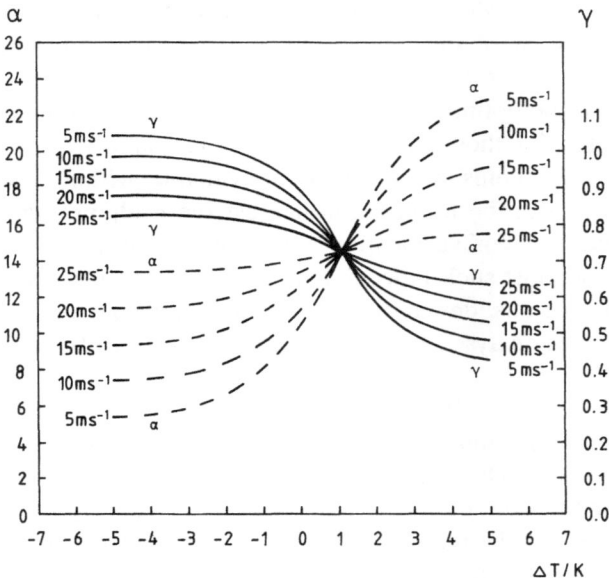

Fig. 3. Boundary-layer parameterisation following the arguments of Brown *et al.* (1986) approximated by smooth functions. The dashed line gives the ageostrophic angle α and the full line the ratio of actual wind speed to geostrophic wind speed γ for surface wind speeds (in m/sec) as indicated. ΔT is air-sea temperature difference.

These boundary-layer models are used in both directions, going from actual winds to pressure gradients and back again.

Model I is the simplest approximation of the relation between actual and geostrophic wind. A constant $\gamma = 0.7$ has often been used in oceanographic applications to derive the surface wind from the pressure field where the cross-isobar angle α was neglected ($\alpha = 0°$). The cross-isobar angle $\alpha = 17°$ used in this study is an approximate mean value taken, e.g., from Hasse (1974).

Model II is an empirical result from the German Bight in the North Sea, where a dense network of light vessels, island and shore stations was available. Note that mesoscale variability appears at low wind speed in a surface wind exceeding the geostrophic wind.

Model III is a Planetary Boundary Layer model based on a Blackadar-type mixing length, modified by Brown (1986) to allow for secondary flow and thermal winds. Thermal wind is neglected in our application. Roughness is represented by a drag coefficient of $C_D = 1.29 \times 10^{-3}$.

4. Results

There are certain variations of the analysis scheme conceivable, which influence its performance. For example, a first-order fit to the pressure field might be used instead of the second-order fit, the boundary-layer formulation may be varied, or the minimum number of observations included can be adjusted. Additionally, preference may be given to wind or pressure information.

Verification of the optimisation is done against independent observations. The latter are selected for those grid points in the test area with rather ample data supply (at least 20 items of information available within a search radius of 600 km). In this case, the observation nearest the gridpoint was extracted from the data set and used for verification only, not for analysis. All results given in this section are against such independent observations. The comparison is made mostly in terms of r.m.s. deviation, but also the wind speed difference of control observations versus analysed field, designated wind speed bias, is considered.

4.1. RELATIVE INFORMATION CONTENT OF WIND OR PRESSURE DATA

Ship observations of wind speed and direction are more liable to errors than pressure observations. Hence it is appropriate to show that the combined analysis performs better than the corresponding analysis of either pressure or wind field alone. In order to test the relative influence of wind and pressure information, the factor W has been introduced in Equation (3). Results are given in Figure 4,

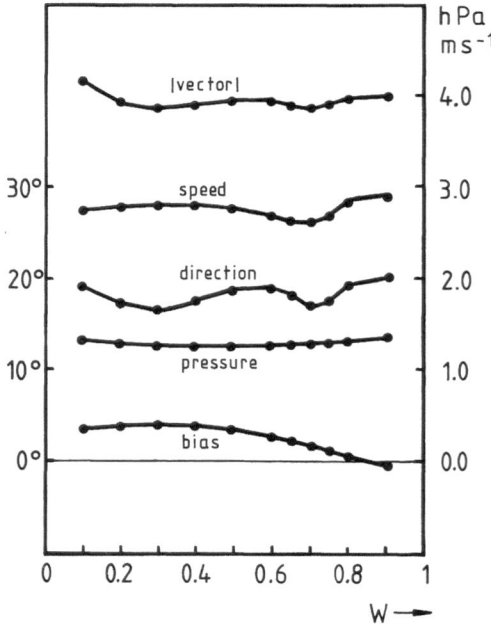

Fig. 4. Performance of the analysis scheme when varying the relative weight W of pressure and wind information ($W = 0$ pressure observations only, $W = 1$ winds only) expressed in r.m.s. deviation and wind speed bias between independent observations and the analysed field, using model III*, at least 12 items of information, 30 cases from 1983.

which shows that a combined analysis gives better results than using either pressures only ($W = 0$) or winds only ($W = 1$). Obviously the increased information content of the combination helps to detect and eliminate erroneous data and provide a better field with the remaining ones. The increase of available information helps to reduce the necessary number of ships to be included. Hence, spatial resolution improves and smoothing decreases; this explains why both the r.m.s. scatter and the wind speed bias are reduced compared to earlier versions of the scheme (Ennenga, 1985; Hasse and Strunk, 1987). The resulting r.m.s. errors may appear rather large but it should be remembered that these "errors" contain contributions by smaller scale meteorological variability. Note also that the r.m.s. deviations represent the variability of single observations. The accuracy of the determined mean wind speed and direction at a grid point is considerably better. As we typically use $n \geq 12$ pieces of information for 6 unknowns, the accuracy of the interpolated grid-point values is better by a factor of 2.5. The resulting error of wind speed is about 1.3 m/s and of direction below 10°, significantly less than the typical value of the ageostrophic angle.

From Figure 4, it is evident that there is no sharp minimum defining an optimum W. At any rate, $W \geq 0.3$ seems to be better than $W \approx 0$. That means that the wind data are at least as useful as the pressure data. In the following, we use a fixed $W = 0.7$ since this provides an acceptable scatter and small bias. We conclude from this experiment that the main advantage of including wind and pressure data is the increase of information that yields a better spatial resolution.

4.2. OPTIMUM AMOUNT OF INFORMATION

Increasing the number of observations increases the reliability of results, but at the same time, it necessitates an extension of the radius of influence and the degree of smoothing. Figure 5 shows the results of varying the minimum amount of information given. It shows that the use of at least 12 items of information provides a relative optimum. Hence we selected 12 items as the minimum amount of information both for routine work and for this study.

4.3. FIRST-ORDER VERSUS SECOND-ORDER FIT

A second-order polynomial fit has more coefficients and needs a larger number of observations than a first-order fit. The flexibility of the second-order polynomial and its ability to follow the synoptic field requires a larger radius of influence at a given data density. With a first-order fit, less observations are necessary, providing for a more local fit. It is not obvious what type of analysis is preferable.

We therefore ran a series of test cases (not reported here) which showed that with the data distribution typically found over the North Atlantic Ocean, the second-order fit gave better results than the first-order one (slightly less r.m.s. deviations and a reduced bias). This is in line with findings from a similar study from the North Sea (Luthardt, 1985). We therefore prefer the second-order fit and revert to first order only in the rare cases where the distribution criterion

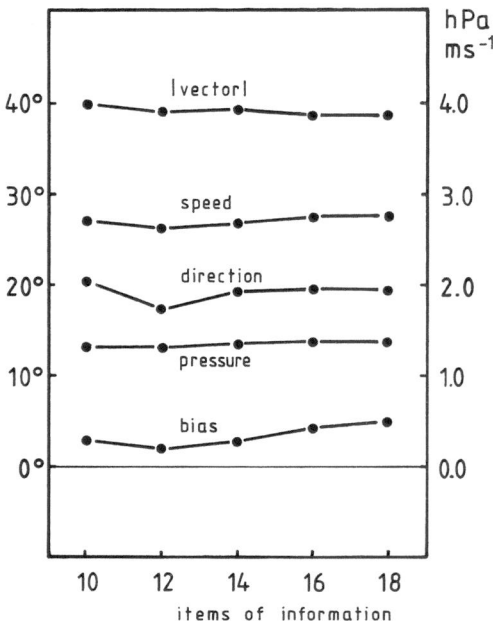

Fig. 5. Variation of r.m.s. deviations and wind speed bias between independent observations and the analysed field with the number of items of information used ($W = 0.7$, boundary-layer model III*, 30 cases from 1983).

cannot be met even with pseudo observations and extending the radius of influence for pressure observations to 1200 km.

4.4. INFLUENCE OF BOUNDARY-LAYER MODEL

Stability influences the surface wind field, as has long been known (Hellmann, 1917). At sea, the actual to geostrophic wind ratio is smaller under stable than under unstable conditions (e.g., Hasse, 1974). Additionally the sea surface roughness may vary with wind speed. The influence of stability can be inferred with the aid of the air-sea temperature difference ΔT. Because of its small value, ΔT apparently scatters considerably; also, there is local variability of wind speed. We therefore tried two approaches: (i) mean wind speed and air-sea temperature difference were determined from all observations in the vicinity of the grid point within 600 km radius and used to obtain α and γ. (ii) individual wind speed and ΔT were used with each observation (indicated below by an asterisk).

Models II and III differ from model I by the inclusion of wind and stability dependence; models II and III differ mainly by the different degree of complexity in the formulation. Figure 6 shows that there is not much difference between II, II*, III and III*. Except for the comparisons in this subsection, all results reported in this paper are obtained with the stability- and wind-speed-dependent model III*.

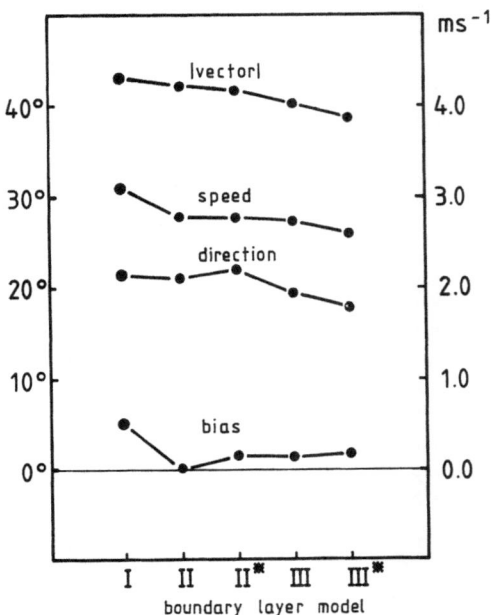

Fig. 6. RMS-deviations and wind speed bias between independent observations and analysed field for three boundary-layer models: Model I, actual to surface wind ratio $\gamma = 0.7$ and ageostrophic angle $\alpha = 17°$. Model II, wind speed and stability dependent, Equation 5. Model III, wind speed and stability dependent, as shown in Figure 3. Model II and III are used in two versions: (i) γ and α depend on mean conditions at each grid point, being taken as constant for all those observations in the vicinity of the grid point that entered into the fit. (ii) Actual stability and wind speed at each observation are used to obtain γ and α; this version is indicated by an asterisk. All calculations for this figure are from 30 cases from all seasons of 1983, using at least 12 items of information at each grid point and weight $W = 0.7$.

4.5. Effective resolution

The spatial resolution depends on the distribution of observations. With respect to oceanographic applications, we interpolate on a $2° \times 2°$ longitude-latitude grid. In the area between $10°$ N and $65°$ N of the North Atlantic Ocean, there are roughly 600 grid points in appropriate agreement with the 300–400 observations available on a typical daytime main observing hour. An impression of an analysed field is given in Figure 7. A more quantitative depiction of the mean spatial resolution in different regions of the North Atlantic Ocean is given in Figure 8. Here, the equivalent filter half-width is used for illustration. Note that the filter half-width is about half the size of the radius of influence, i.e., the distance of the outermost observation used in interpolation onto the grid point.

It is difficult to quantify the degree of improvement compared to other assimilation schemes since different data-handling approaches inhibit direct comparison. Data sets provided by the European Center have found some use in

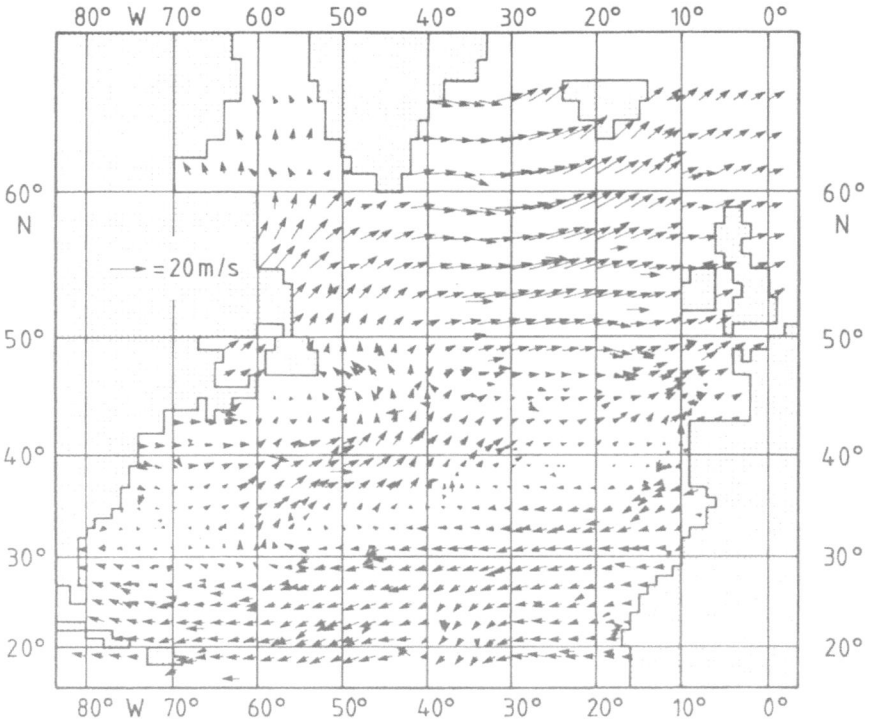

Fig. 7.   Analysed wind and pressure field at 1200 GMT, 1st January, 1983, using at least 12 items of
information, a weighting factor of $W = 0.7$, boundary-layer model III*. Wind speed scale for the
arrows is given in the upper left corner. Both original wind observations and analysed fields are
shown. The latter is on a regular $2° \times 2°$ grid, while the original observations include those that are
eliminated during the analysis.

oceanography and meteorology, e.g., the FGGE data set. MacVeigh *et al.* (1987)
have shown that the resulting fields from ECMWF have a spatial resolution
cut-off at about 1000 km. Our analysis scheme provides a mean resolution of
220 km equivalent filter half-width. This is certainly a considerable improvement
compared to typical resolution of data assimilation schemes used for numerical
weather forecasting.

5. Summary

We have provided an objective numerical scheme to analyse the surface wind
field at sea. Emphasis is on high spatial resolution and retaining the characteristics of surface winds. To achieve this, all available surface information is
used, i.e., pressure *and* wind reports. Simultaneous analysis of pressure and
actual surface winds is possible by converting the actual wind vector into a
geostrophic wind using a boundary-layer formulation. The results indicate that
combining wind and pressure observations provides more information; hence,

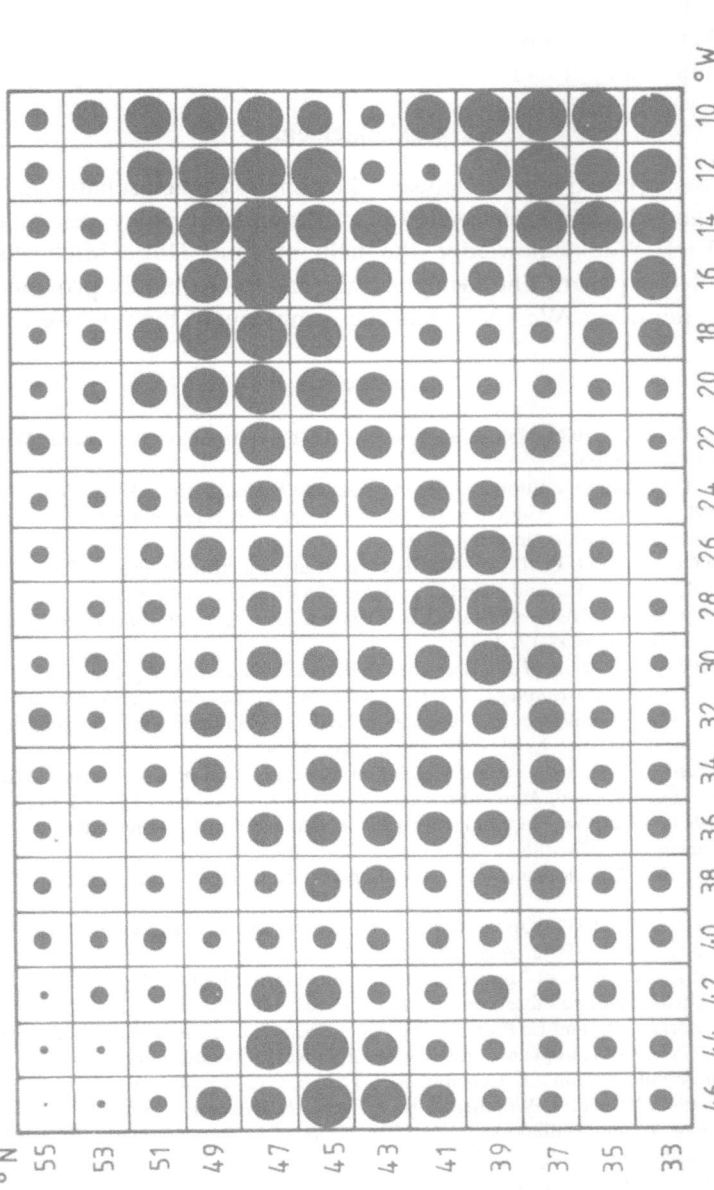

Fig. 8. Illustration of mean obtainable resolution in the test area. The larger the size of the blobs, the better the resolution. Resolution is measured in terms of equivalent filter width as indicated by the scale at the right-hand side. From 30 cases in 1983, 1200 GMT using at least 12 items of information, relative weighting factor of $W = 0.7$, boundary-layer model III*.

interpolation onto a grid point can be made from observations at a smaller distance. This not only yields higher resolution, but also helps to avoid biases in the surface wind speed that are otherwise introduced by the smoothing when interpolating over a large area.

It is tempting to use "surface wind" fields from numerical analyses of meteorological forecast centers. Unfortunately, since their techniques are optimised for operational forecasting, these fields have less desirable properties for air-sea interaction studies or as boundary conditions in oceanographic modelling. It appears that for this application, our approach provides a reasonable, high resolution, unbiased approach.

Acknowledgement

This work was supported by the Deutsche Forschungsgemeinschaft and is part of research of Sonderforschungsbereich 133, "Warmwassersphäre des Atlantik" at Universität Kiel. We acknowledge the help of Frau Doris Michaelis in the preparation of this paper. The synoptic ship data were kindly provided by the Deutscher Wetterdienst. We appreciate the comments of the reviewer.

References

Brown, R. A. and Levy, G.: 1986, 'Ocean Surface Pressure Fields from Satellite-Sensed Winds', *Mon. Wea. Rev.* **114**, 2197–2206.
Cressman, G. P.: 1959, 'An Operational Objective Analysis System', *Mon. Wea. Rev.* **87**, 367–374.
Ennenga, U.: 1985, 'Objektive Analyse aktueller Wind- and Druckfelder über dem Nordatlantik', *Ber. Inst. f. Meereskunde Kiel*, Nr. 142, 1–103.
Gilchrist, B. and Cressman, G. P.: 1954, 'An Experiment in Objective Analysis', *Tellus* **6**, 309–318.
Godshall, F. A., Seguin, W. R., and Sabol, P.: 1976, *GATE Convection Subprogram Data Center: Analysis of Ship Surface Meteorological Data Obtained During GATE Intercomparison Periods*, NOAA Technical Report EDS 17, 1–73.
Gustafsson, N.: 1981, 'A Review of Methods for Objective Analysis', in L. Bengtsson, M. Ghil, and E. Källen (ed.): *Dynamic Meteorology, Data Assimilation Methods*, Springer-Verlag, pp. 17–76.
Hasse, L.: 1974, 'On the Surface to Geostrophic Wind Relationship at Sea and the Stability Dependence of the Resistance Law', *Beitr. Phys. Atm.* **47**, 45–55.
Hasse, L. and Strunk, H.-A.: 1987, 'High Resolution Determination of True Surface Winds at Sea from Wind and Pressure Data', *Mesoscale Analysis and Forecasting*, ESA SP-282, pp. 457–459.
Hellmann, G.: 1917, 'Über die Bewegung der Luft in den unteren Schichten der Atmosphäre, 2. Mitteilung, *Meteorol. Z.*, 273–285.
Kaufeld, L.: 1981, 'The Development of a New Beaufort Equivalent Scale', *Meteorol. Rdsch.* **34**, 17–23.
Luthardt, H. and Hasse, L.: 1981, 'On the Relationship between Surface and Geostrophic Wind in the German Bight Area', *Beitr. Phys. Atm.* **54**, 222–232.
Luthardt, H.: 1985, 'Estimation of Mesoscale Surface Fields from Meteorological Parameters in the North Sea Area from Routine Observations', *Beitr. Phys. Atm.* **58**, 255–272.
MacVeigh, J. P., Barnier, B., and Le Provost, C.: 1987, 'Spectral and Empirical Orthogonal Function Analysis of Four Years of European Center for Medium Range Weather Forecast Wind Stress Curl over the North Atlantic Ocean', *J. Geophys. Res.* **92**, No. C12, 13.141–13.152.
Panofsky, H. A.: 1949, 'Objective Weather Map Analysis', *J. Meteorol.* **6**, 386–392.
Shaw, D., Lönnberg, P. and Hollingworth, A.: 1984, *The 1984 Revision of the ECMWF Analysis System*, Tech. Memorandum No. 92, ECMWF.
WMO: 1970, *The Beaufort Scale of Wind Force*, WMO Commission for Maritime Meteorology, Marine Sciences Affairs Report No. 3, 22.

VELOCITY AND TEMPERATURE SPECTRA
AND COSPECTRA IN AN UNSTABLE SUBURBAN
ATMOSPHERE

M. ROTH, T. R. OKE and D. G. STEYN

*Atmospheric Science Programme, Department of Geography, The University of British Columbia,
Vancouver, B.C., Canada, V6T 1W5*

(Received 24 October, 1988)

Abstract. Boundary-layer flow over very rough surfaces is poorly understood so the applicability of standard micrometeorological theory is uncertain. This study presents observations of the turbulent fluctuations of meteorological parameters over a suburban area. Even though the height of measurement is considered to be close to the junction between the inertial and roughness sub-layers, the wind and temperature spectra and the momentum and sensible heat flux cospectra are in good agreement with reference data from smoother surfaces. Recommendations are made concerning site requirements, height of measurement and averaging times for the study of turbulence and turbulent fluxes over suburban terrain.

1. Introduction

Several experimental results (e.g., Thom *et al.*, 1975; Garratt, 1978 and 1980) show that conventional flux-profile relationships and the validity of the non-dimensional Monin–Obukhov functions for momentum and heat transfer must be questioned over horizontally uniform but very rough natural surfaces such as tall crops and forests. This reinforces the idea that the surface layer over a very rough surface must be considered in two parts: usually referred to as the inertial sub-layer (Tennekes, 1973) and the roughness sub-layer (Raupach, 1979). In the former, under adiabatic conditions, height above the effective surface is the only controlling length scale and the semi-logarithmic profile laws are obeyed. On the other hand, in the roughness sub-layer which is adjacent to the surface itself, the flow is affected by wake motions introduced by individual roughness elements. In this region, length scales defined by the surface characteristics become important.

In order to measure variances and covariances of meteorological variables that are consistent with Monin–Obukhov similarity theory (MOST), it is essential to avoid the roughness sub-layer and to work in the surface layer (inertial sub-layer) wherein the vertical fluxes of entities are assumed to be approximately constant with height, as long as horizontal advection due to changes in surface character in the upwind zone is absent. In this layer, turbulent fluxes can be evaluated using standard micrometeorological approaches such as the aerodynamic, Bowen ratio-energy balance and eddy correlation methods.

Boundary-layer flow over very rough terrain has been investigated mainly for surfaces covered by tall crops or trees with some additional understanding gained from wind tunnel work. Based on such observations, a range of values for the lower height limit, z_* (measured from zero-displacement, d, upwards) for the validity of the Monin–Obukhov functions above large roughness elements is reported. Raupach *et al.* (1980), based on a wind tunnel study under neutral conditions, suggest that (for wind), $(z_* + d) = h + 1.5D$, where h is the height of the roughness elements and D is the inter-element spacing. Under unstable conditions over a forest, Garratt (1980) concludes that (for wind), $z_*/z_0 = 100$ (where z_0 is the surface roughness length) is a useful generalization and the ratio probably decreases (down to 10) with increasing density of roughness elements. For temperature, he finds $z_*/z_0 = 100$ for sparse and about 65 for denser forests.

Much less effort has been spent understanding boundary-layer flow over urbanized areas. This is disappointing considering that an ever increasing number of The World's population is exposed to urban atmospheres. Without doubt, it would be of considerable interest and value to know if and where standard micrometeorologidal theories are applicable in the urban system. Both energy balance and atmospheric diffusion studies would benefit from a better understanding of the structure of turbulence in the urban boundary layer. There are a number of reasons for this lack of understanding. One is the set of practical problems associated with the selection and location of an observation site ensuring the unobstructed exposure of the sensors. Another is the lack of a recognized methodological framework within which the information is to be analysed, although recent attempts have been made to overcome this.

Based on the literature already mentioned, Oke (1984) suggested a possible framework for the complex and high roughness environment of urban terrain. He divided the urban atmosphere into an urban canopy layer (UCL) which is dominated by microscale features associated with individual roughness elements and an overlying urban boundary layer (UBL). The transition between the UCL and UBL is not characterized by a sharp discontinuity; rather, the microscale features of the UCL meld into local or meso-scale ones in the roughness layer (called the 'roughness sub-layer' by Raupach, 1979). The integrated effects of the urban 'surface' thereby form an horizontally-homogeneous turbulent surface layer and the mixed layer of the UBL. In a more recent study, Schmid (1988) develops a model to estimate the two-dimensional surface area contributing to a turbulent flux measurement and hence provides a possible way of assessing if a flux measurement taken at a specific height is representative of the area under investigation. Oke *et al.* (1989) investigate the applicability of standard boundary-layer theory and observation methods to the urban system and conclude that 'despite the physical problems presented by the nature of the 'surface', it is possible to obtain valid areally-averaged fluxes from fixed-point observations provided that careful site selection, height of measurement and temporal sampling procedures are followed'. Analyses to estimate the value of z_* as performed for forests, are not available for urban surfaces; therefore, estimations for 'proper' (within the surface layer) measuring heights rely on the already cited recommendations from other systems.

This study presents the results of a project designed to measure and analyse the turbulent fluctuations of the longitudinal wind component u, the vertical wind component w, air temperature T and the covariances of momentum uw and sensible heat flux wT at one height over an urbanized area. The turbulence measurements are presented in a form so as to arrive at conclusions concerning the appropriateness of the observation height involved and the averaging times used. Thereby the analysis should provide information regarding the suitability of this site for turbulence measurements and their use in energy balance computations.

The approach taken is to compute the energy spectra and cospectra of the turbulence fluctuations and so yield information about the dominant eddy sizes involved in the turbulent transfer through the analysis of the spectral shapes. The question of the proper averaging time to be used (an increase in measuring height requires a longer averaging time) is approached by inspecting the low frequency end of the spectra (cospectra) to see if they conform with MOST. This theory has been shown to apply over surfaces exhibiting the required low roughness and homogeneous fetch requirements. In this respect, the empirical results from the Kansas 'ideal' boundary-layer experiment (Kaimal et al., 1972) will be used as a standard for comparing the results presented in this study. The Kansas results will be referred hereinafter as the 'reference' or 'Kaimal' spectra and data.

If, as suggested by the studies cited earlier, an extra length scale due to the influence of individual roughness elements is introduced, deviations from the shape of 'ideal' reference spectra (cospectra) should be observed. If present, they might indicate that the sensors are at too low a height so that results relate to processes in the roughness layer.

2. Experimental

The data for this study were gathered at the Sunset suburban site in Vancouver, British Columbia, Canada as part of a surface and boundary-layer research programme during the summer of 1986. The general requirements leading to the selection of this particular site are described and discussed in Kalanda et al. (1980) and Steyn (1980) and include as a prime criterion, the fetch required for the atmospheric layer of interest to adjust to a change in surface properties upstream.

The aerodynamic roughness of the site is assessed to be 0.5 m, using Lettau's (1969) land-use/roughness element analysis. Based on estimations from land-use analysis, the zero-plane displacement length d is about 3.5 m. Instruments to probe the surface layer were mounted on the top of a triangular-section, steel lattice, free-standing tower.

Using the dimensions of the Sunset tower site (building height $h = 8.5$ m, building spacing $D = 23$ m) and the formulae of Garratt (1980) and Raupach et al. (1980), the top of the roughness sub-layer, z_*, is calculated to be located between 39 and 50 m for the momentum flux and between 32 and 50 m for the sensible heat flux. However, these values depend on the density of the surface roughness elements and may be lower. Nevertheless, in this study with measurement heights, z', (above d) of only 19 m for sensible heat and 22 m for momentum, there is reason to suspect that the observations may be influenced by wake effects.

Wyngaard (1973) addressed the problem of averaging time. For a sensor height of about 20 m and an uncertainty of 15%, his analysis yields an averaging time of 19 min for the second moments of vertical wind and temperature, 543 min for the momentum flux and 56 min for the sensible heat flux during unstable conditions ($z'/L = -1$). In this study, the averaging time was pre-selected to be 60 min for all components.

The turbulent fluctuations of the longitudinal (u) and vertical (w) wind velocities, the covariance of the kinematic momentum flux (uw) and the mean wind direction were measured with a modified Gill twin propeller-vane anemometer (GTVA) (Pond and Large, 1978). The modifications included a tilt of 60 deg from the horizontal plane for the sensor measuring the vertical wind. Rotation increases the frequency response and introduces about one half of the horizontal wind component into the tilted sensor. The turbulent fluctuations of the vertical velocity (w) and the temperature (T) and hence the covariance of the kinematic heat flux (wT) were measured with a sonic anemometer and fine-wire thermocouple (SAT) (Campbell Scientific, Model CA27T).

Observations were taken during a period of good weather, on eight days at the end of August and the beginning of September 1986. Atmospheric stabilities encountered were $-2.4 < z'/L < 0.06$ with an arithmetic mean value of -0.75. All measurements were recorded with a Campbell Scientific (Model CR21X) data logger. The sampling frequency was 10 Hz for the GTVA and the SAT components and 0.3 Hz for the Gill-vane. The GTVA signals were low-pass filtered at a frequency of 10 Hz to remove commutator noise.

3. Results

3.1. VELOCITY AND TEMPERATURE SPECTRA

The spectra and cospectra presented here are normalized by the respective variances and covariances and hence are a function of non-dimensional frequency ($f = nz'/U$) and stability (z'/L) only. No attempt was made to classify the normalized spectra (cospectra) according to stability, although some stability dependence could be observed.

In Figure 1, the composite w spectrum from the SAT and GTVA systems are presented and compared with the spectrum of Steyn (1982) measured at the same site and with a model spectrum suggested by Højstrup (1981). The agreement between the SAT and the GTVA spectra is excellent; however, it should be

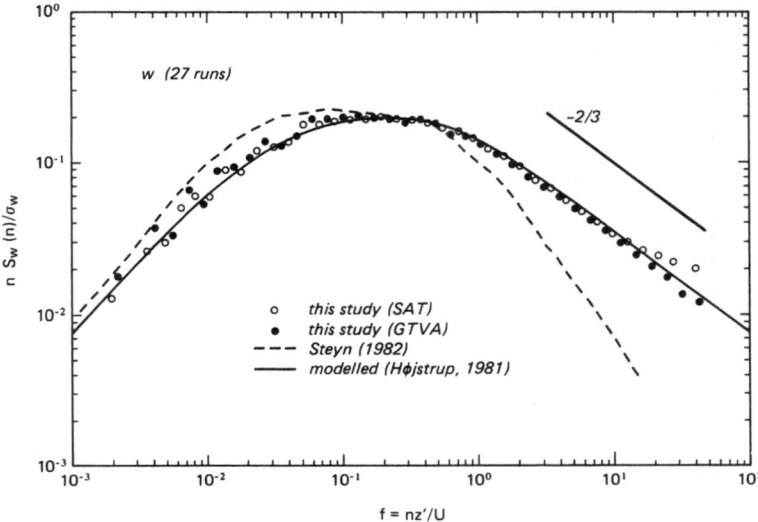

Fig. 1. Composite spectrum of vertical velocity normalized by the variance. Also indicated are spectra from other authors.

noted that the Gill measurements were extended with an artificial $-2/3$ slope in the inertial sub-range. Nevertheless, the fact that the two composite spectra measured with two different systems, follow each other so closely is very promising and gives confidence in the use of the GTVA.

Comparison between the results from this study and that of Steyn (1982) reveals two obvious differences. Firstly, the faster roll-off in Steyn's spectrum is due to the fact that he did not apply a frequency correction to the propellor measurements. On the low frequency side, the composite spectrum of Steyn shows higher energy content than that from this study. However, this 'increase' is relative since the position of the vertical spectrum depends on the stratification of the atmosphere and shifts to lower frequencies with increasing instability (Steyn's observations are from very unstable conditions, z'/L up to -100).

The comparison with the model spectrum (which is based on observations from smoother surfaces and evaluated at $z/L = -0.75$), is very good with only one minor difference: the peak in the measured spectrum from this study is less well defined and at non-dimensional frequencies of $0.05 < f < 0.1$, a 'flat' region with slightly higher energies than the model can be observed. This makes it difficult to decide upon a specific peak frequency. Taking the spectral density with the largest magnitude, the peak frequency (f_m) is at 0.2 with a possible range of $0.08 < f_m < 0.3$. Since $f = nz'/U = z/l_m$, where l_m is the peak wavelength, with $f_m = 0.2$ the dominant eddy scale is computed to be $l_m = 5z'$ or about 100 m.

In Figure 2, the observations of the u component (from the GTVA in this study) are compared with the composite spectrum from Steyn (1982) measured at

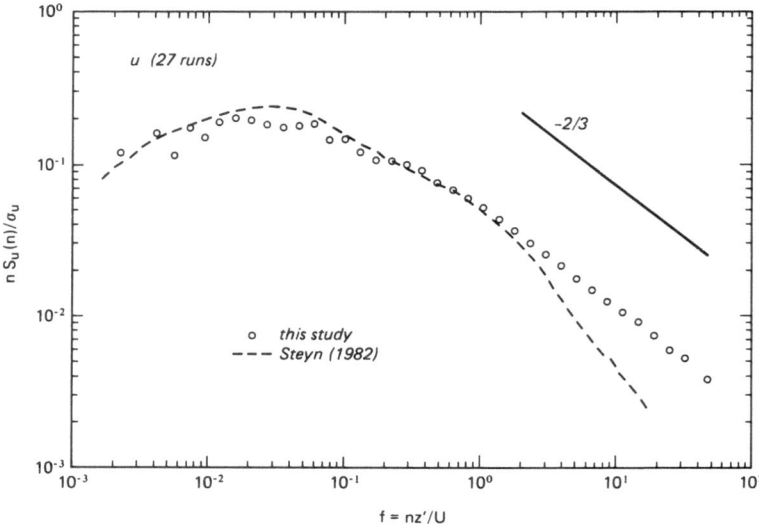

Fig. 2.   Composite spectrum of along-wind velocity normalized by the variance and comparison with
that of Steyn (1982).

the same site. The results are quite similar, the differences at the higher
frequencies again originating in the frequency response deficiencies of the Gill
anemometer used by Steyn whereas the GTVA measurements were corrected for
the frequency response, therefore exhibiting the required $-2/3$ slope. The main
differences between the observations from this field programme and those of
Steyn are a prominent dip at about $f = 0.03$ and a smaller dip at $f = 0.2$ in the
results from this study.

In general, the observed composite along-wind spectrum exhibits a maximum
at low frequencies and a weak inflexion point at higher frequencies (the small dip
mentioned above). These features are shown by Kaimal (1978) to be typical for
unstable spectra in the surface layer over relatively smooth surfaces. The
determination of a peak frequency is made difficult because of the prominent dip
at the low frequencies. The highest spectral value can be found at about
$f_m = 0.017$ ($l_m = 60z' = 1320$ m) with a possible range of $0.01 < f_m < 0.03$. For
both velocity components analysed here, the agreement between the results from
this study and other urban observations as well as reference data is good.

The composite temperature spectrum in Figure 3 very closely follows the $-2/3$
slope in the inertial subrange. The T spectrum is different from the w spectrum
but similar to the u spectrum in that it shows one broad peak and only a slight
roll-off at the low frequency end. The peak frequency (as determined from the
largest spectral density) is at about $f_m = 0.023$ ($l_m = 43z' = 817$ m). However, it is
more realistic to give a range of f_m's as $0.01 < f_m < 0.06$.

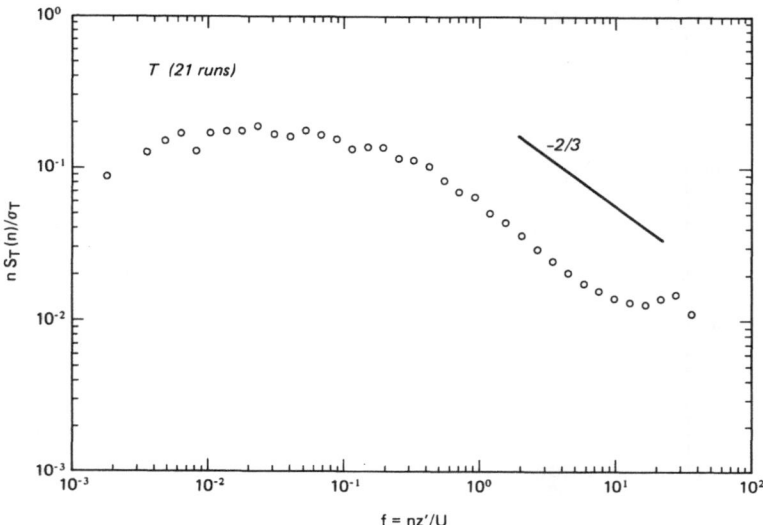

Fig. 3. Composite spectrum of temperature normalized with the variance.

Temperature spectra from other studies utilizing measurements over rough urban surfaces agree quite well with the observations from this study. Clarke *et al.* (1982) report a broad peak at about $f_m = 0.02$, and Coppin (1979) at approximately $f_m = 0.04$. The composite spectrum is also in relatively good agreement with the Kaimal data.

3.2. FLUX COSPECTRA

Figure 4 shows the composite heat flux cospectrum. It shows one broad peak bordered by a sharp roll-off at the low frequency end and another at higher frequencies. At non-dimensional frequencies $2 < f < 8$, hence within the inertial sub-range, the spectrum exhibits a $-4/3$ slope (as required by theory) which is followed by a faster roll-off at the highest frequencies. Theoretically, at very high frequencies, cospectra should become zero because the increasing isotropy of the flow field results in a correlation approaching zero. The increasingly faster roll-off in this composite spectrum confirms this feature (in fact the last two points, not plotted, were slightly negative). The peak frequency, at $f_m = 0.05$ ($l_m = 20z' = 380$ m), has a range of about $0.04 < f_m < 0.1$. Coppin (1979) is the only other author to report heat flux cospectra measured over an urban surface (however, it was not presented in logarithmic co-ordinates). He found the peak to occur at about $0.06 < f_m < 0.1$ for unstable stratification, which is in good agreement with the results from the present study.

Figure 5 displays the composite cospectrum of the momentum flux. It is similar to the heat flux cospectrum (Figure 4) with a fast roll-off at the low frequency end

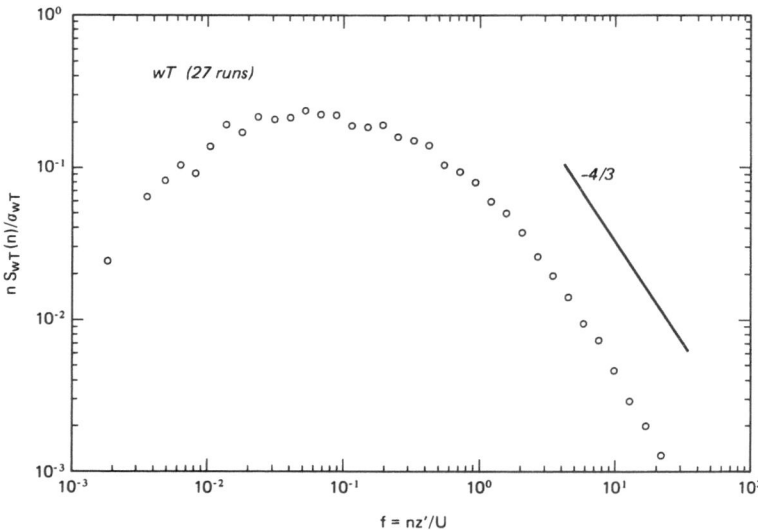

Fig. 4.   Composite cospectrum of sensible heat flux normalized by the covariance.

but the peak is not as broad. The peak frequency, based on the highest cospectral estimate, is at $f_m = 0.04$ ($l_m = 25z' = 550$ m) with a possible range of $0.02 < f_m < 0.06$. Kaimal *et al.* (1972) report that cospectral estimates for the momentum flux at very low frequencies ($f < 0.01$) show a tendency to reverse sign and become negative. Under unstable conditions in particular, the low frequency cospectral estimates fluctuate between large positive and negative values. This is observed

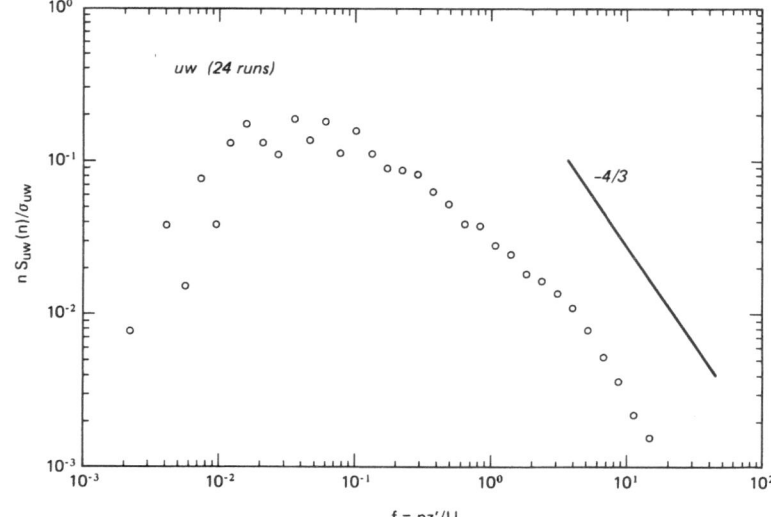

Fig. 5.   Composite cospectrum of momentum flux normalized by the covariance.

in this study; however, the sign reversal occurs up to slightly higher frequencies than for the reference case. In the inertial sub-range, the *uw* cospectra are required (by theory) to follow a $-4/3$ slope. An examination of this behaviour is not possible here because the high frequency estimates are contaminated by noise, and the observations at $0.2 < f < 1.5$ were probably introduced by movements of the tower.

3.3. AVERAGING TIMES

To obtain valid estimates of turbulent variances and covariances, the chosen averaging time has to be long enough to enable measurement of the low frequency contributions by the variables of interest. The recommendations of Wyngaard (1973) are outlined in Section 2. In practice the averaging time should be chosen so that the low frequency energy cut-off is negligible.

Converting the recommended averaging time for w and T to a non-dimensional frequency, by using a typical wind speed of $3 \, \text{m s}^{-1}$ and a height (z') of 20 m, the frequency cut-off is $f_c = 0.006$. Inspecting the low frequency end of the vertical wind (Figure 1) and temperature (Figure 3) spectra shows that about 20 min (or $f_c = 0.006$) might be sufficient to obtain a reasonable estimate for the w variance, but the temperature spectrum still shows a considerable amount of energy at non-dimensional frequencies below 0.006. The low frequency shape of the u spectrum (Figure 2) is similar to that of temperature, suggesting that a longer averaging time (about 60 min; $f_c = 0.002$) would be more appropriate for both. However, the specification of a low frequency cut-off for these two spectra is made difficult since they tend to develop a low frequency peak related to

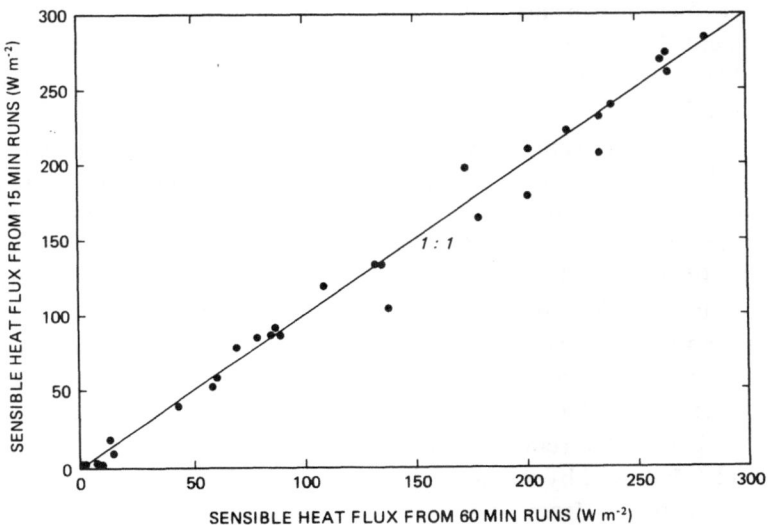

Fig. 6. Comparison of sensible heat fluxes averaged over 60 and 15 min, respectively.

mesoscale phenomena under certain stability conditions. Hence, it is necessary to decide which frequency limits can be regarded as being due to truly microscale turbulence.

The cospectra of heat (Figure 4) and momentum (Figure 5) both show a relatively sharp roll-off on the low frequency side. For the momentum cospectra, an averaging time of between 30 and 40 min ($f_c = 0.004$ and 0.003) might be sufficient, particularly when considering the large statistical variation in the cospectral densities at that end. For the cospectra of heat flux, Wyngaard's suggested averaging time of 56 min seems long enough to include all low frequency contributions. Indeed, the sharp roll-off in Figure 4 suggests that an even shorter averaging time in the order of 30 min or less ($f_c = 0.004$) could be justified here.

To substantiate the fact that for the sensible heat flux an averaging time of less than 60 min is sufficient, the heat flux covariances computed over 60 min were compared with those from within the same record but averaged over only 15 min ($f_c = 0.007$), i.e., four 15 min periods were added and averaged and then compared with the flux estimate of the corresponding 60 min record. According to McBean (1972), a cut-off at $f_c = 0.007$ might result in a flux underestimation of about 16%. However, as seen in Figure 6, there is no significant under- or overestimation over a large range of heat fluxes from about 5 to 300 W m^{-2}.

4. Summary and Conclusions

The results from this study add to the sparse number of observations of turbulence over rough urban surfaces and provide some insight regarding the nature of the turbulent fluxes. The selected runs (about 70% of the entire data set) are found capable of providing some general conclusions and a basis for comparison with other studies.

Considering that the sensors were operated at a height of only about 20 m above zero-plane displacement, which is below the recommended height for this site (between about 32 and 50 m), the results show remarkably good agreement with reference data and with studies from other urban turbulence programmes.

More specifically, on the high frequency side, in the inertial sub-layer, the spectra of w and T follow the theoretically required $-2/3$ slope very closely. No conclusions can be made concerning the high frequency shape of the u spectra, since the signals were corrected for inadequate frequency response by the propellor system on the basis of adjusting the roll-off to a $-2/3$ slope. In the case of the wT cospectra, the required $-4/3$ slope in the inertial subrange is matched very well and is followed by an increasingly steeper roll-off which is in agreement with theory and results from smoother surfaces. Because of the noise problem described earlier, the uw cospectral densities at the high frequency end cannot be analysed. Compared to the momentum cospectra, the inertial sub-range of the heat flux cospectra start at higher non-dimensional frequencies (> 2 vs. about 1).

On the low frequency side, the spectra from this study show the same behaviour as the observations from smoother surfaces. The w spectra exhibit a rapid roll-off whereas the spectra of u and T are characterized by considerable scatter and only drop slightly with decreasing frequency. Large scatter and frequent sign reversal are noticeable at the low frequency end of the two cospectra investigated. Compared to the reference cospectra from smoother sites, the cospectra from this study seem to roll-off slightly faster.

The analysis of the peak frequency, as an indication of the dominant eddy scale involved in the energy transfer, deserves special attention. Comparing the f_m values from this study with the reference data, no differences can be observed. The peak frequencies of all components either coincide with the reference value or are within the range given. The agreement with other urban studies is also good. The only difference is in the temperature spectrum, where Coppin (1979) reports a slightly higher value than observed here.

Minor deviations from the reference spectra observed over smoother surfaces include the flatter appearance of the peak regions in the w and u spectra, the slightly faster roll-off at the low frequency end of the cospectra and a relatively flat region in the momentum cospectra between the peak and the inertial subrange. However, more careful investigation is needed before it can be concluded that these differences are true urban anomalies.

Regarding the objectives of this study, the following conclusions can be drawn:

- The Sunset site, which has been used in several previous micrometeorological studies, seems suitable for turbulence and energy-balance measurements representing suburban terrain.
- To obtain useful estimates of variances and covariances for this site using a z' of about 20 m, the averaging-time recommendations suggested by Wyngaard (1973) seem to be appropriate for the w component (20 min), but should be longer in the case of the u and T components (45 min) and can be relaxed for the heat and momentum fluxes (to 30 min or less).
- The fact that measurements over a rough surface, at a height that is less than is normally assumed to be required, exhibit more concurrence than difference with reference data is somewhat puzzling. It raises the possibilities that either the measurement height constraints over a rough urban surface can be relaxed or (assuming the sensors are placed in the roughness sublayer) that the effects of individual roughness elements is not strong enough to be observed in spectral analyses.

In addition it should be pointed out that this study provides the first cospectra of the fluxes of heat and momentum over a suburban surface that are based on an extended data set. It is also the first evidence that heat flux cospectra over a suburban surface agree well with observations from smoother surfaces.

Acknowledgements

This research was supported by funds from the Natural Science and Engineering Research Council of Canada and the Atmospheric Environment Service of Environment Canada. The field site was made available by B.C. Hydro and Power Authority.

References

Clarke, J. F., Ching, J. K. S., and Godowitch, J. M.: 1982, *An Experimental Study of Turbulence in an Urban Environment*, Tech. Report. U.S. E.P.A., Research Triangle Park, N.C., NMS PB 226085.

Coppin, P. A.: 1979, *Turbulent Fluxes over a Uniform Urban Surface*, Res. Report No. 31, Inst. Atmos. Marine Sci., Flinders Univ., Adelaide.

Garratt, J. R.: 1978, 'Transfer Characteristics for a Heterogeneous Surface of Large Aerodynamic Roughness', *Q. J. R. Meteorol. Soc.* **104**, 491–502.

Garratt, J. R.: 1980, 'Surface Influence upon Vertical Profiles in the Atmospheric Near-Surface Layer', *Q. J. R. Meteorol. Soc.* **106**, 803–819.

Højstrup, J.: 1981, 'A Simple Model for the Adjustment of Velocity Spectra in Unstable Conditions Downstream of an Abrupt Change in Roughness and Heat Flux', *Boundary-Layer Meteorol.* **21**, 341–356.

Kaimal, J. C.: 1978, 'Horizontal Velocity Spectra in an Unstable Surface Layer', *J. Atmos. Sci.* **35**, 18–24.

Kaimal, J. C., Wyngaard, J. C., Izumi, Y., and Coté, O. R.: 1972, 'Spectral Characteristics of Surface-Layer Turbulence', *Q. J. R. Meteorol. Soc.* **98**, 563–589.

Kalanda, B. D., Oke, T. R., and Spittlehouse, D. L.: 1980, 'Suburban Energy Balance Estimates for Vancouver, B.C. Using the Bowen Ratio-Energy Balance Approach', *J. Appl. Meteorol.* **19**, 791–802.

Lettau, H. H.: 1969, 'Note on Aerodynamic Roughness-Parameter Estimation on the Basis of Roughness-Element Description', *J. Appl. Meteorol.* **8**, 828–832.

McBean, G. A.: 1972, 'Instrument Requirements for Eddy Correlation Measurements', *J. Appl. Meteorol.* **11**, 1078–1084.

Oke, T. R.: 1984, 'Methods in Urban Climatology', in *Applied Climatology, 25th International Geographical Congress*, Symposium No. 18, Zürich, 1984.

Oke, T. R., Cleugh, H. A., Grimmond, S., Schmid, H. P., and Roth, M.: 1989, 'Evaluation of Spatially-Averaged Fluxes of Heat, Mass and Momentum in the Urban Boundary Layer', Proc. Symp. Topoclimatology (A. P. Sturman, ed.), Christchurch, N.Z. (in prep.).

Pond, S., and Large, W. G.: 1978, 'A System for Remote Measurements of Air-Sea Fluxes of Momentum, Heat and Moisture During Moderate to Strong Winds', Manuscript Rep. No. 32, Inst. Oceanogr., The University of British Columbia, 55 pp.

Raupach, M. R.: 1979, 'Anomalies in Flux-Gradient Relationships over Forest', *Boundary-Layer Meteorol.* **16**, 467–486.

Raupach, M. R., Thom, A. S., and Edwards, I.: 1980, 'A Wind-Tunnel Study of Turbulent Flow Close to Regularly Arrayed Rough Surfaces', *Boundary-Layer Meteorol.* **18**, 373–397.

Schmid, H. P.: 1988, 'Spatial Scales of Sensible Heat Flux Variability. Representativeness of Flux Measurements and Surface Layer Structure over Suburban Terrain', Ph.D. Thesis, The University of British Columbia, Vancouver.

Steyn, D. G.: 1980, 'Turbulence, Diffusion and the Daytime Mixed Layer Depth over a Coastal City', Ph.D. Thesis, The University of British Columbia, Vancouver.

Steyn, D. G.: 1982, 'Turbulence in an Unstable Surface Layer over Suburban Terrain', *Boundary-Layer Meteorol.* **22**, 183–191.

Tennekes, H.: 1973, 'The Logarithmic Wind Profile', *J. Atmos. Sci.* **30**, 234–238.

Thom, A. S., Stewart, J. B., Oliver, H. R., and Gash, J. H. C.: 1975, 'Comparison of Aerodynamic and Energy Budget Estimates of Fluxes over a Pine Forest', *Q. J. R. Meteorol. Soc.* **101**, 93–105.

Wyngaard, J. C.: 1973, 'On Surface-Layer Turbulence', in D. A. Haugen (ed.), *Workshop on Micrometeorology*, Amer. Meteorol. Soc., Boston, pp. 101–149.

A MICROMETEOROLOGICAL INVESTIGATION OF SURFACE EXCHANGE OF O₃, SO₂ AND NO₂: A CASE STUDY

B. B. HICKS, D. R. MATT and R. T. McMILLEN

NOAA, Atmospheric Turbulence and Diffusion Division, P.O. Box 2456, Oak Ridge, TN 37831, U.S.A.

(Received in final form 2 November, 1988)

Abstract. Data obtained in an intensive field study of the dry deposition of sulfur dioxide, ozone, and nitrogen dioxide, conducted in 1985 in central Pennsylvania, are used to illustrate the factors that must be considered to assure that high quality results are derived. In particular, the quality of the site must be such that flux measurements made above the surface are representative of surface values. For this purpose, tests involving momentum transfer and the surface energy budget are especially useful. In addition, conditions must not be changing rapidly, and the statistical uncertainty associated with flux measurement must be low. For the set of data presented here, conservative quality-assurance guidelines are used to reject potentially erroneous flux data. For ozone, most of the measured fluxes are of use in deriving surface resistances. For SO₂, far fewer data points are available. For NO₂, fluxes appear to lack the order of the O₃ and SO₂ fluxes, and do not enable surface resistances to be computed. The highest-quality SO₂ and O₃ data yield surface resistances in fair average agreement with model predictions for SO₂, but substantially higher than predictions for O₃.

Nomenclature

C	Pollutant concentration in air ($kg\,m^{-3}$)
C_d	Drag coefficient
C_p	Specific heat of air at constant pressure ($cal\,g^{-1}\,°C^{-1}$)
e	Vapor pressure (kPa)
E	Evaporation rate ($kg\,m^{-2}\,s^{-1}$)
F	Pollutant flux ($kg\,m^{-2}\,s^{-1}$)
G	Ground heat transfer ($W\,m^{-2}$)
g	Acceleration due to gravity ($m\,s^{-2}$)
H	Sensible heat flux ($W\,m^{-2}$)
k	von Karman constant ($=0.4$)
L	Latent heat of vaporization of water ($cal\,kg^{-1}$)
L_O	Monin-Obukhov length scale (m)
R_a	Aerodynamic resistance ($s\,m^{-1}$)
R_b	Quasi-laminar boundary-layer resistance ($s\,m^{-1}$)
R_c	Canopy resistance ($s\,m^{-1}$)
R_n	Net radiation ($W\,m^{-2}$)
R_T	Total resistance ($s\,m^{-1}$)
T	Temperature (°C)
t	Time (s)
u	Wind speed ($m\,s^{-1}$)
u_*	Friction velocity ($m\,s^{-1}$)
V_d	Deposition velocity ($m\,s^{-1}$)
w	Vertical velocity component ($m\,s^{-1}$)
z	Height (m)
z_0	Roughness length (m)
θ	Wind direction (deg)
ρ	Air density ($kg\,m^{-3}$)

Boundary-Layer Meteorology **47**: 321–336, 1989.
© 1989 *Kluwer Academic Publishers.*

τ Reynolds stress (kg m^{-1} s^{-2})
χ Pollutant mixing ratio (kg kg^{-1})
ψ_C Stability correction function: concentration
ψ_H Stability correction function: heat
ψ_M Stability correction function: momentum

Following micrometeorological convention, overbars indicate time averages, and primes denote departures from these averages.

1. Introduction

Recent improvements in computer-based data acquisition systems and real-time analytical capabilities have provided an opportunity for more intensive data collection than has previously been possible in micrometeorological studies. At the same time, interest in such studies has tended to shift from the measurement of atmosphere-surface exchanges of various meteorological properties (heat, moisture and momentum) to the development of methods to compute these fluxes in models, especially for the cases of trace gas and particle transfer (see Hales *et al.*, 1987).

A key consideration in all micrometeorological flux-measuring programs is the need to instill confidence that flux determinations made at some height above the surface are indeed representative of the surface itself. Development work and exploratory field studies conducted in the 1950's and 1960's generated a sense of security in the following ways.

(a) In some circumstances, in-air measurements of water vapor flux can be referenced against absolute values of evaporation rate measured by weighing lysimeters.

(b) In some circumstances, in-air measurements of momentum fluxes can be referenced against absolute measurements of surface frictional drag, using drag plates.

(c) In many cases, the surface heat budget imposes an absolute reference for the measurement of the total turbulent heat exchange: $H + LE = R_n - G$.

(d) In all cases, in-air determination of fluxes at height z can only be considered representative of the surrounding surface if the constraints of horizontal homogeneity and time stationarity are satisfied. In both instances, it is required that the flux divergence associated with spatial and temporal changes is small in comparison with the surface flux itself. That is,

$$|\rho \, uz(\partial\chi/\partial x)| \ll |F| \tag{1}$$

and

$$|\rho \, z(\partial\chi/\partial t)| \ll |F| \tag{2}$$

where χ is mixing ratio of the atmospheric property of interest and F is the flux (computed as the covariance with vertical velocity, w: $F = \overline{\rho w' X'}$) of the property of interest. The considerations represented by Equations (1) and (2) lead directly to the familiar micrometeorological specifications for a good site (kilometers of horizontal uniformity) and for stationary conditions (no rapid changes of relevant properties such as wind speed, temperature, or concentration).

As interest in the surface exchange of trace gases has grown, the need has arisen to focus experimental attention on areas where the relevant chemical measurements are feasible, frequently at the expense of the conventional micrometeorological requirements for good fetch and stationarity. The consequences are not always clear; conditions may sometimes be such that surface values of fluxes may be impossible to compute even though measurements at height z are obtained with demonstrably acceptable precision.

Here, the many considerations that arise are explored, using a data set obtained in a field study of trace gas fluxes to crops as an example. The data in question were obtained during a 1985 field study in central Pennsylvania. The purpose is to provide an example of the steps that need to be considered in the interpretation of micrometeorological data involving trace chemical information. The complete data set is available elsewhere (Hicks *et al.*, 1988a).

2. Site Description

The experiment described here was conducted at the Pennsylvania State University Agricultural Research Farm at Rock Springs, PA, about 13 km southwest of State College. The site is described by Duan *et al.* (1988). The topography is not simple; it is in a valley between two of the characteristic parallel ridges of the area, and is planted with a variety of crops on an annual rotational basis. The most uniform upwind surface is for winds from about 230–280 deg, for which the wind blows along the valley floor parallel to the neighboring ridges.

The site was not uniformly vegetated at the time of the experiment considered here (Figure 1). The instruments were located such that the upwind fetch could be dominated either by soybeans or maize, depending on wind direction. The distribution of crops was such that no data representative of only soybeans or only maize could have been obtained. Instead, the fetch ranged from mostly soybeans to mostly maize, with a wide variation in between.

3. Analytical Goals

The goal of the study was to improve the formulation of processes controlling dry deposition using variables that can be measured routinely, or predicted in models. The multiple-resistance analog is used as guidance (see O'Dell *et al.*, 1977;

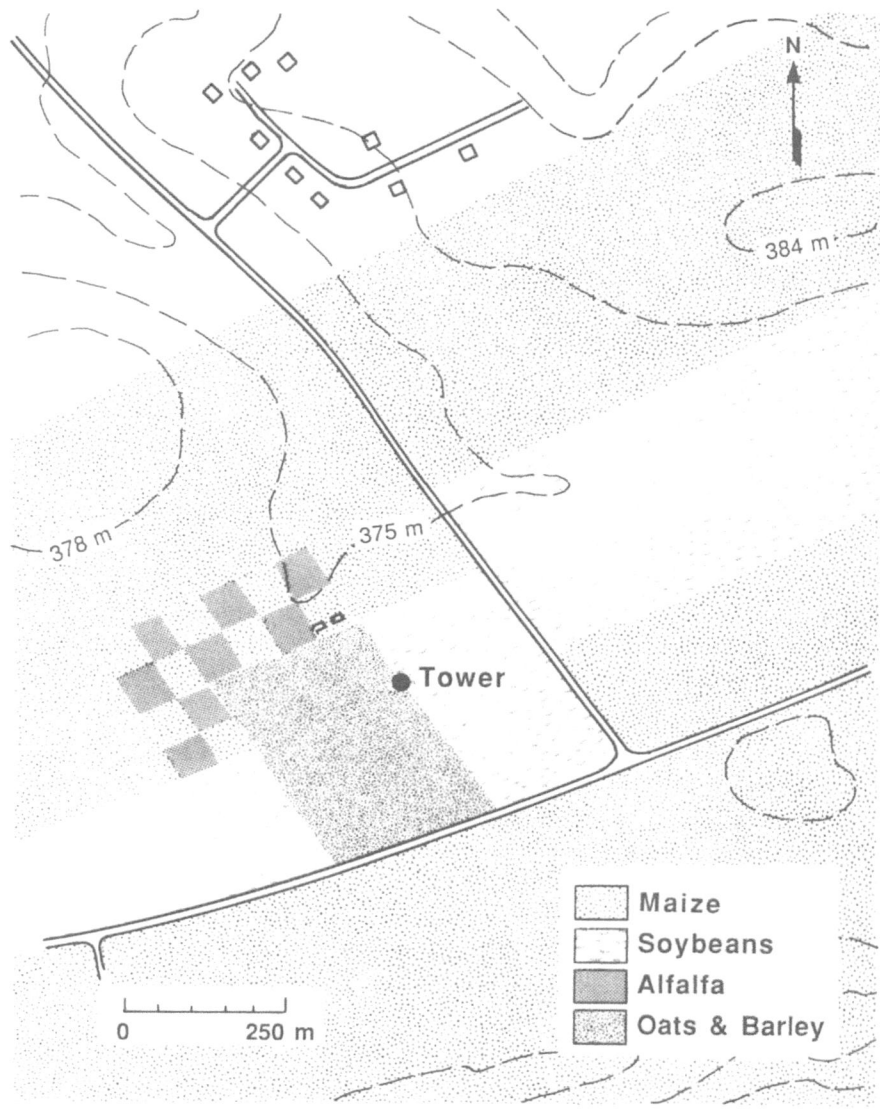

Fig. 1. Topological relief map of the immediate area with plant cover indicated.

Meyers, 1987). In this conceptual framework, resistances to transfer are identified with each contributing process. These individual resistances are then assembled into an overall resistance model, according to the way in which the various processes interact. The validity of the integrated model is then tested by direct comparison against field observations of the net resistance to transfer. There are thus two different objectives of experimental programs (such as that described here) addressing questions of model development:

(a) To provide and test formulations of individual processes contributing to the exchange;

(b) To test the validity of comprehensive models constructed from these process formulations.

High-quality observations are required of trace gas and particle fluxes. Hence, a further supporting objective needs to be taken into account:

(c) To quantify uncertainties associated with flux measurements and with the consequent process formulation.

4. Theoretical Background

Existing knowledge is largely limited to circumstances in which the surface flux through the air is the consequence of vertical turbulent exchange. The standard approach is based on the multiple-resistance model of agricultural meteorology (see O'Dell *et al.*, 1977; Baldocchi *et al.*, 1987; Baldocchi, 1987; Hicks and Matt, 1988). In this model, turbulent exchange is represented by an aerodynamic resistance R_a, transfer across the quasi-laminar layer in contact with the surface is expressed in terms of a boundary-layer resistance R_b, and the various surface processes contributing to the capture and uptake of materials are combined into a surface (or canopy) resistance R_c. Then, the usual expression relating the flux F to concentration C via a deposition velocity V_d,

$$F = V_d \cdot C ,\tag{3}$$

is generalized to yield

$$F = C/(R_T)$$

$$= C/(R_a + R_b + R_c) .\tag{4}$$

A basic intention of field programs is therefore to relate R_a, R_b, and R_c to underlying variables capable of direct measurement. The key question concerns the canopy resistance R_c, since this is influenced by many chemical, biological, and physical mechanisms associated with the nature of the surface itself, and which are still relatively poorly understood.

The quantity R_c can be investigated in two distinctly different ways, either by intensive studies of individual surfaces, and then synthesizing canopy behavior by assembling the individual formulations of component resistances into a comprehensive multiple-resistance model, or by solving for R_c in Equation (4), given field measurements of F, C, R_a and R_b. The first approach constitutes a goal that has yet to be attained for a vegetated canopy. The second approach presents difficulties, but has yielded success in some instances. The following discussion is intended to present a step-by-step explanation of the philosophies involved in such an analysis. The sequence of questions that must be answered in order to generate a valid set of values for R_c is as follows:

(a) Is the site suitable?

Response: • Study surface roughness as a function of wind direction.
• Study energy balance as a function of wind direction.
• Select wind direction "windows", for acceptable data.

(b) Are atmospheric conditions adequate?

Response: • Identify cases violating the stationarity constraint, and exclude offenders or apply appropriate corrections.

(c) What are R_a and R_b for each case?

Response: • Apply site "calibrations": derived from micrometeorological results to quantify both terms.

(d) What are R_c values?

Response: • Solve for R_c using Equation (4).

(e) What are the uncertainties associated with R_c determinations?

Response: • Analyze for error propagation, considering statistical uncertainty in F, errors in C, etc.

5. The Raw Data

Table I shows an example of raw data derived from a field study. The variables reported are as follows:

Identification variables:

Julian date of measurement; hour and minute of end of 30-min run; number of individual measurements used to construct averages, covariances, etc.

Ambient meteorological variables (with dimensions and units as in the table of nomenclature):

Wind speed, wind direction, temperature, vapor pressure, sensible heat flux, latent heat flux, and Reynolds stress.

Flux variables:

Mean concentration; covariance; standard error on the covariance.

These variables are given for each chemical species: SO_2, O_3, and NO_2.

6. Momentum Fluxes, Drag Coefficients, Roughness Lengths, and Heat Balance

Inspection of the data reveals occasions for which the covariance $\overline{u'w'}$ is positive, counter to expectations. Such occasions cannot be rejected unilaterally from analysis, since to do so would bias the results. Several considerations lead to the

TABLE I

Raw data obtained on Julian day No. 201

Hr	Min	Num	u	θ	T	e	H	LE	$\overline{w'u'}$	SO₂ [C]	$\overline{w'C'}$	Flux error
4	0	17465	0.2	107	16.1		−2.8		0.00	16.1	−0.020	0.001
4	30	17475	0.2	121	15.9		−1.4		0.00	12.2	0.001	0.001
5	0	17457	0.2	348	16.2		−9.5		0.01	10.8	−0.048	0.002
5	30	17476	0.9	236	15.7		−0.2		0.00	7.0	−0.011	0.002
6	0	17473	0.3	286	15.6		−2.4		0.00	7.7	0.000	0.001
6	30	17474	0.2	232	15.6		2.3		0.00	4.5	0.007	0.002
7	0	17467	1.0	254	15.8		1.3		−0.00	3.9	−0.005	0.001
7	30	17467	1.1	247	16.8		3.8		−0.01	4.8	−0.024	0.002
8	0	17474	1.6	257	18.0		6.8		−0.02	8.1	−0.056	0.004
9	0	10537	2.3	253	19.6	8.2	0.8	1.2	−0.01	12.6	0.004	0.004
10	0	8345	2.4	251	20.9	8.5	4.6	1.2	−0.01	8.9	−0.001	0.005
10	30	17462	2.1	251	21.0	8.5	44.2	14.9	−0.05	8.3	−0.042	0.006
11	0	17472	2.4	250	21.0	8.7	34.9	11.5	−0.05	8.6	−0.037	0.005
11	30	11248	2.9	251	20.7	9.0	27.0	8.5	−0.10	35.9	−0.109	0.016
12	0	17465	3.4	246	21.1	8.9	32.8	9.5	−0.11	35.6	−0.079	0.013
12	30	17461	3.2	254	21.9	8.8	41.5	16.7	−0.10	32.2	−0.075	0.014
13	0	17461	3.2	284	22.3	8.5	45.4	14.8	−0.11	30.2	−0.080	0.014
13	30	16590	3.1	281	22.4	8.5	23.5	16.8	−0.09	5.2	−0.040	0.006
14	0	17460	3.0	259	22.2	8.7	17.8	9.4	−0.08	4.4	−0.029	0.006
14	30	17463	2.9	277	22.5	8.8	23.7	10.0	−0.10	4.3	−0.037	0.006
15	0	14985	3.7	279	22.5	8.9	18.9	7.8	−0.14	4.6	−0.029	0.007
15	30	17459	4.0	277	22.7	9.0	15.5	11.6	−0.14	7.1	−0.049	0.008
16	0	17465	4.0	279	23.0	9.0	17.9	9.4	−0.13	10.6	−0.083	0.009
16	30	17467	3.9	294	22.8	9.0	10.2	7.8	−0.14	8.1	−0.034	0.010
17	0	17457	3.9	277	22.6	9.3	13.2	7.1	−0.11	8.4	−0.055	0.008
17	30	17467	3.8	272	22.4	9.3	12.3	4.7	−0.12	7.3	−0.038	0.008
18	0	17462	3.3	274	22.0	9.4	−2.4	5.3	−0.07	8.1	−0.034	0.007
18	30	17469	2.5	261	21.4	9.4	−2.8	4.3	−0.04	7.3	−0.024	0.004
19	0	17460	1.7	234	20.8	9.4	−0.1	4.2	−0.01	7.2	−0.013	0.003
19	30	17458	1.7	228	20.4	9.4	−1.9	3.4	−0.02	9.0	−0.018	0.004
20	0	17459	2.3	252	20.0	9.4	−3.9	3.0	−0.03	11.8	−0.025	0.005
20	30	17460	1.1	253	18.5	9.6	−0.6	2.1	−0.01	10.7	−0.013	0.002
22	0	17470	0.3	236	16.1		0.8		0.00	12.1	0.003	0.001
22	30	17479	0.4	258	16.2		0.6		0.00	13.5	0.009	0.001
23	0	17471	0.4	290	15.8		−4.0		0.00	11.9	−0.007	0.001
23	30	17469	0.3	131	15.4		1.8		0.00	10.9	0.003	0.000

exclusion of large amounts of data, however. First, the anemometry will not resolve Reynolds stresses well in light winds. Since the purpose here is to search for a most-acceptable direction sector for subsequent analysis (bearing in mind the characteristics of the site that favor fetches to the west and southwest), no loss in generality is incurred by excluding light-wind cases, for the present.

Figure 2 is a plot of drag coefficient as a function of wind direction, in which all data are plotted except for ten values for which $|C_d| > 1.0$. The data are widely

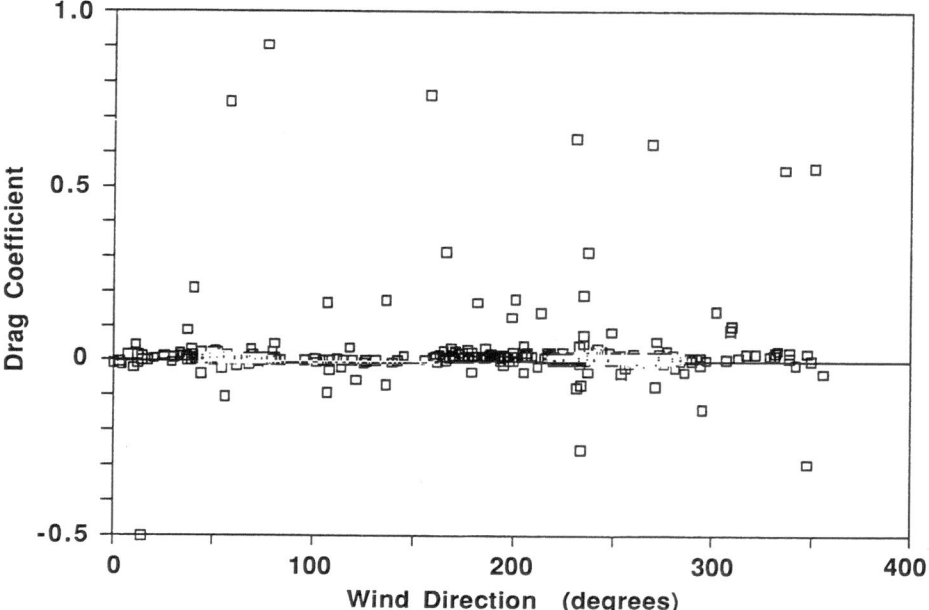

Fig. 2.   Drag coefficient as a function of wind direction for the 1985 study at State College, PA. All
data are plotted, except for a few outliers yielding $|C_d| > 1$.

scattered. Figure 3(a) shows the improvement that results if data sets with
$u < 1$ m s^{-1} have been excluded.

To investigate surface roughness as a function of wind direction, atmospheric
stability must be taken into account. Stability is normally expressed in terms of
$z/L_0 \equiv (-\mathrm{kg\ Hz}/\rho c_p) u_*^3 T$. In practice, it is often adequate to consider near-
neutral occasions, defined on the basis of net radiation ($-1 < R_n < 10$ W m^{-2}, say)
or heat flux ($-0.5 < H < 5$ W m^{-2}, say). Figure 3(b) shows the near-neutral subset
of the data shown in Figure 3(a), selected on this basis.

Using near-neutral data, it is usually straightforward to produce a plot of
surface roughness length z_0 versus direction. The intent is to define a wind
direction sector in which z_0 does not vary; within this sector, subsequent analysis
can then be based on the momentum "calibration" represented by z_0. Figure 4 is
a plot of roughness length versus wind direction for the present data set. After
consideration of the surroundings of the experimental location, four acceptable
"windows" can be identified, as represented in the diagram. For two "windows"
(A and C), the fetch was over relatively short crops (soybeans and sorghum)
which can be characterized by a roughness length averaging about 7 cm. For
another two windows, fields of maize dominated the upwind fetch; a roughness
length of about 25 cm then appears appropriate. Fetch section C corresponds to

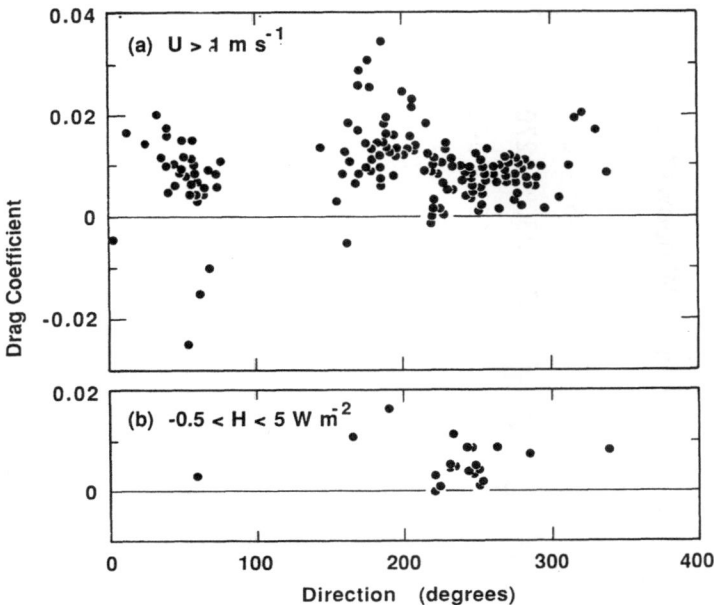

Fig. 3. (a) Drag coefficient versus wind direction, for wind speeds greater than 1 m/s. (b) Near-neutral drag coefficients versus wind direction.

the southwest-to-west sector identified above as the best fetch, based on visual observation.

As a further test, surface heat budget considerations can be applied. In essence, it is required that the measured sensible and latent heat fluxes differ from the net radiation by an amount which is explicable in terms of heat storage either by vegetation or the ground. This constraint presents a further step of site quality control. If the surface heat balance criterion is not satisfied, then it means that the site is affected by advection (or flux divergences) and that therefore fluxes measured by micrometeorological methods distant from the surface cannot be considered representative of surface values. It is obvious that the purpose of studies of the kind considered here is precisely to study exchange at the surface. Hence the surface energy budget test is an important step.

7. Resistance Analysis

Although measurements of the friction velocity (u_*) were made, and are available, the analysis that follows does not rely heavily on these direct measurements.

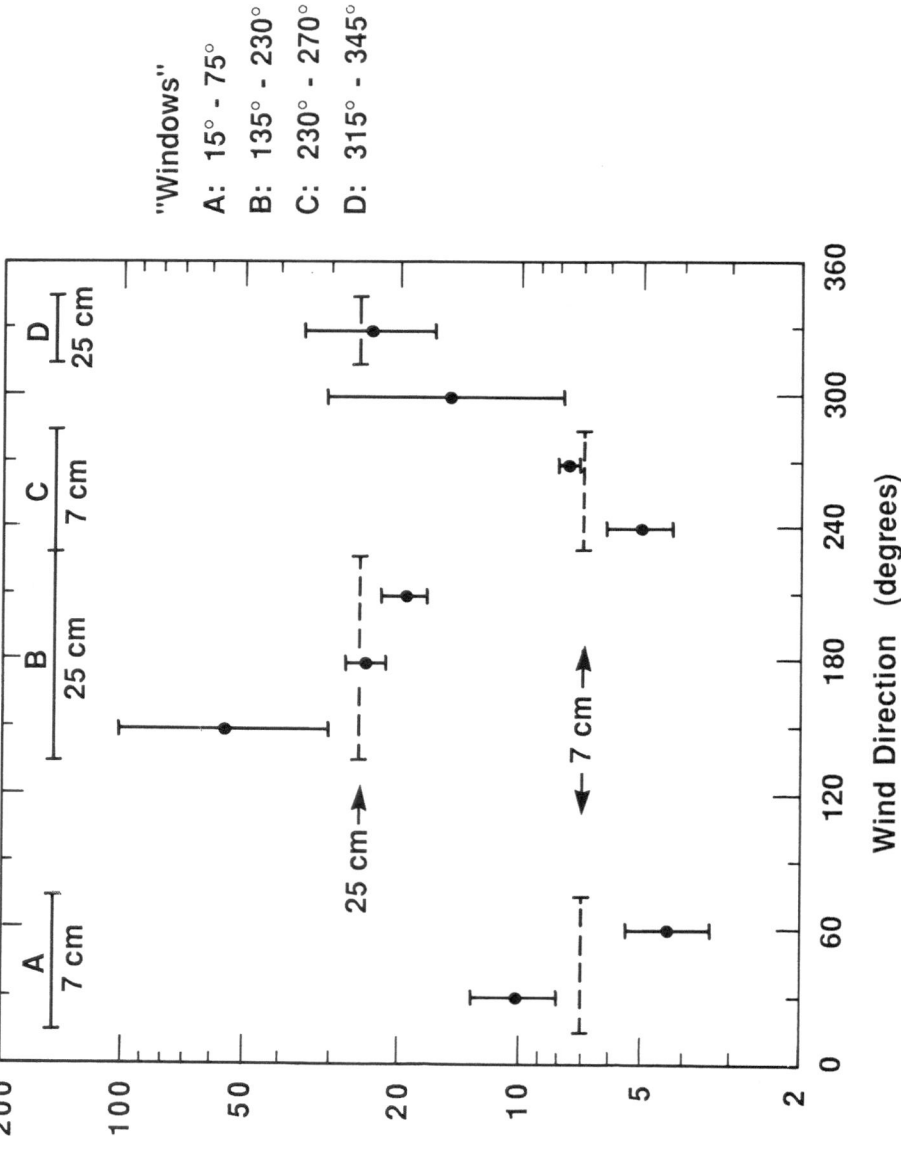

Fig. 4. Roughness length versus wind direction, showing the four "windows" selected for special attention.

Instead, smoothed values of u_* have been derived using the site calibrations illustrated in Figure 4, and these indirect measurements have been used to evaluate the resistance components R_a and R_b in the subsequent analysis. In this way, the unwelcome effects of propagating measurement errors in u_* and in other eddy flux quantities through the analysis are reduced.

The specific data selection steps that are taken for each sequence of chemical observations are as follows.

(a) For the entire time sequence, evaluate the rate of change of concentration with time and eliminate all data points for which the resulting flux divergence term imposes an uncertainty exceeding 0.1 cm/s in the corresponding deposition velocity. That is, eliminate all records for which

$$z(\partial C/\partial t)/C > 10^{-3} \text{ m/s}. \tag{5}$$

(b) Sort the remaining data by wind direction, and eliminate all runs with fetch outside the acceptable "windows" identified in Figure 4.

(c) Eliminate all runs for which $u < 1$ m/s, since these will yield flux data of doubtful quality.

(d) Eliminate all runs for which the statistical error associated with the computation of the average covariance exceeds 20% of the covariance itself. For this purpose, the standard error on the average covariance is computed directly from the raw data as the covariance is computed, in real-time.†

An example of an acceptable data set is given in Table II, for the case of SO_2. The body of available information has been reduced substantially, from an initial set of more than 350 data points to only 48 that satisfy all constraints.

A common sequence of analytical steps is employed for each of the chemical species under scrutiny (in this case, O_3, SO_2, and NO_2). It is first necessary to evaluate the aerodynamic resistance terms R_a and R_b. For this purpose, a smoothed value of the friction velocity is required. This is derived using the recorded wind speeds in conjunction with heat flux covariances, assuming the roughness lengths appropriate for the fetch "window" in question. Standard flux-gradient relations are used to quantify the integrated stability correction function ψ_m, for use in the relation

$$u_* = ku/[\ln(z/z_0) - \psi_m]. \tag{6}$$

The value 0.41 is assumed for the von Karman constant k, and ψ_m is taken to be

$$\psi_m(u) = \exp(0.032 + 0.448(\ln(-z/L)) - 0.132(\ln(-z/L))^2) \tag{7}$$

$$\psi_m(s) = -5 \, z/L$$

---

† It should be noted that this quantity is different from the statistical error associated with the determination of the eddy flux due to turbulent variability. The present purpose is to test for sensor inadequacies, not to assess the certainty of determination of the geophysical flux (see Fairall, 1984).

TABLE II

Raw data obtained on several days

Day	Hr	Min	u	θ	H	SO_2 ppb	$\overline{w'c'}$	Error	z_0	Err/Covar	z/L	ψ_M	u_*	ψ_H	R_a	R_b	R_c
199	10	30	1	54	175.2	3.9	-0.0442	0.0053	7	-0.120	-1.138	1.09	0.12	1.91	53.6	42.0	-8.3
199	11	0	1.3	68	130.4	4.3	-0.0439	0.0058	7	-0.132	-0.521	0.73	0.14	1.36	55.2	35.7	5.9
199	15	0	1.4	62	76.2	0.7	-0.0226	0.0045	7	-0.199	-0.303	0.50	0.14	1.00	61.3	35.6	-64.1
200	11	0	1.7	266	103.8	13.9	-0.1082	0.0071	7	-0.066	-0.246	0.43	0.17	0.88	53.4	30.0	44.7
199	18	30	1.1	248	19.5	1.8	-0.112	0.0022	7	-0.196	-0.196	0.35	0.10	0.76	89.3	48.6	19.7
200	11	30	2.1	265	118.6	21.4	-0.1219	0.0117	7	-0.096	-0.162	0.30	0.20	0.66	47.1	25.0	103.3
199	18	0	1.1	220	38.7	1.7	-0.017	0.0032	7	-0.188	-0.131	0.24	0.15	0.57	44.3	33.8	20.4
201	10	30	2.1	251	44.2	8.3	-0.0421	0.0057	7	-0.135	-0.036	0.11	0.20	0.31	52.1	25.3	119.6
200	13	0	2.7	255	84.4	23.3	-0.1508	0.014	7	-0.093	-0.063	0.11	0.25	0.31	41.9	20.4	92.4
202	9	30	2.0	216	95.1	23.0	-0.223	0.0144	25	-0.065	-0.060	0.10	0.26	0.30	27.9	19.3	56.0
203	11	30	2.6	264	77.3	3.4	-0.0503	0.0064	7	-0.127	-0.062	0.11	0.24	0.30	43.0	20.8	3.1
200	13	30	2.9	255	87.5	25.4	-0.1971	0.0152	7	-0.077	-0.048	0.08	0.27	0.24	38.6	18.5	71.8
200	12	30	2.9	255	74.2	23.2	-0.1082	0.0101	7	-0.093	-0.044	0.07	0.26	0.23	39.8	19.0	155.4
201	7	30	1.1	247	3.8	4.8	-0.0241	0.0023	7	-0.095	-0.042	0.17	0.10	0.21	105.2	50.0	42.1
201	11	0	2.4	250	34.9	8.6	-0.0367	0.0054	7	-0.147	-0.035	0.05	0.22	0.18	47.7	22.5	164.0
203	12	0	3.0	255	54.8	3.9	-0.0358	0.0064	7	-0.179	-0.030	0.04	0.27	0.16	39.5	18.5	51.7
202	11	30	2.5	217	93.0	29.1	-0.1674	0.0229	25	-0.137	-0.030	0.04	0.32	0.15	23.4	15.4	135.1
203	12	30	3.1	266	58.1	4.5	-0.047	0.0067	7	-0.143	-0.029	0.04	0.28	0.15	38.3	17.9	40.2

where $\psi_m(u)$ is for unstable and $\psi_m(s)$ is for stable stratification (see Kanemasu *et al.*, 1979).

The aerodynamic resistance is then computed as

$$R_a = (1/ku_*) \cdot [\ln(z/z_0) - \psi_c] \tag{8}$$

where ψ_c is the appropriate stability correction function corresponding to ψ_m for momentum. Available evidence indicates that $\psi_c \simeq \psi_H$ (for heat transfer). Note that R_a differs slightly from the quantity u/u_*^2, since atmospheric instability influences the transfer of heat (and trace species) differently from momentum.

For purposes of computing ψ_c, once stability z/L is known, an equation similar to (7) is used:

$$\psi_c(u) = \exp(0.589 + 0.39 \ln(-z/L) - 0.09(\ln(-z/L))^2)$$
$$\psi_c(s) = -5 \, z/L \, . \tag{9}$$

The quantity R_b is not well specified, but is not large in the case of the trace gases considered here. Having evaluated u_*, the value of R_b is derived as

$$R_b = 5/u_* \, , \tag{10}$$

which is an extension of results obtained previously for heat exchange (see Garratt and Hicks, 1973). In practice, R_b is best known and best understood as a property influencing thermal exchange. Its relevance to trace gas transfer (and especially for particles) is less well understood.

The quantity R_c is then derived as a residual, using Equation (4).

8. Results

Figures 5, 6, 7 illustrate the diurnal variability of R_c derived for O_3, SO_2, and NO_2, respectively. Averages of the experimental results are plotted with model predictions. Error bars indicate standard errors on the mean values. Inspection indicates that the O_3 results are the most well behaved, that the SO_2 data appear to have somewhat different general characteristics, and that the NO_2 data are completely dissimilar.

The lines in Figures 5 and 6 present predictions of R_c produced by an inferential model (see Hicks *et al.*, 1988), in which stomatal resistance, cuticular resistance, etc., are parameterized in terms of variables that can be easily measured. The variables used are those that influence photosynthesis: photosynthetically active radiation, air temperature, humidity, etc.

The ozone canopy resistances of Figure 5 display the overall characteristics expected on the basis of stomatal control. The average canopy resistance reaches a minimum value of about 400 s/m during the middle of the day, after dropping from relatively high overnight values soon after dawn. In the late afternoon, the

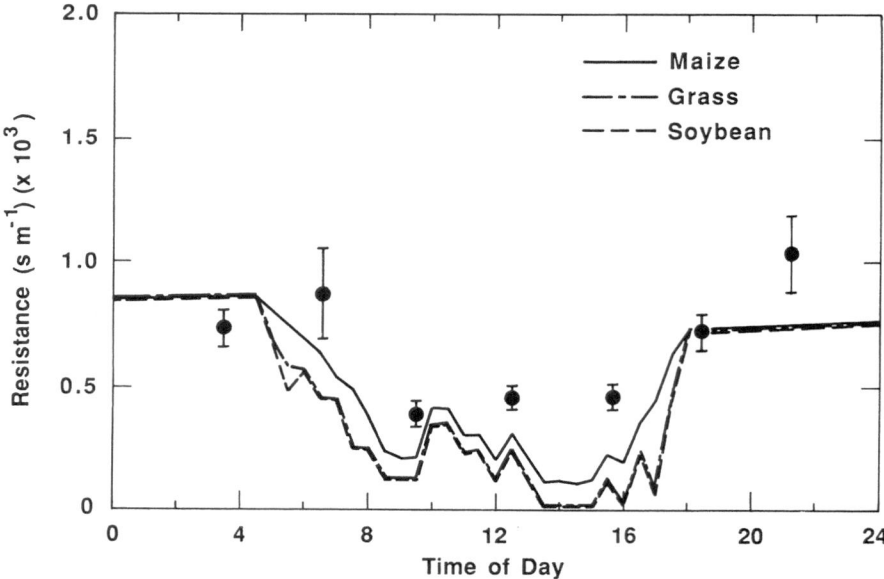

Fig. 5.   Predicted and measured canopy resistance versus time of day, for ozone.

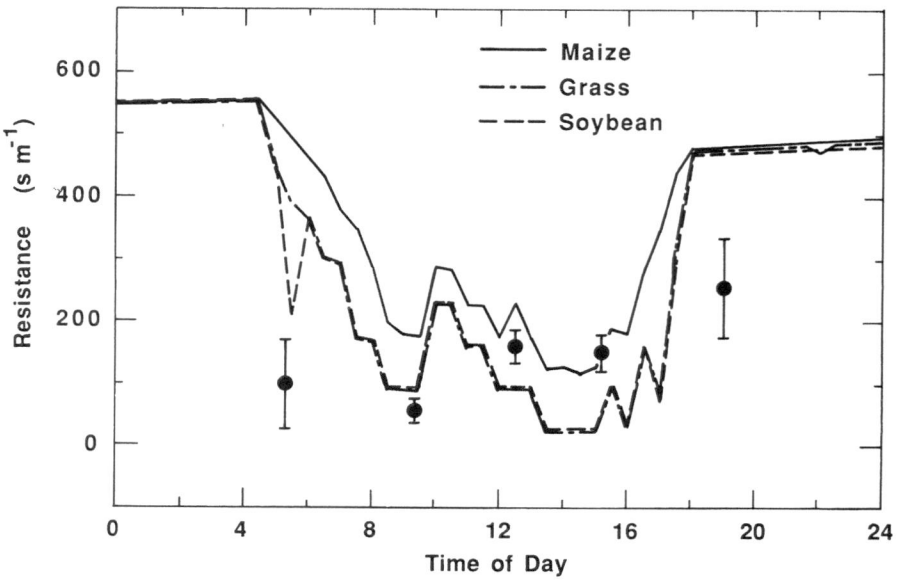

Fig. 6.   Predicted and measured canopy resistance versus time of day. for sulfur dioxide.

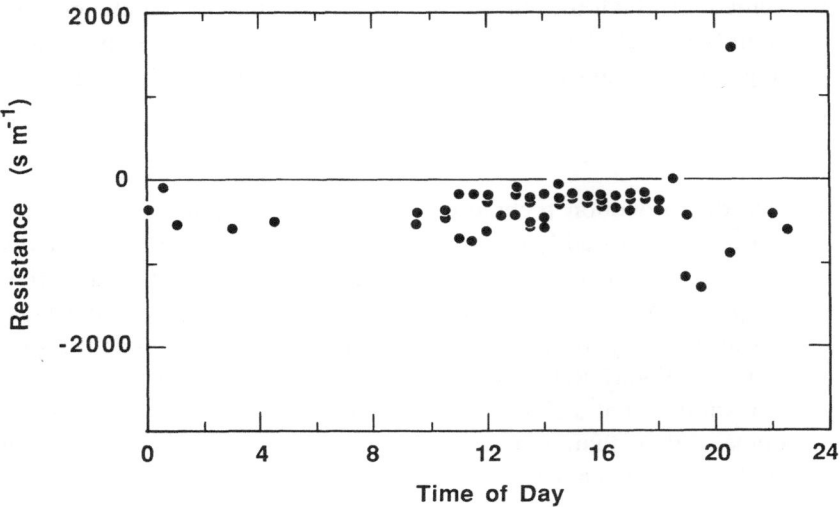

Fig. 7.   Measured canopy resistance versus time of day, for nitrogen dioxide.

resistances again climb, presumably because diminishing radiation results in gradual stomatal closure.

Comparison with the model results indicates some considerable disagreement, however. In particular, the measured daytime values of R_c are substantially more than are computed using the model. The cause for this disagreement remains to be identified.

Much of the same general behavior is evident in the SO_2 data of Figure 6, although in this case the near-dawn canopy resistance is very low. The possibility of a surface-wetness influence cannot be discounted; if dewfall were occurring, then an exceedingly low canopy resistance to SO_2 would be expected, with a correspondingly high resistance for O_3. Comparison between Figures 5 and 6 provides some support for this possibility.

In general, the canopy resistances for SO_2 in daytime are lower than for O_3. At the time of this experiment, the canopy was dry (any overnight dewfall evaporated quickly). For SO_2, the experimental data are in closer agreement with model predictions than for O_3. It is possible that the O_3 data represent true stomatal resistance, whereas the SO_2 results are also influenced by the con-sequences of uptake to leaf cuticle. If subsequent studies support the results of Figures 5 and 6, then an intriguing and important possibility arises: that some of the SO_2 transfer is to leaf cuticle rather than to internal leaf tissue via stomatal openings. Such behavior has been suggested elsewhere (Taylor *et al.*, 1988).

The continually negative resistances evident for NO_2 transfer as illustrated in Figure 8 are a consequence of the frequent finding of eddy fluxes directed upwards from the surface. This set of data is incompatible with the situation usually assumed, in which the surface is an effective sink for the chemical species in question and in which deposition to the surface dominates the atmosphere/surface exchange. This matter is discussed elsewhere (Hicks and Matt, 1988).

9. Conclusions

Conservative data-selection techniques result in the elimination of most of the original measurements of eddy fluxes of O_3, SO_2, and NO_2 fluxes at a field site in central Pennsylvania. The remaining data reveal the expected diurnal variation in surface canopy resistance, R_c, for O_3, with some evidence of a dewfall-retardation effect in the early morning. (At this time, SO_2 deposition appears to have been promoted.) Midday SO_2 canopy resistances are about 50% of O_3 values, raising the possibility that a significant fraction of SO_2 exchange may be directly to leaf cuticle rather than entirely through stomata. NO_2 fluxes are highly scattered and primarily such that fluxes are mostly away from the surface, not towards it.

References

Baldocchi, D. D., Hicks, B. B., and Camara, P.: 1987, 'A Canopy Stomatal Resistance Model for Gaseous Deposition to Vegetated Surfaces', *Atmos. Environ.* **21**, 91–101.

Baldocchi, D. D.: 1988, 'A Multi-Layer Resistance Analog Model for Estimating Sulphur Dioxide Deposition to a Deciduous Oak Forest Canopy', *Atmos. Environ.* **22**, 869–884.

Duan, B., Fairall C. W., and Thomson, D. W.: 1988, 'Eddy Correlation Measurements of the Dry Deposition of Particles in Wintertime', *J. Appl. Meteorol.* **27**, 642–652.

Fairall, C. W.: 1984, 'Interpretation of Eddy Correlation Measurements of Particulate Deposition and Aerosol Flux', *Atmos. Environ.* **18**, 1329–1337.

Garratt, J. R. and Hicks, B. B.: 1973, 'Momentum, Heat and Water Vapour Transfer to and from Natural and Artificial Surfaces', *Quart. J. Roy. Meteorol. Soc.* **99**, 680–687.

Hales, J. M., Hicks, B. B., and Miller, J. M.: 1987, 'The Role of Research Measurements Networks as Contributors to Federal Assessments of Acid Deposition', *Bull. Amer. Meteorol. Soc.* **68**, 216–225.

Hicks, B. B. and Matt, D. R.: 1988, 'Combining Biology, Chemistry, and Meteorology in Modelling and Measuring Dry Deposition', *J. Atmos. Chem.* **6**, 117–131.

Hicks, B. B., Matt, D. R., and McMillen, R. T.: 1988a, *A Micrometeorological Investigation of Surface Exchange of Trace Gases: A Case Study*, NOAA Technical Memorandum (in press).

Hicks, B. B., Baldocchi, D. D., Meyers, T. P., Matt, D. R., and Hosker, R. P.: 1988b, 'A Multiple Preliminary Resistance Routine for Deriving Dry Deposition Velocities from Measured Quantities', *Water, Air & Soil Pollut.* **36**, 311–330.

Kanemasu, E. T., Wesely, M. L., Hicks, B. B., and Heilman, J. L.: 1979, 'Techniques for Calculating Energy and Mass Fluxes', in B. J. Barfield and J. F. Gerber (eds.), *Modification of the Aerial Environment of Crops*, ASAE, St. Joseph, Michigan, pp. 156–182.

Meyers, T. P.: 1987, 'The Sensitivity of Modeled SO_2 Fluxes and Profiles to Stomatal and Boundary Layer Resistances', *Water, Air, and Soil Pollution* **35**, 261–278.

Meyers, T. P. and Yuen T. S.: 1987, 'An Assessment of Averaging Strategies Associated with Day/Night Sampling of Dry Deposition Fluxes of SO_2 and O_3', *J. Geophys. Res.* **92**, 6705–6712.

O'Dell, R. A., Taheri, M. and Kabel, R. L.: 1977, 'A Model for Uptake of Pollutants by Vegetation', *J. Air Pollut. Contr. Assoc.* **27**, 1104–1109.

Taylor, G. E., Hansen, P. J., and Baldocchi, D. D.: 1989, 'Pollutant Deposition to Individual Leaves and Plant Canopies: Sites of Regulation and Relationship to Injury', in W. W. Heck, D. T. Tingey and O. C. Taylor (eds.), *Assessment of Crop Loss from Air Pollutants*, Elsevier Press, in press.

EFFECT OF FINITE SAMPLING ON ATMOSPHERIC SPECTRA

J. C. KAIMAL, S. F. CLIFFORD and R. J. LATAITIS

Wave Propagation Laboratory/NOAA/ERL, 325 Broadway, Boulder, Colorado 80303, U.S.A.

(Received in final form 2 November, 1988)

Abstract. The effect of a finite averaging time on variances is well known, but its effect on power spectra is less clearly understood. We present numerical solutions for the spectral distortion arising from sampling over a finite time interval T and show that the commonly used filter function $(1 - \text{sinc}^2 \pi f T)$, valid for variances, is a reasonable approximation for power spectra only when $T \geq 10\tau_m$, where f is the cyclic frequency, and τ_m is the dominant time scale of the process. Our results exhibit an increasingly steeper low-frequency roll-off as T decreases relative to τ_m, indicating that the measured spectrum is subject to a greater suppression of the lower frequencies ($f < 1/T$) than predicted by $(1 - \text{sinc}^2 \pi f T)$. This suppression is, in a sense, compensated by an overestimation of spectral estimates in the frequency range $f \geq 1/T$.

1. Introduction

In studies of geophysical turbulence, one often deals with records collected over periods shorter than the scales of motion important to the study. This is particularly true in boundary-layer work where analysts have had to infer ensemble properties of the flow from data sets and statistics sampled over periods ranging from several minutes (diffusion experiments) to several hours (turbulence experiments). Convenience of processing and recording can also influence the choice of sampling duration, as at the Boulder Atmospheric Observatory (BAO) operated by the Wave Propagation Laboratory (Kaimal and Gaynor, 1983), where the averaging period for data archival is fixed at 20 min. The effect of imposing a finite sampling window on wind velocity statistics, especially the variance, has been examined by a number of investigators (Ogura, 1959; Lumley and Panofsky, 1964; Wyngaard, 1973; Pasquill and Smith, 1983; Panofsky, 1988; Wollenweber and Panofsky, 1989). Hans Panofsky, in his last years, became interested in this problem and worked on developing models for vertical wind and wind direction variance dependence on averaging interval. One of the authors (JCK) had long discussions with Hans on the types of data he needed from the BAO to verify his model. Wollenweber and Panofsky (1989 – this issue), described the results of their comparisons of model predictions with data from both the BAO and Vandenberg Air Force Base.

In this paper we examine the effects of finite sampling on power spectra and show that the simple filter function which correctly represents the effect of finite sampling on the variances (Pasquill and Smith, 1983) cannnot be applied to power spectra to represent energy lost at the low-frequency end. An exact numerical solution is presented and the condition for approach to the simple form is defined. These results would have been of great interest to Hans, whose creative insights and methods of analysis inspired a generation of boundary-layer meteorologists.

2. Effect of Finite Averaging on Variance

Adopting the notations used by Pasquill and Smith (1983), we define a variance $\sigma_{\infty,0}^2$ of a zero-mean random variable x extending over infinite time (Figure 1), sampled with essentially zero time resolution. We assume that this time series is subdivided into sequential samples of duration T and define a variance $\sigma_{T,0}^2$ for each sample. We also define the variance $\sigma_{\infty,T}^2$ of the mean values $\bar{x}$ of each sequential sample. If we denote the average over all samples by $\langle \ \rangle$, we have (Pasquill and Smith, 1983)

$$\langle \sigma_{T,0}^2 \rangle = \sigma_{\infty,0}^2 - \sigma_{\infty,T}^2 \, . \tag{1}$$

In other words, the mean square departure from the long-term mean is the average of the individual variances plus the variance of the individual means. This relationship has been used extensively to reconstruct long-term variances of atmospheric properties such as wind and temperature from short-term records.

It can be shown that the variance of $\bar{x}$ is simply the variance of $x(t)$, subjected to a rectangular running mean filter of width T:

$$\sigma_{\infty,T}^2 = \int_0^\infty [S(f)]_\infty \, \text{sinc}^2(\pi f T) \, df \, , \tag{2}$$

where $[S(f)]_\infty$ is the power spectrum of the infinite time series and sinc $\pi f T$ is the abbreviated notation (Bracewell, 1986) for $(\sin \pi f T)/\pi f T$.

The effect of finite sampling on the "ensemble" variance can then be expressed as

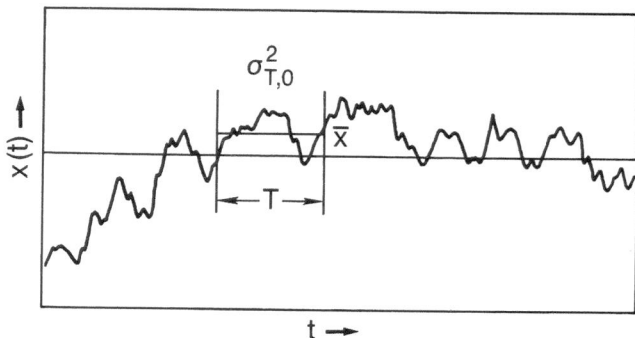

Fig. 1.   Segment of infinite time series $x(t)$ showing variance and mean value during a sampling interval T.

$$\langle \sigma_{T,0}^2 \rangle = \int_0^\infty [S(f)]_\infty (1 - \mathrm{sinc}^2 \pi f T)\, df. \tag{3}$$

In place of $[S(f)]_\infty$, one can now substitute smooth spectral forms derived from field experiments, as Panofsky (1988) did, and construct models for variance attenuation as functions of z/L and z_i/L (z being the height above ground, z_i the depth of the convective boundary layer, and L the Obukhov length).

From (3) one could argue (incorrectly) that the left side is the integral over the averaged finite time spectrum $[S(f)]_T$ of the windowed time series so that

$$\langle [S(f)]_T \rangle = [S(f)]_\infty (1 - \mathrm{sinc}^2 \pi f T), \tag{4}$$

where, by definition, $[S(f)]_\infty = \lim_{T \to \infty} \langle [S(f)]_T \rangle$.

The idea of such a simple filter function (Figure 2) defining the low-frequency roll-off is an attractive one. It complements the shape of a moving-average smoothing filter, it is less oscillatory than the $(1 - \mathrm{sinc}\, \pi f T)^2$ filter function (Figure 2) appropriate for a high-pass moving average filter, and it can be replaced in practice by a rectangular function that cuts off at $0.44/T$ (Smith, 1961) or $0.5/T$ (Panofsky, 1988). That such a simple filter function cannot be automatically assumed for the effect of finite sampling on $[S(f)]_\infty$ becomes clear when one realizes that the windowing and mean-removal operations involved in computing $[S(f)]_T$ are nonlinear operations.

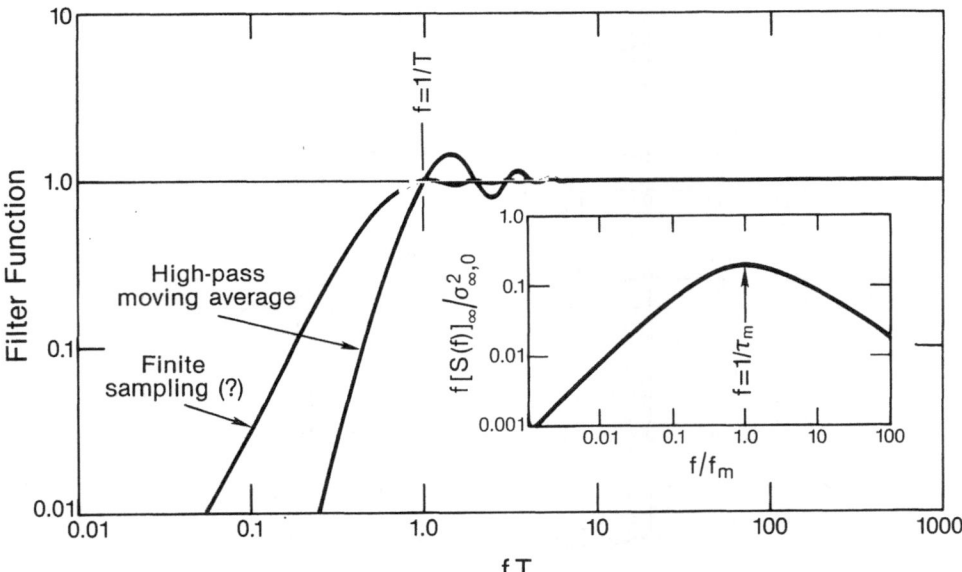

Fig. 2.   Two common high-pass filter functions compared. The insert shows the spectral form chosen for the numerical computations described here.

In the sections to follow, we examine the effect of these processes on the estimation of atmospheric spectra. It should be pointed out that our analysis is not limited to just consecutive samples from long records. The samples could be separated in either time or space, or both, but they would have in common some quality, be it stability, height above ground, or time of day, that defines them as part of an ensemble.

3. Fourier Transform Sequence

The effects of windowing and mean removal are demonstrated in Figure 3 with graphic representations of those operations in both the time and frequency domains. If $x(t)$ is the infinite time series and $X(f)$ its complex transform in the frequency domain, windowing can be represented in the time domain by multiplication of $x(t)$ with window $h(t)$, where

$$h(t) = \begin{bmatrix} 1; & |t| \leq T/2 \\ 0; & |t| > T/2. \end{bmatrix} \tag{5}$$

Its transform $H(f) = T \operatorname{sinc}(\pi f T)$ is shown in Figure 3. In the frequency domain, the process translates to one of convolution (denoted by $*$)

$$x(t) \cdot h(t) \leftrightarrow X(f) * H(f), \tag{6a}$$

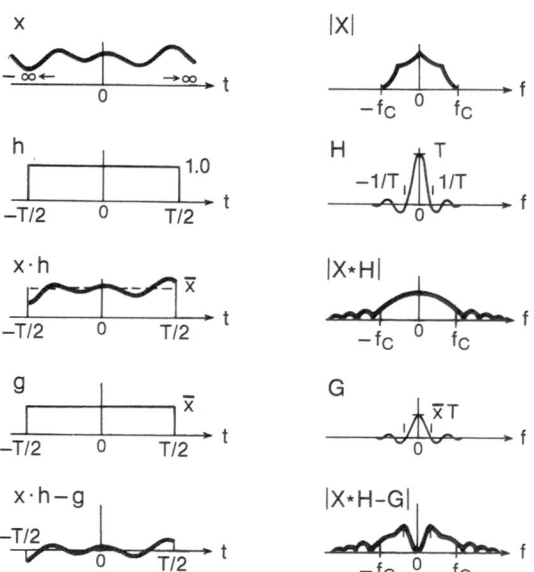

Fig. 3. Fourier transform sequence showing effect of finite sampling and mean removal in time (left) and frequency (right) domains.

where

$$X(f) * H(f) \equiv \int\limits_{-\infty}^{\infty} X(f-f')H(f')\,df'. \tag{6b}$$

The effect of convolution is to smooth out the details in $X(f)$ and spread the energy beyond the cut-off frequency $\pm f_c$. (The shape of $|X(f)|$ in Figure 3 is designed to illustrate this point.) The shorter the value of T, the greater the width of $H(f)$ and the larger the smearing and leakage in $|X(f) * H(f)|$. The larger the T, the more closely will $|X(f) \circ H(f)|$ resemble $|X(f)|$.

The next step, mean removal, is the process that suppresses frequencies below $f = 1/T$. Subtraction of $\bar{x}$ in the time domain is a subtraction also in the frequency domain. Representing the mean by a function $g(t)$ and its transform by $G(f)$,

$$g(t) = \begin{bmatrix} \bar{x}; & |t| \leqslant T/2 \\ 0; & |t| > T/2, \end{bmatrix} \tag{7}$$

and

$$G(f) = \bar{x}T \text{ sinc } \pi fT , \tag{8}$$

we have

$$[x(t) \cdot h(t)] - g(t) \leftrightarrow [X(f) * H(f)] - G(f) . \tag{9}$$

Note that for an infinite time series, mean removal amounts to subtraction of a delta function at $f = 0$.

4. Modification of Power Spectrum

The power spectrum of the infinite, zero-mean time series x, represented by $[S(f)]_\infty$, is the limit as $T \to \infty$ of $(1/T)\langle|X(f)|^2\rangle$. For a short-term sample, $-T/2 < t < T/2$, with mean removed as in (9), the spectrum assumes the form

$$[S(f)]_T = \frac{1}{T}|X(f) * T \text{ sinc } \pi fT - \bar{x}T \text{ sinc } \pi fT|^2 , \tag{10}$$

and the average of all such short samples is

$$\langle[S(f)]_T\rangle = T\langle|X(f) * \text{ sinc } \pi fT - \bar{x} \text{ sinc } \pi fT|^2\rangle. \tag{11}$$

To proceed further with (11), we define

$$X(f) \equiv \int\limits_{-\infty}^{\infty} e^{-i2\pi ft} x(t)\,dt \tag{12a}$$

which has an inverse

$$x(t) \equiv \int_{-\infty}^{\infty} e^{+i2\pi ft} X(f) \, df \,, \tag{12b}$$

so that

$$\bar{x} \equiv \frac{1}{T} \int_{-T/2}^{T/2} x(t) \, dt = \int_{-\infty}^{\infty} X(f) \, \text{sinc} \, \pi fT \, df \,. \tag{12c}$$

Substituting (12c) into (11) and collecting terms, (11) reduces to the form (see Appendix)

$$\langle [S(f)]_T \rangle = T \int_{-\infty}^{\infty} [S(f')]_{\infty} [\text{sinc} \, \pi(f-f')T - \text{sinc} \, \pi fT \cdot \text{sinc} \, \pi f'T]^2 \, df' \,, \tag{13}$$

and not the form suggested in (4). It is interesting to note that despite the fact that the spectrum has a more complicated form, if we square the terms in the bracket and integrate (13), we get the result

$$\langle \sigma^2_{T,0} \rangle = \int_0^{\infty} \langle [S(f)]_T \rangle \, df$$

$$= \int_0^{\infty} [S(f)]_{\infty} [1 - \text{sinc}^2 \, \pi fT] \, df \tag{14}$$

that agrees with (3). Note also that, from (13), we have $[S(f)]_{\infty} = \lim_{T \to \infty} \langle [S(f)]_T \rangle$, as it should.

Finally, it is interesting to consider the level of approximation that would give the relationship in (4). After much manipulation and rearrangement of terms, (13) can be expressed in the form:

$$\langle [S(f)]_T \rangle = [S(f)]_{\infty} (1 - \text{sinc}^2 \, \pi fT) + R(f) \,, \tag{15}$$

where $R(f)$ represents the remainder (see Appendix) which integrates to zero, consistent with the relationship in (14). To illustrate the effect of finite sampling and mean removal, we recast (15) in the form of a transfer function $K(f)$, where

$$K(f) = \frac{\langle [S(f)]_T \rangle}{[S(f)]_{\infty}} \tag{16a}$$

$$= (1 - \text{sinc}^2 \, \pi fT) + R(f)/[S(f)]_{\infty} \,. \tag{16b}$$

$K(f)$ is not a true transfer function since it depends on the choice of $[S(f)]_\infty$. For convenience, we choose a form similar to the one recommended by Panofsky (1988) for w spectra. In its frequency-weighted form, it can be expressed as

$$\frac{f[S(f)]_\infty}{\sigma_{\infty,0}^2} = \frac{(f/f_m)}{(1 + 1.5f/f_m)^{5/3}},$$ (17)

where f_m is the frequency at the spectral peak. From the peak frequency, we define the dominant eddy time scale† $\tau_m (= 1/f_m)$ for the process. Thus, we have

$$[S(f)]_\infty = \sigma_{\infty,0}^2 \tau_m (1 + 1.5f\tau_m)^{-5/3},$$ (18)

a form which fits most boundary-layer spectra and satisfies the expected relationship between the integral of the spectrum and variance.

Functional forms for $K(f)$ and $\langle [S(f)]_T \rangle$ for a range of T/τ_m values are presented in Figures 4 and 5. The functions were evaluated numerically since no easy analytic solution exists for our choice of $[S(f)]_\infty$.

We see in Figure 4 that $K(f)$ approaches $(1 - \text{sinc}^2 \pi fT)$ as $T \to \infty$. When T drops below $10\tau_m$, the curve departs rapidly from the above form. At $f > 1/T$, the energy is systematically overestimated, while at $f < 1/T$, it is underestimated relative to $(1 - \text{sinc}^2 \pi fT)$. The overshoot at $f \approx 1/T$ and the oscillations which damp out slowly with increasing frequency are significant. The magnitudes of overshoots can be very large: 10% at $T = \tau_m$ and 50% at $T = 0.01\tau_m$.

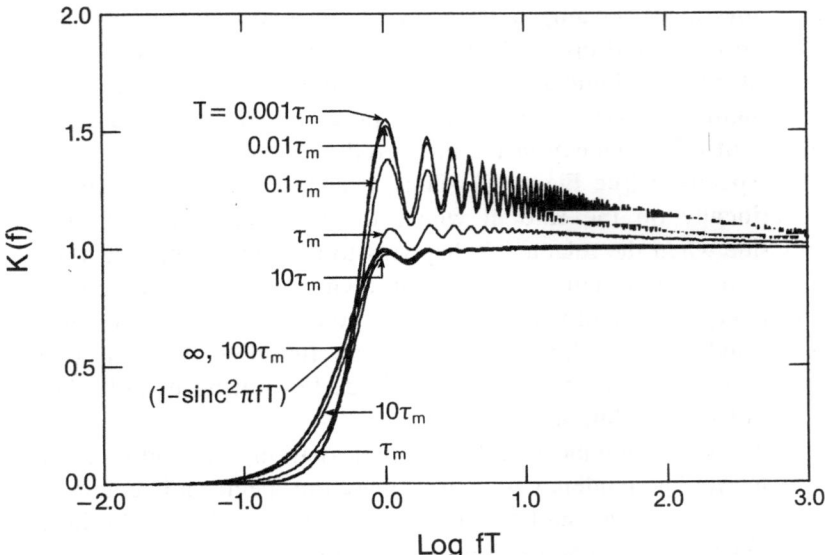

Fig. 4. $K(f)$ shown as a function of fT for a range of T/τ_m values. The curves depart significantly from $(1 - \text{sinc}^2 \pi fT)$ for $T < 10\tau_m$.

† Typically $\tau_m \simeq 2\pi\tau_o$, where τ_o is the integral time scale of the process.

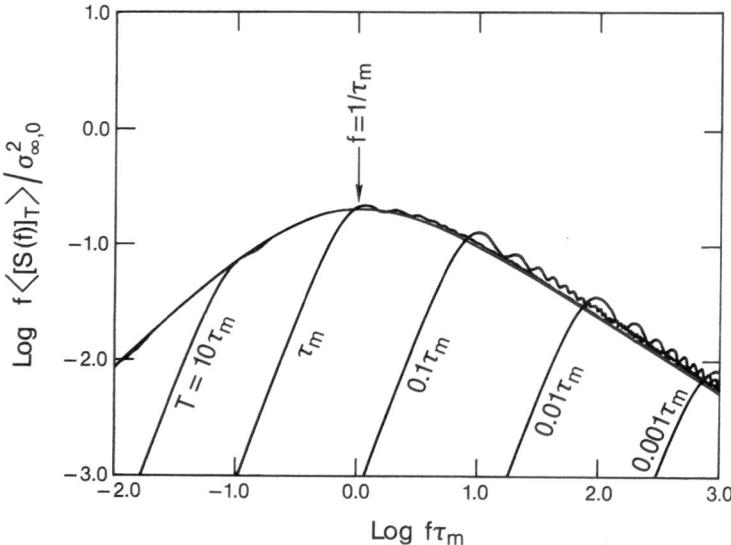

Fig. 5. Plots of observed spectrum as a function of $f\tau_m$ for different values of T relative to τ_m.

The amplitudes of the oscillations and their tendency to damp out seem strongly influenced by the choice of $[S(f)]_\infty$. Leif Kristensen (private communication, 1988) points out that selecting a less realistic, but analytically tractable form for the spectrum, $\tau_m/\{1+(f\tau_m)^2\}$, results in larger oscillations that neither damp out nor drop gradually towards unity, as in Figure 4. The oscillations can also have infinite amplitude if the spectrum is band-limited (or falls off steeply enough), because the convolution process leaks energy into the region beyond the cut-off frequency in the measured spectrum, where no energy exists in the true spectrum (see Figure 3). Thus, it appears that the behavior of $K(f)$ depends critically on the rate at which the spectrum rolls off on the high frequency side when the selected T happens to be small compared with τ_m. The tails of the true spectrum and the convolving sinc function determine the behavior of $K(f)$ for small values of T. We note that the behavior of $K(f)$ also depends strongly on the choice of the window function. A cosine or Gaussian taper on the window will reduce the amplitude but not necessarily the elevated mean value of the oscillations.

In Figure 5, we have a more realistic representation of the effect of decreasing T relative to τ_m. The tendency to overestimate the spectrum becomes noticeable only for $T < 10\tau_m$. The subtle departures from $(1 - \text{sinc}^2\ \pi fT)$ in their low frequency roll-offs are not so obvious in this plot as in Figure 4.

5. Conclusions

We have shown that the effect of finite sampling on the power spectrum can be approximated by the function $(1 - \text{sinc}^2\ \pi fT)$ only when $T \geq 10\tau_m$. However, it correctly represents the effect of the variance, allowing accurate recovery of long-term or ensemble variances from short-term estimates.

In the atmospheric boundary layer, τ_m ranges from a few seconds to several minutes (varying with stability, height above ground and the parameter in question). If we require that T be larger than $10\tau_m$, we find that the sampling duration should be 60 min or more to include the worst case scenario. The implication here is that neither the low-frequency roll-off nor the magnitude of the spectrum at $f > 1/T$ can be simply defined if the sampling time is not significantly larger than the dominant time scale of the process.

In boundary-layer work, it is common practice to compute the high- and low-frequency segments separately and to combine them, splicing in the region where the two overlap. The high-frequency spectrum is obtained by dividing the full record (~ 1 h) into shorter records (~ 2 min) and averaging their spectra incoherently to obtain a single smooth spectrum. For the low-frequency spectrum, a new time series is created by block averaging the original series over consecutive nonoverlapping blocks (~ 10 sec). The two spectra should agree over a broad frequency range, and the splicing is seldom apparent (Kaimal and Gaynor, 1983). With the perspective gained from this study, we can now see that the first few spectral estimates at the low end of the smoothed high-frequency spectrum show better agreement with the spectrum of the block-averaged time series when f_m occurs at a frequency higher than $1/T$ (as with vertical velocity near the ground), than when it is lower than $1/T$.

For most applications, the effect of finite sampling is not particularly serious, provided reasonable care is taken in choosing T. However, its implications are clear for studies where spectra of differing sampling intervals are combined, or where assumptions are made about spectral shapes from records that are too short.

6. Acknowledgments

This paper evolved from discussions with J. J. Finnigan (CSIRO, Canberra, Australia), Leif Kristensen (Risø National Laboratory, Roskilde, Denmark) and D. H. Lenschow (National Center for Atmospheric Research, Boulder, Colorado). Their contribution to the development of this study is gratefully acknowledged. We also thank R. S. Lawrence for his expert handling of the numerical analysis and for the computer plots presented here.

Appendix: Derivation of Spectral Forms

Using definitions in (12), we can rewrite (11) in the form

$$\langle [S(f)]_T \rangle = T \left\langle \left| \int_{-\infty}^{\infty} X(f')\{\operatorname{sinc} \pi(f-f')T - \operatorname{sinc} \pi f'T \cdot \operatorname{sinc} \pi fT\} \, df' \right|^2 \right\rangle$$

$$= T \int\limits_{-\infty}^{\infty} \int\limits_{-\infty}^{\infty} \langle X(f')X^*(f'')\rangle$$

$$\times [\text{sinc } \pi(f-f')T - \text{sinc } \pi f'T \cdot \text{sinc } \pi fT]$$

$$\times [\text{sinc } \pi(f-f'')T - \text{sinc } \pi f''T \cdot \sin \pi fT] df' df''. \qquad \text{(A1)}$$

For a stationary random process,

$$\langle X(f')X^*(f'')\rangle = \delta(f' - f'')[S(f')]_\infty \qquad \text{(A2)}$$

where X^* represents the complex conjugate of X, and δ is the Dirac delta function. From these definitions, it follows that

$$\langle [S(f)]_T \rangle = T \int\limits_{-\infty}^{\infty} [S(f')]_\infty [\text{sinc } \pi(f-f')T - \text{sinc } \pi f'T \cdot \text{sinc } \pi fT]^2 df'.$$

$$\text{(A3)}$$

After considerable manipulation and rearrangement of terms, (A3) can be expressed as

$$\langle [S(f)]_T \rangle = [S(f)]_\infty (1 - \text{sinc}^2 \pi fT) + R(f), \qquad \text{(A4)}$$

where $R(f)$ represents the remainder terms given below

$$R(f) = -(1 - \text{sinc}^2 \pi fT)\left\{ [S(f)]_\infty - 2T \int\limits_{-\infty}^{\infty} [S(f')]_\infty \text{ sinc } 2\pi(f-f')T \, df' \right\}$$

$$-2T \int\limits_{-\infty}^{\infty} [S(f')]_\infty \left\{ \text{sinc } 2\pi(f-f')T - \frac{1}{2}\text{sinc}^2 \pi(f-f')T \right.$$

$$+ \text{sinc}^2 \pi fT \left[\text{sinc } 2\pi f'T - \frac{1}{2}\text{sinc}^2 \pi f'T \right]$$

$$\left. - \frac{1}{\pi fT} \text{sinc } 2\pi fT \cdot \text{sinc } \pi(f-f')T \cdot [\sin \pi(f-f')T] \right\} df'.$$

$$\text{(A5)}$$

In order that the variance relationship in (3) be satisfied, we must have

$$\int\limits_{0}^{\infty} R(f) \, df = 0, \qquad \text{(A6)}$$

which is easily verified by integrating (A5).

References

Bracewell, R. N.: 1986, *The Fourier Transform and Its Applications* (2nd Edition), McGraw Hill, New York, 477 pp.

Kaimal, J. C. and Gaynor, J. E.: 1983, 'The Boulder Atmospheric Observatory', *J. Clim. Appl. Meteorol.* **22**, 863–880.

Lumley, J. L. and Panofsky, H. A.: 1964, *The Structure of Atmospheric Turbulence*, Wiley Interscience, New York, 239 pp.

Ogura, Y.: 1959, 'Diffusion from a Continuous Source in Relation to a Finite Observational Interval', *Advances in Geophysics* **6**, 149–159.

Panofsky, H. A.: 1988, 'The Effect of Averaging Time on Velocity Variances', *Meteorol. Atmos. Phys.* **38**, 70–88.

Pasquill, F. and Smith, F. B.: 1983, *Atmospheric Diffusion* (3rd Edition), Ellis Horwood Ltd., Chichester, England, 437 pp.

Wollenweber, G. C. and Panofsky, H. A.: 1989, Dependence of Velocity Variance on Averaging Time', *Boundary-Layer Meteorol.* **47**, 205–215 (this volume).

Wyngaard, J. C.: 1973, 'On Surface Layer Turbulence', in D. A. Haugen (ed.), *Workshop in Micrometeorology*, American Meteorological Society, Boston, 437 pp.

OBSERVATION OF ORGANIZED STRUCTURE IN TURBULENT FLOW WITHIN AND ABOVE A FOREST CANOPY

W. GAO, R. H. SHAW, and K. T. PAW U

Department of Land, Air and Water Resources, University of California, Davis, California 95616, U.S.A.

(Received in final form 10 November, 1988)

Abstract. Ramp patterns of temperature and humidity occur coherently at several levels within and above a deciduous forest as shown by data gathered with up to seven triaxial sonic anemometer/thermometers and three Lyman-alpha hygrometers at an experimental site in Ontario, Canada. The ramps appear most clearly in the middle and upper portion of the forest. Time/height cross-sections of scalar contours and velocity vectors, developed from both single events and ensemble averages of several events, portray details of the flow structures associated with the scalar ramps. Near the top of the forest they are composed of a weak ejecting motion transporting warm and/or moist air out of the forest followed by strong sweeps of cool and/or dry air penetrating into the canopy. The sweep is separated from the ejecting air by a sharp scalar microfront. At approximately twice the height of the forest, ejections and sweeps are of about equal strength.

In the middle and upper parts of the canopy, sweeps conduct a large proportion of the overall transfer between the forest and the lower atmosphere, with a lesser contribution from ejections. Ejections become equally important aloft. During one 30-min run, identified structures were responsible for more than 75% of the total fluxes of heat and momentum at mid-canopy height. Near the canopy top, the transition from ejection of slow moving fluid to sweep bringing fast moving air from above is very rapid but, at both higher and lower levels, brief periods of upward momentum transfer occur at or immediately before the microfront.

1. Introduction

The mechanism by which the atmospheric surface layer and a vegetation canopy such as a forest are coupled aerodynamically is understood in only rudimentary fashion, although the exchanges of heat, mass, and momentum are of fundamental and practical importance. Most work on the turbulent flow within and immediately above a plant canopy has been based on gradient diffusion theory, but theoretical and observational studies (Legg and Monteith, 1975; Raupach and Thom, 1981; Denmead and Bradley, 1987) have shown the inadequacy of this approach. Although higher order closure and Lagrangian models have been developed to describe turbulent transport more accurately (Wilson and Shaw, 1977; Meyers and Paw U, 1986, 1987; Raupach, 1987), these models are based on statistical descriptions of turbulence and must employ semi-empirical relationships of questionable accuracy. Further advances in the simulation of canopy microclimate will only result from a more comprehensive description and understanding of the turbulent motions responsible for the diffusion of scalar and vector quantities within and above plant canopies.

Boundary-Layer Meteorology **47**: 349–377, 1989.
© 1989 *Kluwer Academic Publishers.*

Organized structures have been identified and studied in various turbulent flows (Kline *et al.*, 1967; Chen and Blackwelder, 1978; Praturi and Brodkey, 1978; Antonia *et al.*, 1979; Thomas and Bull, 1983; Schols, 1984). For example, in the wall region of laboratory boundary layers, ejection-sweep cycles have been commonly observed through flow visualization techniques and instrumental measurement (Kline *et al.*, 1967; Thomas and Bull, 1983). In the lower atmosphere, extensive analysis has been carried out on temperature ramps and thermal plume structure, as exemplified by Wilczak (1984) and Wilczak and Businger (1984), although Antonia *et al.* (1979) emphasized that scalar ramps in the atmospheric boundary layer should be considered as indications of organized structures but not necessarily as thermal plume signatures. Little detailed work has been performed on coherent structures within the atmosphere under neutral stability conditions.

Various quantitative methods for the identification of turbulent structures based on flow field measurements have shown reasonable correlation with visually identified structures (Bogard and Tiederman, 1986; Talmon *et al.*, 1986), although in some cases poor correlation was found (Offen and Kline, 1975; Subramanian *et al.*, 1982). One method is quadrant analysis (Wallace *et al.*, 1972; Bogard and Tiederman, 1986; Talmon *et al.*, 1986), and another is that of using scalar ramps (Rajagopalan and Antonia, 1980). In quadrant analysis, $u'w'$ is divided into four categories, with quadrant 2 ($u' < 0$, $w' > 0$) identified as 'ejections', quadrant 4 ($u' > 0$, $w' < 0$) identified as 'sweeps', and the remaining two quadrants identified as 'interactions' between 'ejections' and 'sweeps' (Wallace *et al.*, 1972). Here, u and w are the longitudinal and vertical velocities, respectively, and the primes indicate departures from the mean. In our paper, we revert to the original definition of an ejection based on the visual analysis of passive tracers (Kline *et al.*, 1967; Corino and Brodkey, 1969); we use the term **sweep** to indicate the region of downward motion, and **ejection** the region of upward motion in a coherent structure, without making assumptions as to the horizontal momentum of the air.

Studies of real and artificial plant canopies using quadrant analysis show that relatively short duration quadrant 4 'sweeps' and quadrant 2 'ejections' transport most of the fluxes (Finnigan, 1979a, b; Raupach, 1981; Shaw *et al.*, 1983; Baldocchi and Meyers, 1988). Quadrant 4 events dominate flux transport within the canopy, while quadrant 2 events become increasingly important above the canopy and become the dominant process at several times the canopy height. In the present study, data from sonic anemometer/thermometers and Lyman alpha hygrometers within and above a deciduous forest, reveal periods with a large number of ramp patterns appearing repeatedly in time series of temperature and humidity. The ramps occur nearly simultaneously at several levels both within and above the canopy. Two-dimensional flow fields associated with these ramps, displayed by a time/height cross-section of scalar contours and velocity vectors, show strong organization. The cross-sectional analysis is similar to that of

Wilczak (1984), but we add velocity vectors to aid depiction of the structure kinematics. Several structures are examined in terms of their two-dimensional flow and scalar fields, associated flux and higher order product distributions, and their contributions to heat and momentum transfer.

2. Experiment and Data Acquisition

The data used for this study are from an experiment at Camp Borden, Ontario, Canada during 1986 and 1987. The forest, of approximate mean height $h = 18\,m$, is composed primarily of aspen and maple, with other species including eastern white pine, red pine, black cherry, white ash, eastern hemlock, eastern white cedar, white birch, yellow birch, and shagbark hickory (see Neumann *et al.*, 1988 for further details). Measurements were made at a location downwind of about 4 km of uniformly rough canopy surface on reasonably level terrain.

Seven triaxial sonic anemometer/thermometers (Kaijo-Denki Co., Ltd.) were operated in 1986 at above-ground heights of 5.9, 10.5, 15.4, and 17.6 m on an 18 m tall tower, and at heights of 17.9, 34.2, and 43.1 m on a taller tower 25 m to the west. In 1987, the instrument at 15.4 m was moved to 14.2 m and, on the taller tower, the instrument at 17.9 m was relocated at a height of 25.6 m. Also for summer 1987, no data were taken at 43.1 m. Lyman alpha hygrometers provided water vapor density at 14.2, 17.6, and 34.2 m. All fast-response sensors were sampled at 10 Hz. Other measurements included mean temperature using ventilated thermometers and wind speed from cup anemometers, at 12 levels up to 48 m. Details of the instruments and site are described more fully by Shaw *et al.* (1988).

The data presented here are primarily from two 30-min periods: Run A from 1400 to 1430 EDT on October 2, 1986, with unstable conditions and when partial leaf fall had occurred; Run B from 1600 to 1630 EDT on August 13, 1987, during near neutral stability and with full summer foliage. In addition, temperature traces are presented from Run C during stable conditions from 0300 to 0330 EDT on August 12, 1987. Statistics from these runs are shown in Table I.

TABLE I

30-min means for selected periods

Run	Date	Time (EDT)	U (48 m) (m/s)	u^* (18 m) (m/s)	T^* (18 m) (K)	$\overline{w'T'}$ (18 m) (K m/s)	Correction[†] (18 m) (K m/s)	L (m)
A	10/2/86	1400–1430	4.92	0.74	−0.288	0.213	0.004	−138
B	8/13/87	1600–1630	5.68	0.74	−0.024	0.018	0.011	−1063
C	8/12/87	0300–0330	2.59	0.13	0.062	−0.008	0.000	+18

† Correction = $0.61\,T\overline{w'q'}$.

The Monin–Obukhov length was calculated based on eddy correlation measurements of the sensible heat, water vapor, and momentum fluxes at the canopy top, according to

$$L = - u^{*3}/(k(g/T)(\overline{w'T'} + 0.61\, T\overline{w'q'}))\,,\tag{1}$$

which includes the buoyancy effects of water vapor (Busch, 1973). Here, u^* is the friction velocity, g the acceleration due to gravity, k the von Kármán constant, T the absolute temperature, and q the specific humidity. The water vapor component in the denominator of this expression was small in all cases but, in the near neutral run (Run B), it was of comparable magnitude to the sensible heat flux term.

3. Results and Analysis

3.1. OBSERVATION OF RAMP PATTERNS UNDER VARIOUS STABILITY CONDITIONS

Figure 1a shows a 10-min time series of temperature fluctuations from Run A, with $L = -138$ m, at six heights from 0.33 to 2.39h, where h is the canopy height. The variation of temperature follows distinct ramp patterns which are characterized by a gradual rise terminated by a sharp drop of about 1.5 °C over 1 to 3 s. Most of the ramps were observed at all heights, but were most clearly evident in the middle and upper levels of the forest. The drop at each level precedes that at the next lower level. The magnitude of the temperature drop suggests that the cold air originates well above the canopy because the mean temperature profile (Figure 2) shows only a slight vertical temperature gradient in the above-canopy atmosphere.

Temperature ramps have also been observed in unstable surface layers over smoother surfaces (Taylor, 1958; Priestley, 1959; Kaimal and Businger, 1970; Kaimal, 1974; Wilczak, 1984). In general, thermal convective plumes have been identified as the cause of the ramps. In this study, data collected under near neutral stability (Run B) include ramp patterns in humidity (Figure 1b). No temperature ramps were found in this run because the mean temperature gradient was too weak (Figure 2). Although the mean humidity gradient was apparently sufficient to allow ramp production, the moisture flux did not contribute significantly to destablizing the air, as indicated by the relatively large magnitude of the Monin–Obukhov length ($L = -1063$ m). Therefore, without significant buoyancy effects, shear must be the major factor in the dynamics of the structures associated with the ramps.

Inverse temperature ramps also occurred during stable conditions, when $L = 18$ m (Figure 1c). These ramps are composed of a gradual temperature decrease, followed by a sharp temperature rise, and are essentially mirror images of the ramps described earlier. Because of the stable mean temperature profile, air sweeping down from above the canopy will be warm, creating the steep tem-

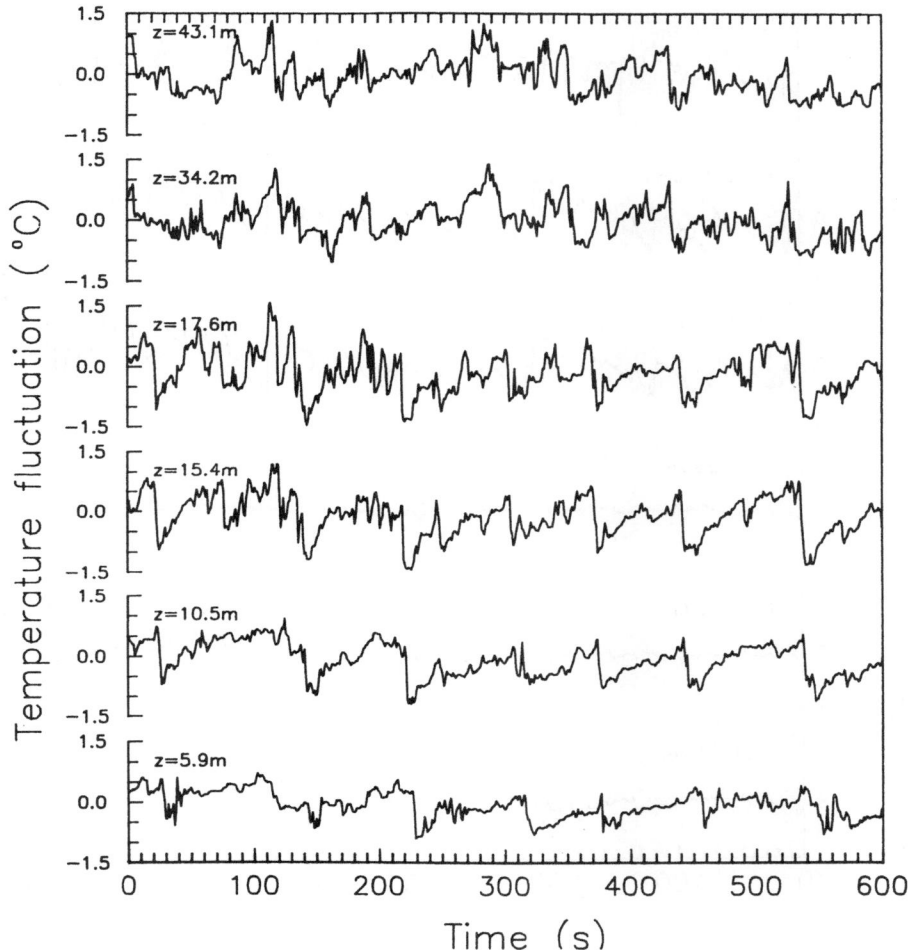

Fig. 1a. Time series of temperature at each of six heights during a 10-min portion of Run A ($L = -138$ m). Ramp patterns can be seen at each height.

perature rise; preceding the sweep, the air in the canopy gradually cools by sensible heat transfer to the radiatively cooled canopy elements.

3.2. Two-dimensional flow and scalar fields of ramp structures

The flow and thermal fields associated with the ramp structures are revealed by plotting the data as time/height cross-sections, in a manner similar to that of Wilczak (1984). These cross-sections were formed by visually identifying the ramps in the time traces at several heights, and plotting contours of temperature deviations T' from the 2-min mean temperature at each level, centered on the time of the temperature drop at the top of the forest. The temperature deviations are defined by

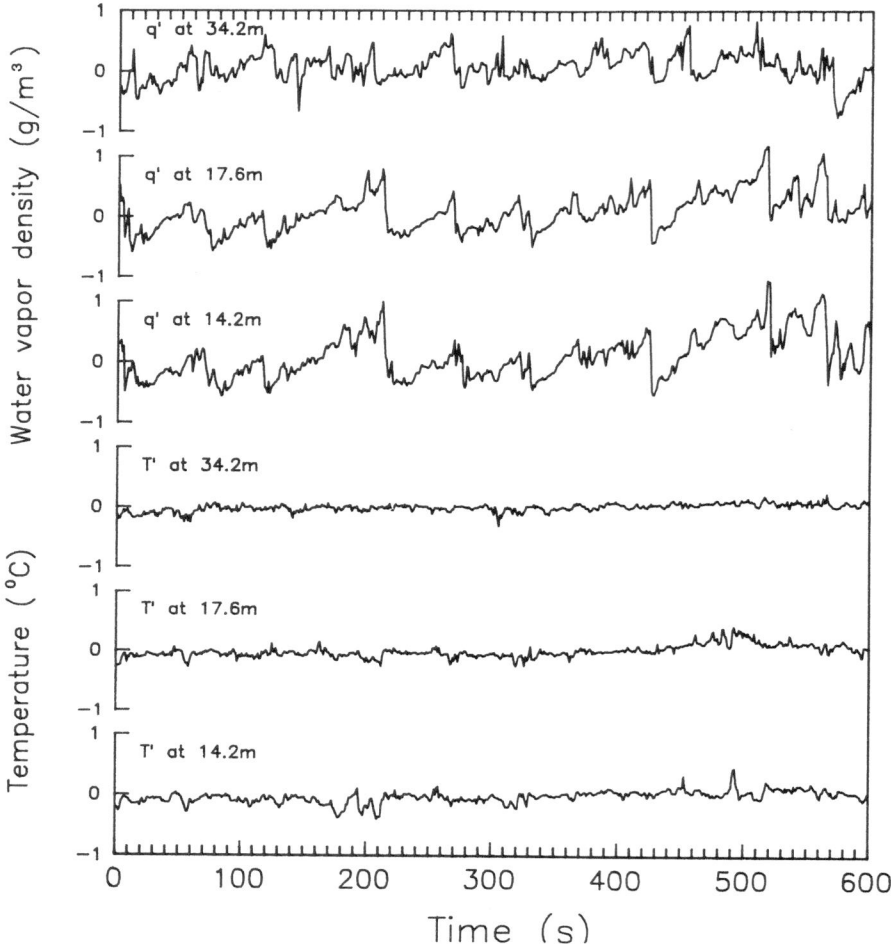

Fig. 1b. Time series of temperature and water vapor density at each of three heights during a 10-min portion of Run B ($L = -1063$ m). Ramps of water vapor density are shown, but no temperature ramp is observed.

$$T'(z, t) = T(z, t) - \bar{T}(z),\qquad(2)$$

where the mean temperature $\bar{T}(z)$ applies to the period surrounding a specific event and is defined as

$$\bar{T}(z) = \frac{1}{N} \sum_{n=1}^{N} T(z, t),\qquad(3)$$

with $N = 120$ to obtain a 2-min average from the 1-s values. An example from unstable conditions during Run A is shown in Figure 3. Wind vectors are also plotted as arrows to depict the physical transport processes associated with the

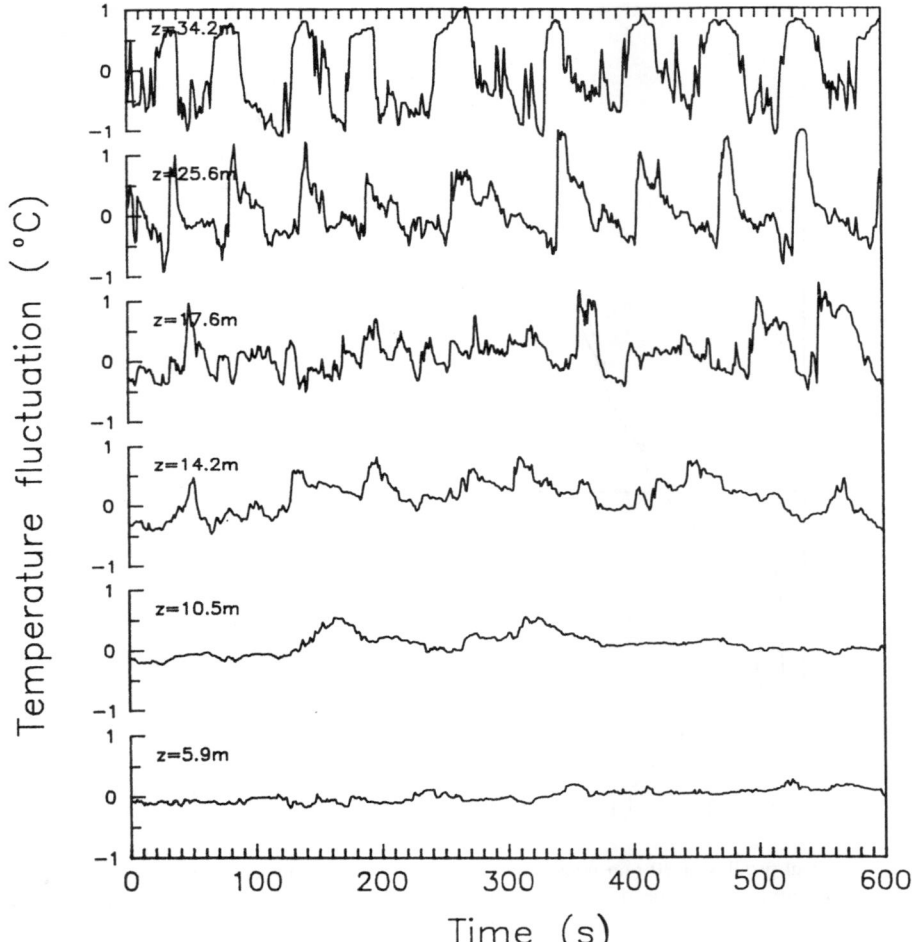

Fig. 1c.   Time series of temperature at each of six heights during a 10-min portion of Run C under stable conditions with $L = 18$ m. Inverse temperature ramps are observed close to the canopy surface.

scalar ramps. A coordinate rotation to set the 30-min mean lateral velocity (v-component) to zero was applied, and the diagram shows the two-dimensional u, w velocity field. The thermal field is composed of warm and cold regions separated by a narrow microfront with a dramatic temperature decrease of 1–2 °C occurring over 2–4 s. The microfront passage at each level lags behind that at upper levels.

Near the top of the forest, the vector flow field shows a weak upward motion before the microfront arrives. Close to the frontal region, the wind rapidly shifts to a strong downward motion. The dramatic sweep preceded by the relatively gentle ejection is in marked contrast with thermal plume structures described by

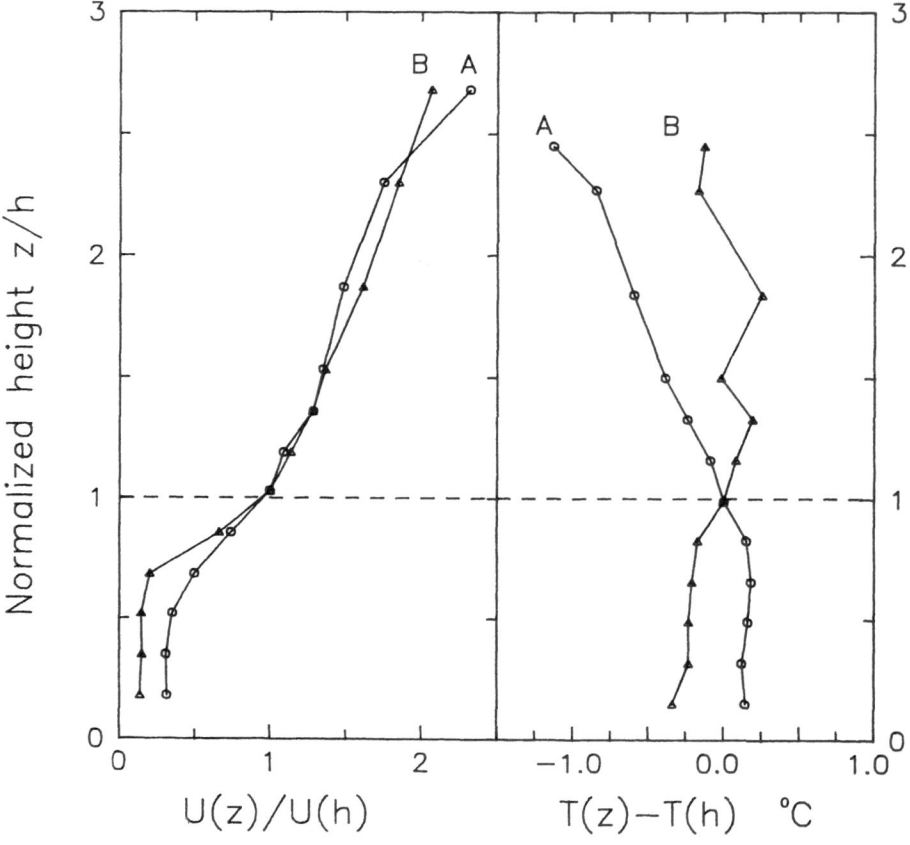

Fig. 2. Vertical profiles of 30-min mean temperature (T) and wind speed (U). $T(h)$ and $U(h)$ are mean temperature and wind speed at canopy top. The circles denote Run A and the triangles denote Run B.

Wilczak (1984), which are characterized by rising motion much stronger than the longer duration, gentle sinking motion.

The shift in the sign of the vertical velocity occurs almost simultaneously at all levels, in contrast with the passage of the thermal microfront for which the lower levels lag behind the upper ones. The effect of this difference is that, at the lower levels, the switch to downward motion occurs a few seconds before the micro-front (see the lower portion of Figure 3). This is consistent with a cooler air parcel sweeping downward through the previously warmer canopy; at lower levels, the air which initially moves down is from the upper, relatively warm canopy layers. As the cool air sweeps farther down, it reaches the lower levels several seconds after the initial wind shift.

Cross-correlations of temperature between two heights and of vertical velocity, calculated from the 30 minutes of data encompassing a single run (Figure 4),

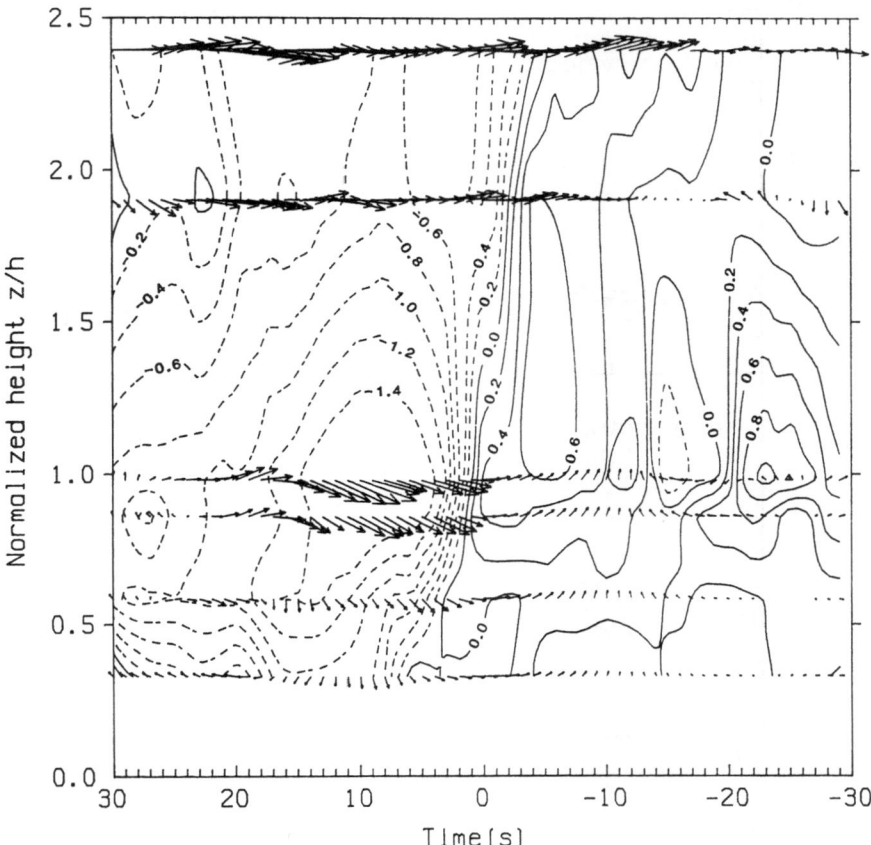

Fig. 3. Vertical cross-section of temperature and velocity fields for a single ramp structure. Positive temperature deviations are represented by solid lines, and negative deviations by dashed lines. Contour intervals are 0.2 °C. Arrows represent the two-dimensional (u, w) wind velocities with the maximum arrow length representing a magnitude of 5.5 m/s. Data are from Run A with $L = -138$ m.

support the spatial relationship noted above. The correlation of the temperature signal at 43 m against that at the top of the forest is a maximum at a negative time lag, implying that, on the average, temperature variations at 43 m occur about 6 s before those at the canopy top. The temperature signal at 1/3 the canopy height lags that at the canopy top by about 10 s. Vertical velocity cross-correlations showed no appreciable time lag between any of the heights. We attribute this difference to the mechanisms for the production of temperature perturbations and those of vertical velocity. Temperature fluctuations are associated with the physical transport of air from regions of different mean temperature by fluid motions. On the other hand, velocity fluctuations are dynamically related to pressure perturbations which propagate very rapidly. It should be remembered that the correlations are for a 30-min period which includes turbulence not directly connected with the ramps, and therefore the time lags

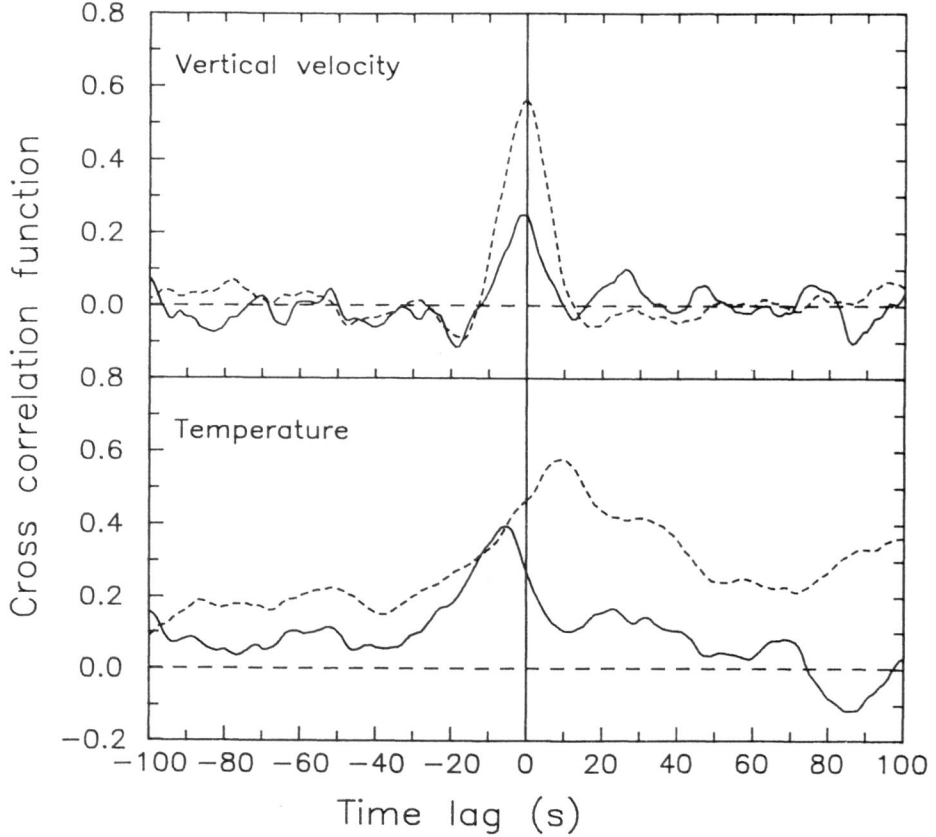

Fig. 4.   Spatial cross-correlations of vertical velocity for two sets of two heights, and of temperature for the same two sets. Solid lines: 43.1 m versus 17.9 m on the tall tower; dashed lines: 5.9 m versus 17.6 m on the short tower. Data are from Run A with $L = -138$ m.

calculated by the correlation analyses may not be identical to the ramp structure time lags.

In an attempt to clarify the origins of air at different levels, we calculated relative potential temperatures $\theta_r(z, t)$ by calibrating the temperature signals from each sonic anemometer/thermometer against the mean profile from the ventilated thermometers, and adjusting for the adiabatic lapse rate. Equation (4) defines the relative potential temperature as

$$\theta_r(z, t) = T(z, t) + \delta \bar{T}(z) + \Gamma \, \delta z \,, \tag{4}$$

where $\delta \bar{T}(z)$ is the calibration offset for each of the sonic thermometers, determined by comparison of the 30-min mean temperatures from the sonics and from the vertical array of ventilated, slow-response thermometers. Γ is the dry adiabatic lapse rate, and $\delta z = z - z_r$, where z_r is the reference height taken to be 5.88 m, the height of the lowest sonic anemometer/thermometer.

Two examples from Run A are presented in Figures 5a and 5b. Note that in these figures and others to follow, the vectors are the fluctuating component of the wind. The effect of subtracting the mean wind is seen by comparing Figures 5a and 3, which are for the same event. In Figure 3 the vectors describe the total velocity field in which the longitudinal velocity is sufficiently strong that the vertical motion is not clearly seen at the higher levels. An apparent change in the flow field arising from subtracting the mean flow is that the rising motion ahead of the microfront is directed backwards and upwind. This is because the upward motion brings fluid of a small longitudinal velocity from low levels, causing the difference between the instantaneous longitudinal velocity and the local mean velocity to be negative.

Figures 5a and 5b appear to show structures at different stages of development. The steep vertical gradient of potential temperature ahead of the microfront in

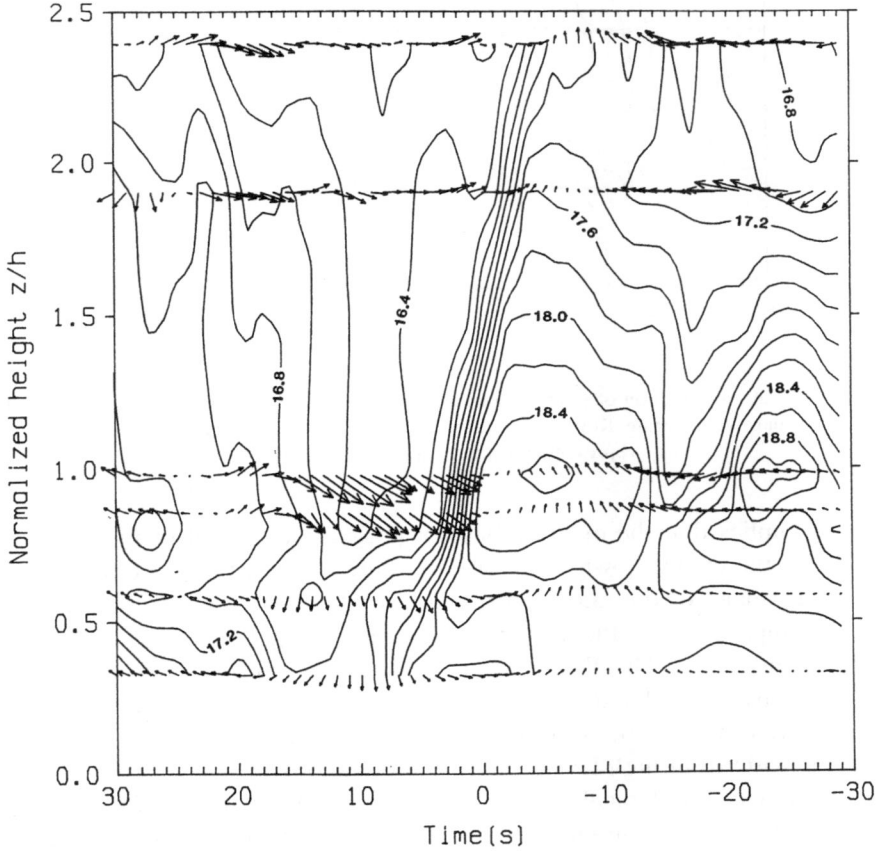

Fig. 5a.   Vertical cross-section of relative potential temperature and fluctuating velocity fields for the same case as Figure 3. Contour interval is 0.2 °C and the maximum arrow length represents a wind magnitude of 3.6 m/s.

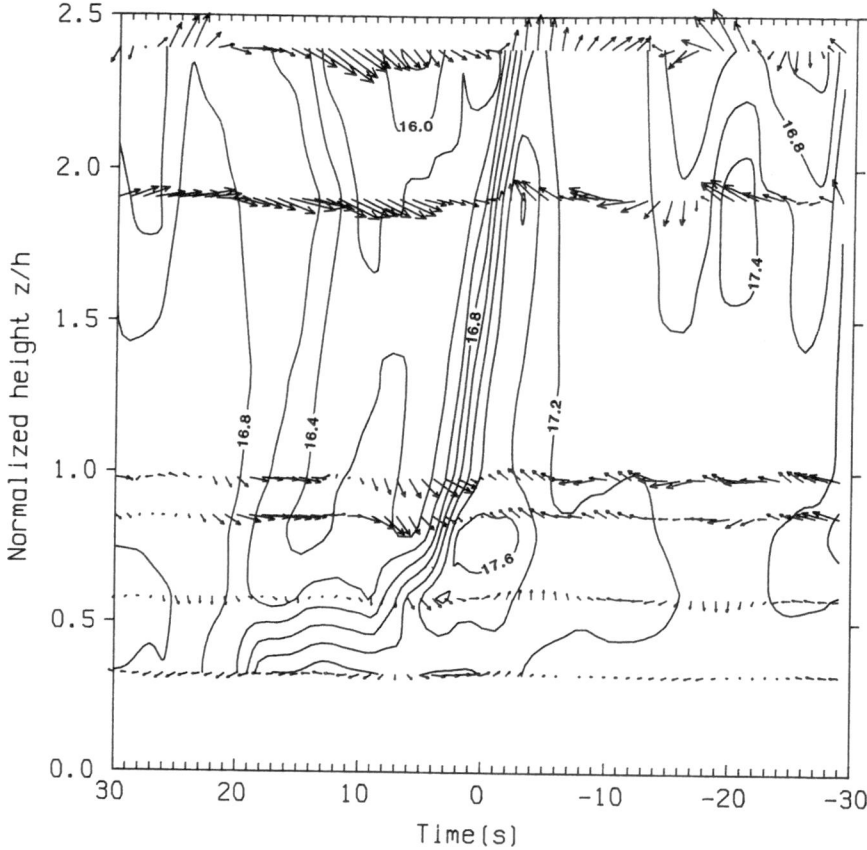

Fig. 5b. Vertical cross-section of relative potential temperature and fluctuating velocity fields for a second individual event during Run A. Contour interval is 0.2 °C and the maximum arrow length represents a wind magnitude of 3.3 m/s.

Figure 5a implies that the ejection has not progressed far out of the canopy. In contrast, for the event shown in Figure 5b, there is air at twice the forest height immediately ahead of the microfront, and also about 20 s earlier, with the same potential temperature as the canopy layer, suggesting that forest air has been ejected to this level. The lifting of the ejected air, and the existence of two ejected regions, is similar to visual observations of coherent structures (Corino and Brodkey, 1969) in laboratory flows. In both diagrams the strong downdraft is closely adiabatic from high levels down to mid-canopy.

In order to examine typical features of the flow and scalar fields for a relatively large number of events, an ensemble average was applied to velocity and scalar fields during and after those ramps that could be visually identified and distinguished from background turbulence by the following two requirements: (a) a gradual temperature or humidity increase terminated by a steep drop; and (b) near-simultaneous presence at several levels. A total of 20 ramp events have been analyzed, 10 of which are from Run A and 10 from Run B.

We assumed that the flow could be divided into two forms of motion: organized coherent structures; and generally smaller scale background turbulence. By ensemble averaging a number of ramps, the effect of smaller scale random turbulence was diminished and the organized motions more easily seen.

Our procedure for ensemble averaging a series of ramp structures involved, first of all, subjectively timing the rapid temperature or humidity drop at each level and referencing this time to that from the instrument located near the top of the forest. At each level, the average time delay or advance was determined from all ramps examined. Next, and for each level separately, we combined and averaged the temperature or humidity, and velocity data centered on the sharp

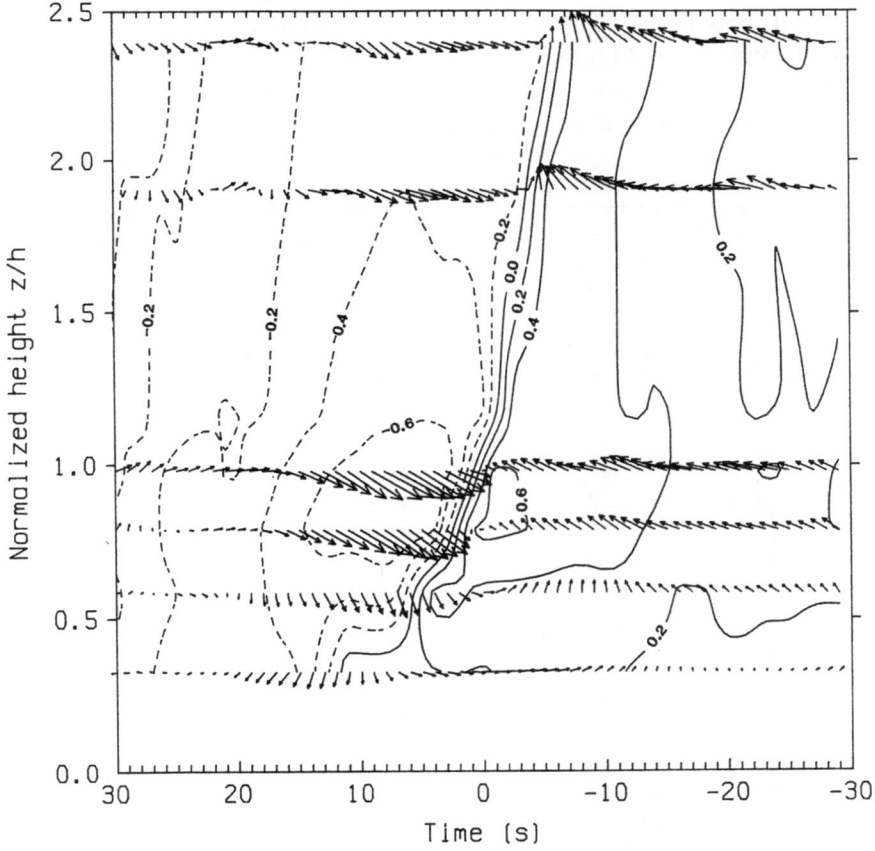

Fig. 6a. Vertical cross-section of ensemble-averaged temperature and fluctuating velocity fields, under unstable conditions ($L = -138$ m), during Run A. Dashed lines are isotherms below the mean, and solid lines are isotherms above the mean. Contour interval is 0.2 °C and the maximum arrow length represents a wind magnitude of 1.9 m/s.

temperature or humidity decrease, to form ensemble average traces. No attempt was made to adjust time scales to match ramps of different duration. For each level, the mean value of the scalar variable for the 60-s period centered on the rapid drop was subtracted from each ensemble trace but, as in Figures 5a and 5b, the wind vectors represent departures from the 30-min means. Finally, the vertical structure of the events was reconstructed by applying the appropriate mean time delay or advance to each level. These data were then plotted in the same manner as Figure 3, to depict the velocity and scalar fields typical of the ramp structures (Figures 6a and 6b), for unstable conditions (Run A, $L = -138$ m) and near-neutral conditions (Run B, $L = -1063$ m).

Compared with the single structures presented in Figures 3, 5a, and 5b, the ensemble-averaged flow and thermal fields of Figure 6a are smoother. The microfront in the thermal field is not quite as sharp, with about a 1 °C change over a 3 s period, because of the averaging process. Ensemble averaging in near-neutral conditions ($L = -1063$ m), with humidity as tracer, yields an equivalent microfront (Figure 6b). The humidity field is divided into dry and moist regions with a 0.8 g/m^3 drop over 3 s across the microfront.

Ensemble-averaged vertical velocity contours are plotted in Figures 7a and 7b, in which a zero line is seen to separate the field into regions of updraft and

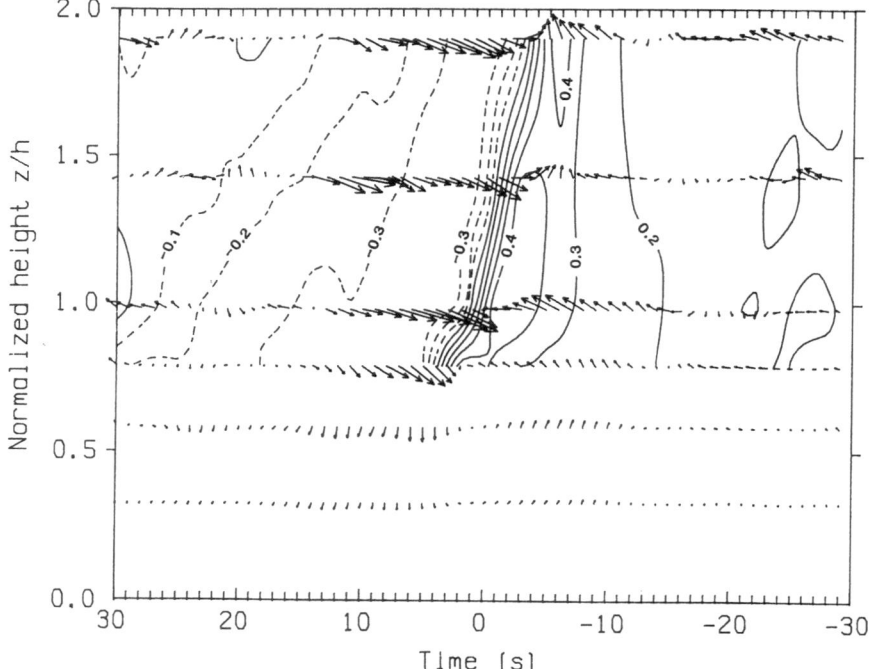

Fig. 6b. Same as in Figure 6a, but contours represent water vapor density deviations at intervals of 0.1 g/m^3, for near-neutral conditions ($L = -1063$ m), during Run B. The maximum arrow length represents a wind magnitude of 2.0 m/s.

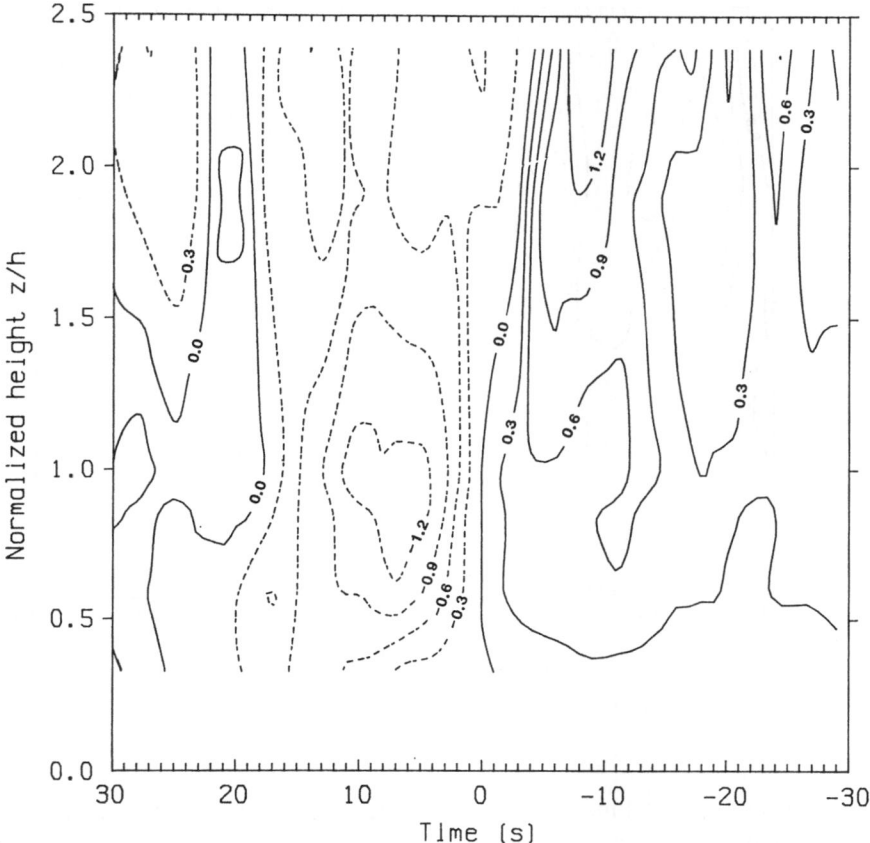

Fig. 7a. Vertical cross-section of ensemble-averaged vertical velocity field normalized by friction velocity at the top of the forest during Run A ($L = -138$ m). Contour interval is 0.3. Dashed lines depict negative vertical velocities and solid lines depict positive vertical velocities.

downdraft. In the sweep region, the downward flow is accelerated to a maximum value near the top of the canopy, and decelerated as the ground surface is approached. In the ejection region, the fluid moves upward at a low speed within the canopy, and is slightly accelerated away from the canopy surface. In both unstable (Figure 7a) and near-neutral conditions (Figure 7b), the magnitude of the downdraft is consistently larger than that of the updraft near the canopy surface, while they approach equivalency at 1.5 to $2h$. We note that the acceleration of the downward flow occurs in the region where shear, therefore vorticity, has maximum value (see mean velocity profile in Figure 2).

The translation speeds of the thermal fronts for the 10 events from Run A were calculated by timing the passage of the sharp temperature drops observed at tree top level between the two towers, which were horizontally separated by 25 m; the wind during this period was in approximate alignment with the towers. The mean

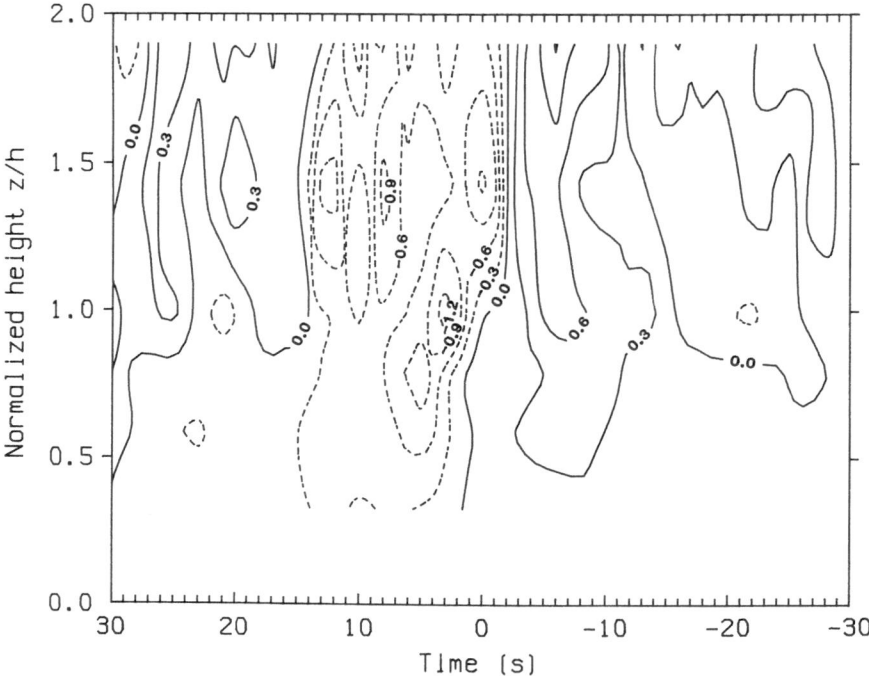

Fig. 7b. Same as in Figure 7a, but data are for near-neutral conditions during Run B, with
$L = -1063$ m.

translation speed was 2.4 m/s which is 3.2 times the friction velocity at canopy surface (0.74 m/s as shown in Table I). This ratio is much smaller than the value 15.7–34.5 calculated by Antonia *et al.* (1979) for temperature ramps over a smooth surface. The translation speed is close to the mean wind speed at the canopy height (2.1 m/s), and is 0.49 times the mean wind speed at 48 m, which is similar to previous results for thermal plumes (Kaimal, 1974; Wilczak and Tillman, 1980) and ramps in the marine surface boundary layer (Antonia *et al.*, 1979). Therefore, our translation speeds are consistent with previous work if scaled by mean wind, but are inconsistent if scaled by friction velocity.

The slope of the microfront was estimated on the assumption that the structures move with the same translation speed U_t at all levels. From the time difference Δt in the occurrence of the microfront at two levels, and from their vertical separation Δz, the slope β is estimated according to

$$\beta = \arctan(U_t \, \Delta t / \Delta z) . \tag{5}$$

As the diagrams show, the slope of the microfront is not constant with height. Using the data of Figure 6a for the upper portion of the forest, a slope was calculated of 48° from the vertical, in the downwind direction. This slope, coincidentally, is quite close to the average for several thermal plumes examined by Kaimal (1974).

3.3. MOMENTUM AND HEAT FLUX DISTRIBUTIONS

Patterns of instantaneous momentum and heat flux are plotted using the same height versus time contour techniques as for the scalar and velocity fields, for individual events (Figures 8a and 8b) and for ensemble averages of ten events (Figures 9a, 9b and 10). The ramp structures are primarily associated with an upward heat flux and a downward momentum flux in both the ejection and sweep phases. This is because the sweep behind the microfront decreases local temperature and increases local momentum to give positive $w'T'$ and negative $u'w'$. Preceding the microfront, the ejection carries warm air with low momentum to higher levels, thus also creating positive heat flux and negative momentum flux.

The momentum flux in the sweep phase is much larger than that in the ejection phase near the canopy surface (Figures 8a, 9a and 10). At the higher levels above

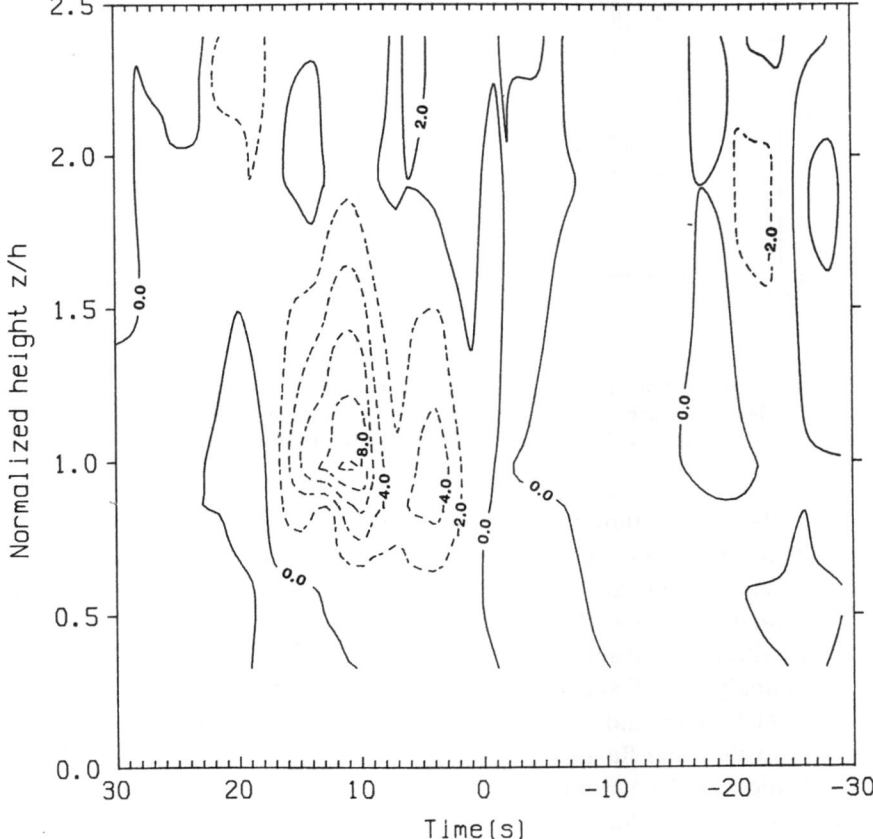

Fig. 8a. Vertical cross-section of 1-s average momentum flux normalized by the square of the friction velocity calculated at the top of the forest, for the single structure during Run A shown in Figure 3. Contour interval is 2.0. Dashed lines indicate negative $u'w'$ and solid lines indicate positive $u'w'$.

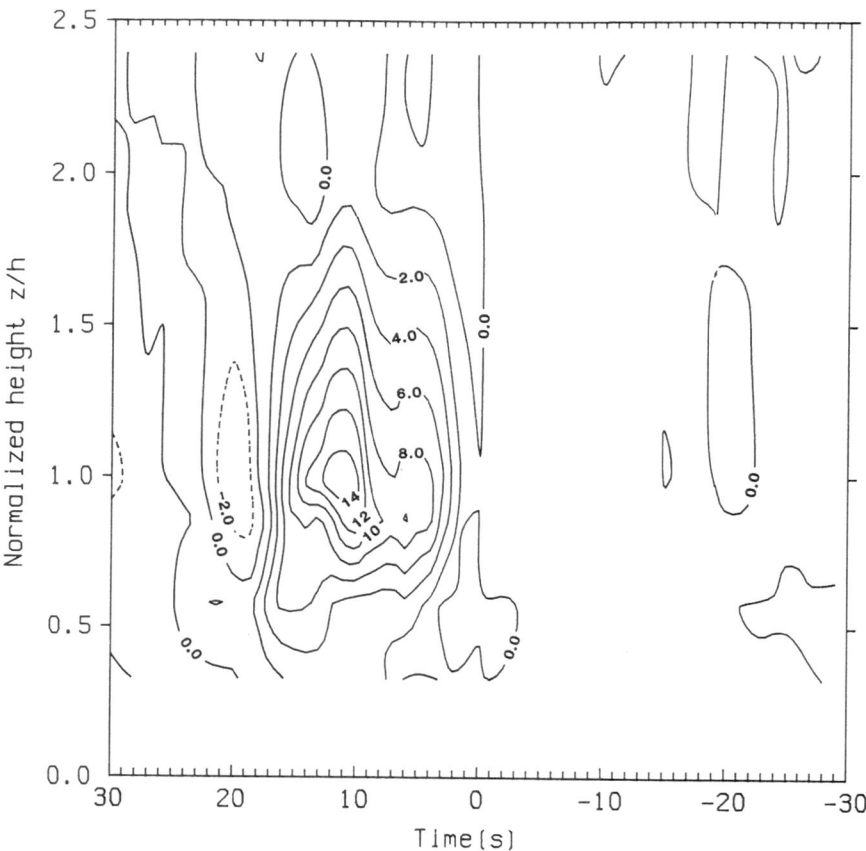

Fig. 8b. Vertical cross-section of 1-s average heat flux normalized by $-T^*u^*$ calculated at the top of the forest, for the same single structure as in Figure 3. Contour interval is 2.0. Solid lines indicate positive $w'T'$ and dashed lines represent negative $w'T'$.

the canopy, the momentum fluxes transferred by downward and upward flow become equally important. This can consistently be seen in both unstable and near neutral cases. The heat flux magnitude is also larger in the sweep phase than in the ejection phase near the canopy surface (Figures 8b and 9b). The flux distributions within the structures found here may provide a basis for explaining the quadrant analyses of Raupach (1981) for rough surfaces in a wind tunnel, and Shaw et al. (1983) in and above an agricultural crop, which show that the contribution to the total fluxes by quadrant 4 'sweeps' is dominant over that by quadrant 2 'ejections' near the rough surface. Our results are consistent with quadrant analysis if the flux contribution of non-coherent-structure turbulence was of minor importance for the data gathered by Raupach (1981) and Shaw et al. (1983).

The instantaneous fluxes reach magnitudes which are larger than the 30-min

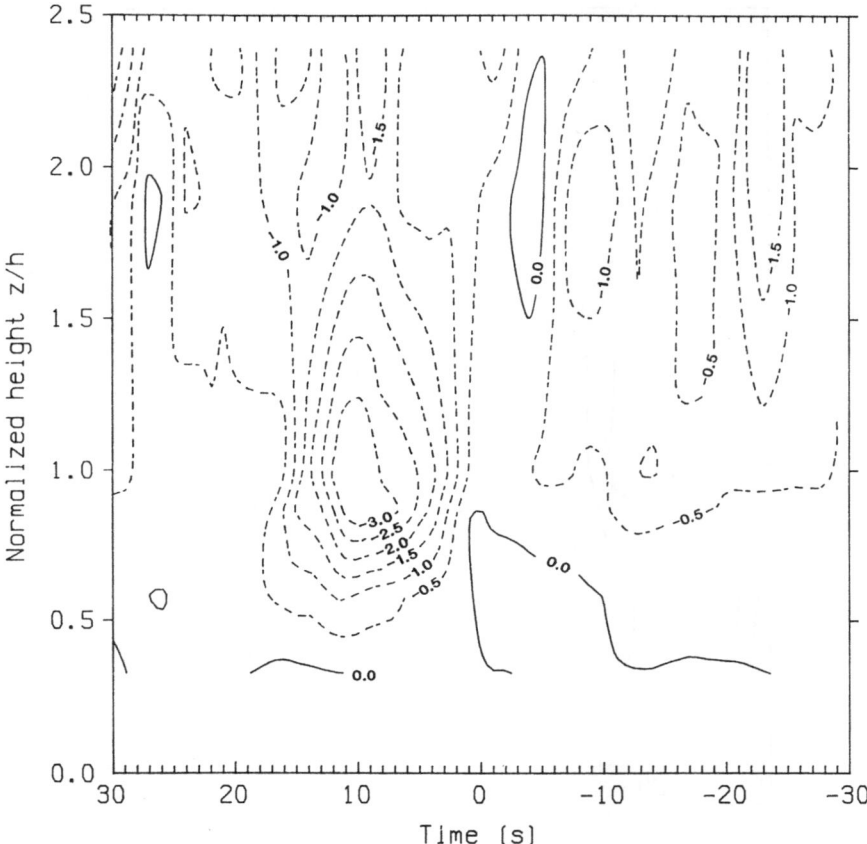

Fig. 9a.   Same as in Figure 8a, but for an ensemble average of 10 events identified during Run A
($L = -138$ m). Contour interval is 0.5.

mean fluxes by a factor of 10 for momentum and 14 for heat. Ensemble averaging of the fluxes resulted in the relative high and low values being attenuated by a factor of 2 to 3 (see Figures 8a, b and 9a, b), when compared with the individual event fluxes. This was caused by differences in the timing patterns of the individual events close to the tight gradients near the microfront.

In both the unstable and near-neutral runs (Figures 9a and 10), there are two regions of brief upward momentum transfer immediately preceding or at the microfront. One is located in the middle and lower layers of the forest, and the other is at about 1.5- to 2-times the treetop height. With sufficient resolution, these time/height regions would form a continuous separation of the 'ejection' and 'sweep' portions of the structures, as appears in the individual structure shown in Figure 8a, but the switch from 'ejection' to 'sweep' was so rapid near the top of the forest that the 1-s smoothing applied to the data did not reveal such details of the transition. These brief reversals in the direction of the

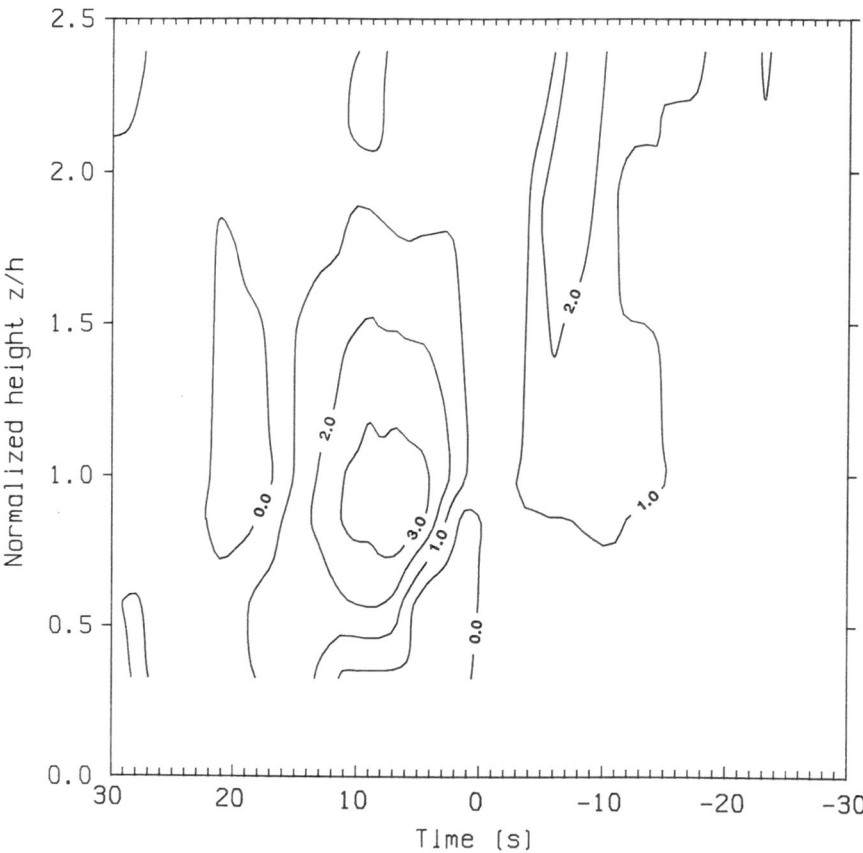

Fig. 9b. Same as in Figure 8b, but for an ensemble average of 10 events identified during Run A
($L = -138$ m). Contour interval is 1.0.

momentum flux may also be identified in Figures 6a and 6b by assigning the
fluctuating velocity vectors to the appropriate quadrants. In this way it is seen
that both upper and lower regions of flux reversal occur in quadrant 1 'outward
interaction' ($u' > 0$, $w' > 0$). We hypothesize that the upward momentum fluxes
are dynamically created by the sweep's sharp microfront, which accelerates the
air ahead of it in both the upward and longitudinal direction. This process implies
that large horizontal momentum from higher levels is directly transported
downward in the sweeps, and indirectly upwards in the area ahead of the sweeps
in the ejection phase. The upward transport is much weaker than the downward
transport, probably because of the general dissipation of the sweep microfront.
These observations and interpretations are consistent with early observations of
coherent structures in laboratory flows (Wallace *et al.*, 1972).

The overall similarity in the spatial and temporal distribution of momentum
flux associated with structures from the two different runs suggests that the

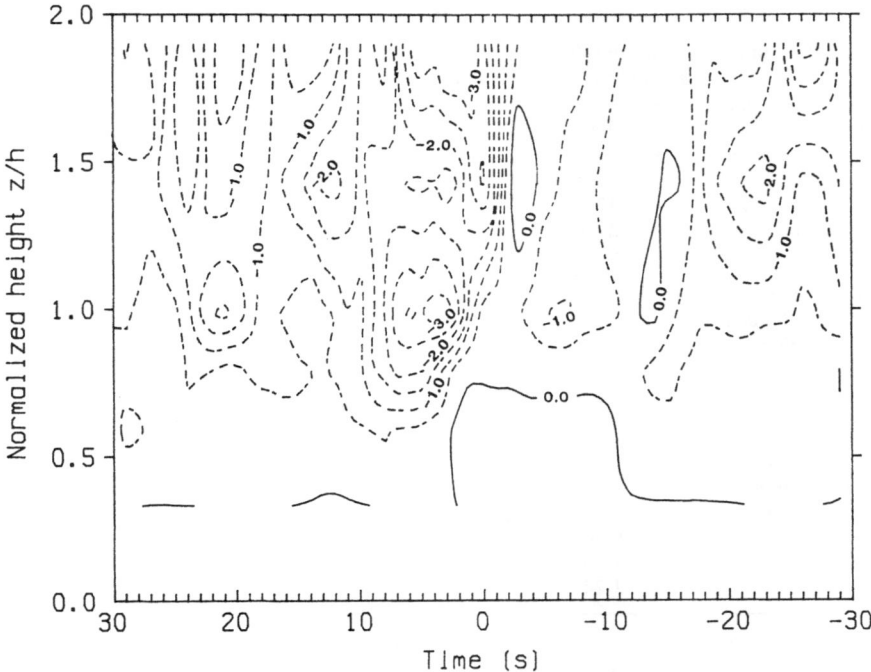

Fig. 10. Same as in Figure 8a, but for an ensemble average of 10 events identified during Run B ($L = -1036$ m). Contour interval is 0.5.

structures have similar dynamics, despite stability differences. The patterns in Figures 9a and 10 are quite similar, even to the extent that the ratios of the momentum flux at the local maximum ahead of the front ($t \approx -5$ s, $z/h \approx 0.5$) and the minimum behind the front ($t = 5$ to 10 s, $z/h \approx 1$) are equal to about 0.05. The momentum fluxes are small in the middle to lower canopy, however, and this result should be viewed with caution.

3.4. CONTRIBUTIONS TO THE TOTAL TRANSFER PROCESSES

In order to estimate the contribution of the structures to the total momentum and heat fluxes during a 30-min sampling period, we conditionally sampled the downdrafts and updrafts associated with 18 temperature ramps identified from Run A using the criteria stated in Section 3.2. To reduce the influence of background turbulence, we took a 10-s running average of the vertical velocity to calculate the duration of each part of each structure as the period occupied by the continuous updraft or downdraft ahead of or behind the microfront, respectively. The results are shown in Table II. The mean duration of the combined updrafts and downdrafts associated with structures was greater within the forest than above (63.9 s at mid canopy versus 41.9 s at twice treetop height). Since

TABLE II

Mean duration in seconds of updrafts and downdrafts, and of ejection/sweep structures, and ratio of normalized flux contribution to the relative duration

Height (z/h)	0.33	0.58	0.86	0.99	1.90	2.39
Duration (s):						
Updrafts	32.4	38.6	27.8	27.4	21.7	23.1
Downdrafts	29.7	25.3	26.2	24.3	20.2	22.3
Structures	60.1	63.9	54.0	51.6	41.9	45.4
Normalized $u'w'$ contribution divided by relative duration:						
Updrafts		0.57	0.53	0.80	1.10	1.06
Downdrafts		2.11	2.25	1.93	1.38	1.56
Structures		1.18	1.37	1.34	1.23	1.30
Normalized $w'T'$ contribution divided by relative duration:						
Updrafts	0.58	0.55	0.71	1.04	1.37	1.32
Downdrafts	1.86	2.21	2.12	1.83	1.36	1.38
Structures	1.24	1.21	1.39	1.42	1.37	1.35

there were 18 events counted during the 30-min run (except at $z/h = 0.33$ where we counted only 15), the mean durations in seconds are numerically equal to the percentages of the time occupied by ejections, sweeps, or structures. That is, at canopy top for example, structures occupied 51.6% of the total 30-min period. Ejections tended to be of slightly longer duration than sweeps.

For Run A, the rates of vertical flux of momentum and heat were integrated over the periods occupied by the sustained updrafts and downdrafts associated with the 18 ramp structures, as defined in the preceding paragraph. This was performed at each level, and the resulting fluxes were divided by the corresponding 30-min mean fluxes to obtain the fractional contribution by coherent structures. The results are presented in Figure 11, which shows that the ramp structures contributed 50–60% of the momentum and heat flux at the highest levels and 70–80% of the fluxes within the canopy. No data are shown for momentum flux at the lowest level because the 30-min mean flux was very small.

A ratio of the normalized flux contribution to the fractional duration of the structure gives a measure of the relative efficiency and significance of the structures; for example, if the structures only occurred 25% of the time but contributed 75% of the fluxes, the ratio would be 3. Table II shows these calculations for the complete structures and for their ejection and sweep components. For complete structures, there is only a small trend with height, the largest values appearing near the top of the canopy in both cases. The average ratios are 1.28 for momentum and 1.33 for heat.

Within the canopy, sweeps transferred considerably more momentum and heat

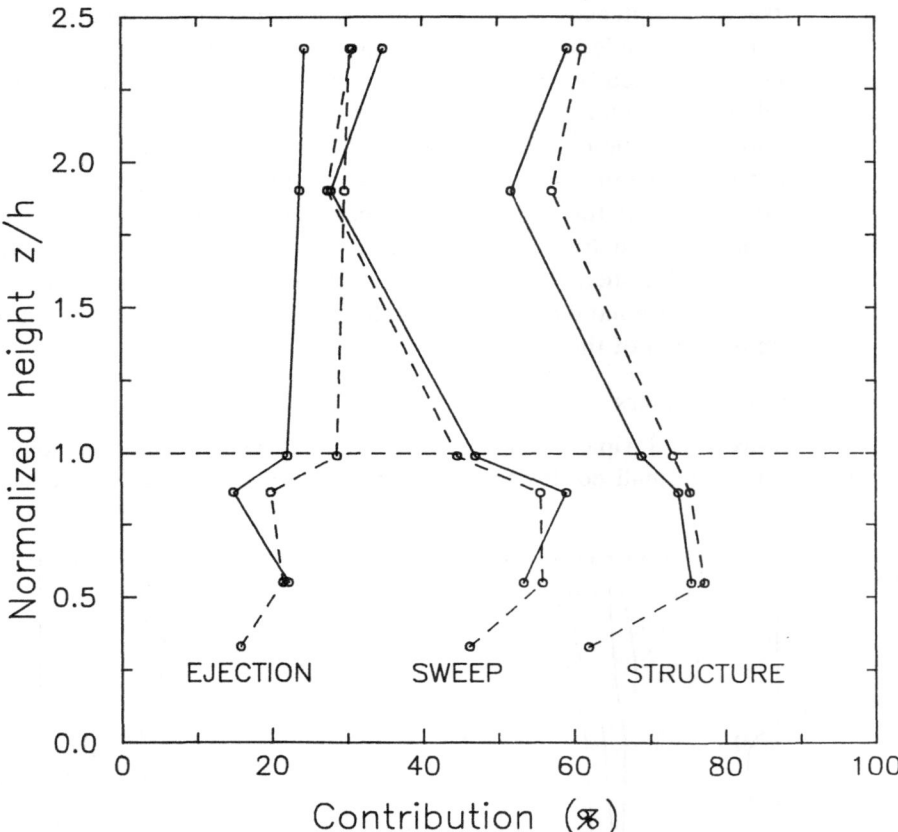

Fig. 11. Vertical profiles of contributions to the 30-min mean fluxes of momentum (solid lines) and heat (dashed lines) from identified structures, and from their ejection and sweep components. Data are from Run A.

than ejections, and were also more efficient at transferring these quantities (Figure 11 and Table II). Ejections were less efficient than average turbulence within the canopy, considering the flux magnitudes and the time required (Table II). At the higher levels above the canopy, however, sweeps and ejections contributed about equally, which is compatible with previous results from traditional quadrant analysis (Raupach, 1981; Shaw *et al.*, 1983). Above the forest, sweeps were still more efficient at transporting momentum but, for heat, ejections and sweeps were equally efficient.

It should be noted that our analysis is based on the sweep and ejection contributions from coherent structures identified by scalar ramps, in contrast to previous quadrant analyses based on conditional sampling without an independent identification technique. Such conditional sampling cannot distinguish between randomly occurring sweep-like and ejection-like quadrant 4 and quadrant 2 turbulence events and real sweeps and ejections clearly connected in a coherent structure sequence. A problem with traditional conditional quadrant analysis using non-zero hole sizes arises partially from the asymmetry shown by our results. Sweeps are clearly dominant near the canopy, so that for a given hole size, many coherent structures would have ejections too weak to be included. The contributions from the ejection portion of the coherent structures are thus underestimated by conventional quadrant analysis (unless a hole size of zero is used, which results in all turbulence being analyzed, instead of just the flow attributed to coherent structures). One way to rectify this situation is to modify the hole size constraint such that the hole is smaller for quadrant 2 than for quadrant 4; the determination of how much smaller should rely on an independent identification of the coherent structures.

3.5. THE THIRD MOMENTS

The ensemble-averaged triple products $w'u'w'$, $w'w'T'$, and $w'w'w'$ are shown in Figures 12a, b, c, normalized by the friction velocity and temperature scale

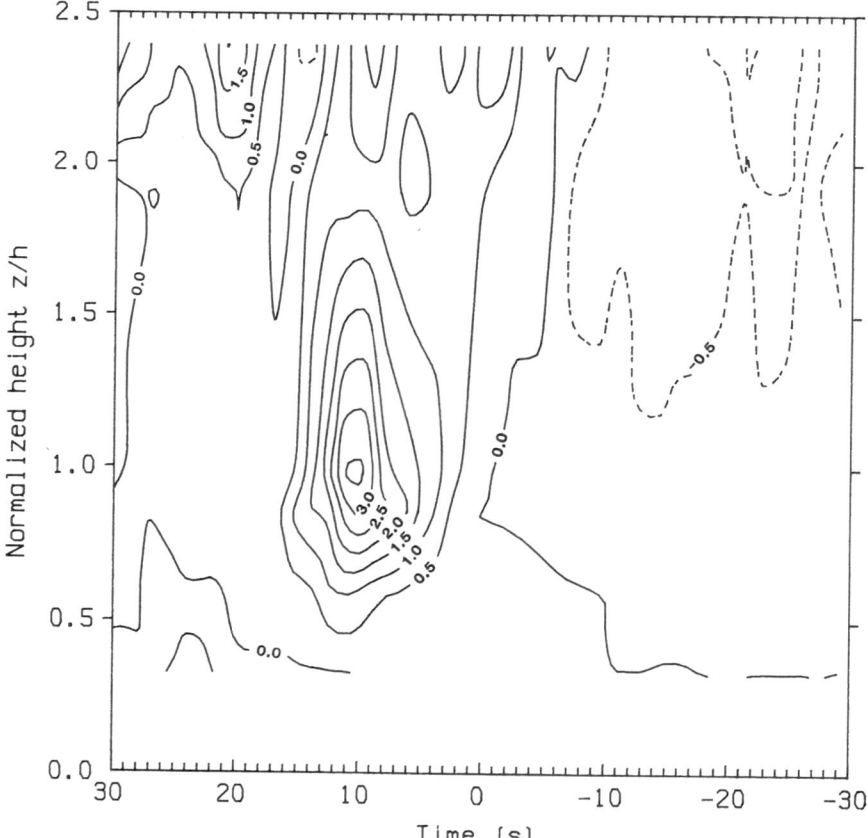

Fig. 12a.   Same as in Figure 6a, but for $w'u'w'$ normalized by u^{*3}, with contour intervals of 0.5. Solid lines represent positive $w'u'w'$ and dashed lines represent negative $w'u'w'$.

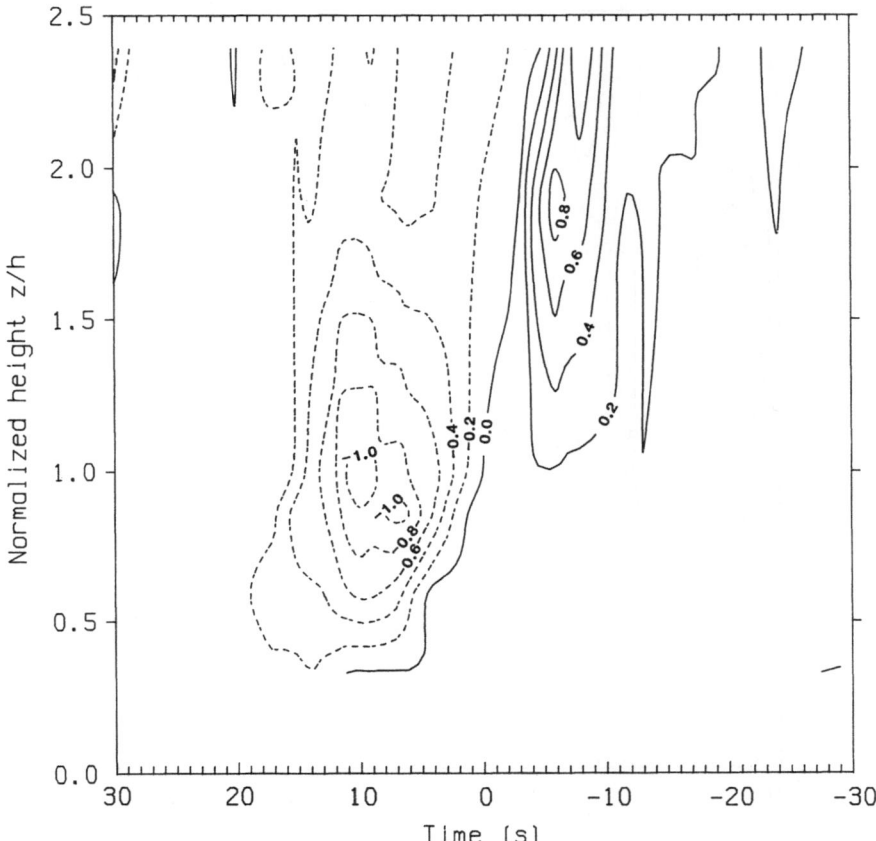

Fig. 12b.   Same as in Figure 12a, but for $w'w'T'$ normalized by $-T^*u^{*2}$, with contour intervals of 0.2.

$(T^* = -\overline{w'T'}/u^*)$ evaluated at the canopy surface, for the unstable case (Run A). These quantities appear in the conservation equations for, respectively, Reynolds stress $\overline{u'w'}$, heat flux $\overline{w'T'}$, and vertical velocity variance $\overline{w'^2}$, and may be interpreted as vertical transport terms. A large downward transport of momentum flux occurs in the sweep behind the front, with a maximum near the canopy top, while the ejection preceding the front contributes only a little (Figure 12a). The region of positive momentum flux shown in Figure 9a at $t = -10$ to 0 s centered at about $z/h = 0.6$ is reflected here as a bulge in the region of positive $w'u'w'$ (Figure 12a).

The transport of heat flux $(w'w'T')$ is also asymmetrical (Figure 12b), with an import into the forest corresponding to $w'T' > 0$ and $w' < 0$ occurring in the sweep region, and an export corresponding to $w'T' > 0$ and $w' > 0$ appearing in the ejection region. The transport in the sweep is much stronger than the export during the ejection, particularly near the canopy surface.

The sweep is also associated with a large negative $w'w'w'$, with a maximum

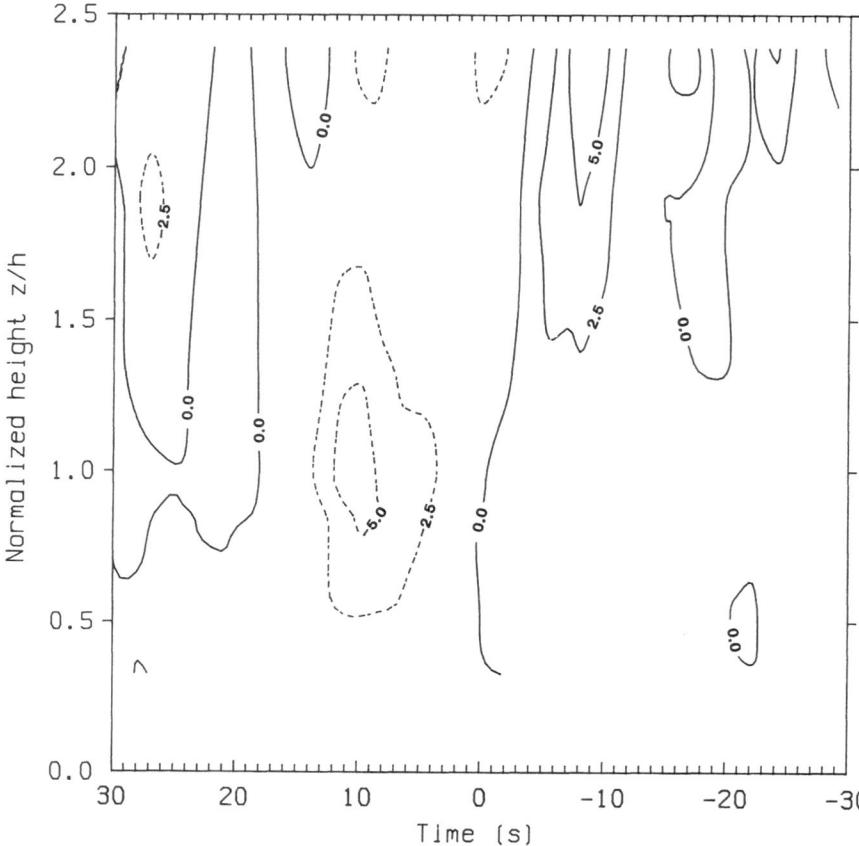

Fig. 12c.   Same as in Figure 12a, but for $w'w'w'$ normalized by u^{*3}, with contour intervals of 2.5.

near the canopy surface (Figure 12c). The sweeps are probably largely responsible for the average negative skewness of vertical velocity observed over rough
surfaces (Raupach, 1981; Shaw *et al.*, 1983; Shaw and Seginer, 1987). In
contrast, over smooth surfaces in unstable conditions, $w'w'w'$ averages as a
positive value, indicating a positive skew of a non-Gaussian distribution (Chiba,
1978).

4. Summary and Conclusions

Organized characteristics of turbulence in the layer within and above a forest
canopy have been examined. Ramp patterns appearing in the scalar temperature
and humidity fields were used to identify coherent structures under both unstable
and near-neutral conditions. The structural characteristics, including the general
pattern of scalars, and the distribution of covariances representing heat and
momentum fluxes, when plotted as height versus time cross-sections, show close

similarity between the unstable and near-neutral coherent structures. This indicates that vertical wind shear is the major factor in the creation of these structures.

In general, a coherent structure consists of a weak ejection from the canopy top followed by a strong sweep into the canopy. Under lapse conditions, the sweep transports cooler air from some distance above the canopy to within the canopy, forming a cold microfront; under near-neutral conditions and with a transpiring canopy, the sweep transports dry air into the canopy. In both cases the coherent structures produce the ramp patterns seen in the time traces of temperature and/or humidity. Within and near the canopy top, the sweeps transport more heat, humidity and momentum than the ejections, but at higher levels the ejections are equally important. Near the canopy surface and within the canopy, the coherent structures may contribute more than 75% of the fluxes; this decreases to between 50 and 60% at twice the forest height.

5. Acknowledgements

The Camp Borden field experiment was funded by Environment Canada. Participants included G. den Hartog and H. H. Neumann of Atmospheric Environment Service, who had overall responsibility for the field experiment, G. Kidd, M. Y. Leclerc, G. Shi, and G. W. Thurtell, University of Guelph, Ontario, and R. H. Shaw, University of California, Davis. One of us (RHS) is grateful to the Atmospheric Environment Service, Environment Canada for support during a sabbatical leave that allowed him to take part in the Camp Borden study. The analysis of the data leading to this report was conducted with the support of the National Science Foundation under Grant No. ATM-8703832. Some graduate student support was also provided by National Aeronautics and Space Administration Grant No. NAG-5-892.

References

Antonia, R. A., Chambers, A. J., Friehe, C. A., and Van Atta, C. W.: 1979, 'Temperature Ramps in the Atmospheric Surface Layer', *J. Atmos. Sci.* **36**, 99–108.
Baldocchi, D. D. and Meyers, T. P.: 1988, 'Turbulence Structure in a Deciduous Forest', *Boundary-Layer Meteorol.* **43**, 345–364.
Bogard, D. G. and Tiederman, W. G.: 1986, 'Burst Detection with Single-point Velocity Measurements', *J. Fluid Mech.* **162**, 389–413.
Busch, N. E.: 1973, 'On the Mechanics of Atmospheric Turbulence', in D. A. Haugen (ed.), *Workshop on Micrometeorology*, Amer. Meteorol. Soc., Boston, pp. 1–65.
Chen, C. P. and Blackwelder, R. F.: 1978, 'Large-scale Motion in a Turbulent Boundary Layer: a Study Using Temperature Contamination', *J. Fluid Mech.* **89**, 1–31.
Chiba, O.: 1978, 'Stability Dependence of the Vertical Velocity Skewness in the Atmospheric Surface Layer', *J. Meteorol. Soc. Japan.* **56**, 140–142.
Corino, E. R. and Brodkey, R. S.: 1969, 'A Visual Investigation of the Wall Region in Turbulent Flow', *J. Fluid Mech.* **37**, 1–30.

Denmead, O. T. and Bradley, E. F.: 1987, 'On Scalar Transport in Plant Canopies', *Irrigation Sci.* **8**, 131–149.

Finnigan, J. J.: 1979a, 'Turbulence in Waving Wheat. I Mean Statistics and Honami', *Boundary-Layer Meteorol.* **16**, 181–211.

Finnigan, J. J.: 1979b, 'Turbulence in Waving Wheat. II Structure of Momentum Transfer', *Boundary-Layer Meteorol.* **16**, 213–236.

Kaimal, J. C.: 1974, 'Translation Speed of Convective Plumes in the Atmospheric Surface Layer', *Quart. J. Roy. Meteorol. Soc.* **100**, 46–52.

Kaimal, J. C. and Businger, J. A.: 1970, 'Case Studies of a Convective Plume and a Dust Devil', *J. Appl. Meteorol.* **9**, 612–620.

Kline, S. J., Reynolds, W. C., Schraub, F. A., and Rundstadler, P. W.: 1967, 'The Structure of Turbulent Boundary Layers', *J. Fluid Mech.* **30**, 741–773.

Legg, B. J. and Monteith, J. L.: 1975, 'Heat and Mass Transfer in Plant Canopies', in D. A. De Vries and N. H. Afgan (eds.), *Heat and Mass Transfer in the Biosphere*, Wiley, New York, pp. 167–186.

Meyers, T. P. and Paw U, K. T.: 1986, 'Testing of a Higher-order Closure Model for Airflow within and above Plant Canopies', *Boundary-Layer Meteorol.* **37**, 297–311.

Meyers, T. P. and Paw U, K. T.: 1987, 'Modelling the Plant Canopy Micrometeorology with Higher-order Closure Principles', *Agric. Forest. Meteorol.* **41**, 143–163.

Neumann, H. H., den Hartog, G., and Shaw, R. H.: 1988, 'Leaf Area Measurements During Leaf-fall for a Deciduous Forest Based on Hemispheric Photographs and Leaf-litter Collection', *Agric. Forest Meteorol.*, in press.

Offen, G. R. and Kline, S. J.: 1975, 'A Proposed Model of the Bursting Process in Turbulent Boundary Layers', *J. Fluid Mech.* **70**, 209–228.

Praturi, A. K. and Brodkey, R. S.: 1978, 'A Stereoscopic Visual Study of Coherent Structures in Turbulent Shear Flow', *J. Fluid Mech.* **89**, 251–272.

Priestley, C. H. B.: 1959, *Turbulent Transfer in the Lower Atmosphere*, University of Chicago Press, Chicago, pp. 53–72.

Rajagopalan, S. and Antonia, R. A.: 1980, 'Interaction between Large and Small Scale Motions in a Two-dimensional Turbulent Flow Duct', *Phys. Fluids* **23**, 1101–1110.

Raupach, M. R.: 1981, 'Conditional Statistics of Reynolds Stress in Rough-wall and Smooth-wall Turbulent Boundary Layers', *J. Fluid Mech.* **108**, 363–382.

Raupach, M. R.: 1987, 'A Lagrangian Analysis of Scalar Transfer in Vegetation Canopies', *Q. J. Roy. Meteorol. Soc.* **113**, 107–120.

Raupach, M. R. and Thom, A. S.: 1981, 'Turbulence in and above Plant Canopies', *Ann. Rev. Fluid Mech.* **13**, 97–129.

Schols, J. L. J.: 1984, 'The Detection and Measurement of Turbulent Structures in the Atmospheric Surface Layer', *Boundary-Layer Meteorol.* **29**, 39–58.

Shaw, R. H., den Hartog, G., and Neumann, H. H.: 1988, 'Influence on Foliar Density and Thermal Stability on Profiles of Reynolds Stress and Turbulence Intensity in a Deciduous Forest', *Boundary-Layer Meteorol.* **45**, 391–409.

Shaw, R. H., Tavangar, J., and Ward, D. P.: 1983, 'Structure of the Reynolds Stress in a Canopy Layer', *J. Clim. Appl. Meteorol.* **22**, 1922–1931.

Shaw, R. H. and Seginer, I.: 1987, 'Calculation of Velocity Skewness in Real and Artificial Plant Canopies', *Boundary-Layer Meteorol.* **39**, 315–332.

Subramanian, C. S., Rajagopalan, S., Antonia, R. A., and Chambers, A. J.: 1982, 'Comparison of Conditional Sampling and Averaging Techniques in a Turbulent Boundary Layer', *J. Fluid Mech.* **123**, 335–362.

Talmon, A. M., Kunen, J. M. G., and Ooms, G.: 1986, 'Simultaneous Flow Visualization and Reynolds-stress Measurement in a Turbulent Boundary Layer', *J. Fluid Mech.* **163**, 459–478.

Taylor, R. J.: 1958, 'Thermal Structures in the Lowest Layer of the Atmosphere', *Australian J. Phys.* **11**, 168–176.

Thomas, A. S. and Bull, M. K.: 1983, 'On the Role of the Wall-Pressure Fluctuations in Deterministic Motions in the Turbulent Boundary Layer', *J. Fluid Mech.* **128**, 283–322.

Wallace, J. M., Eckelmann, H., and Brodkey, R. S.: 1972, 'The Wall Region in Turbulent Shear Flow', *J. Fluid Mech.* **54**, 39–48.

Wilczak, J. M.: 1984, 'Large-scale Eddies in the Unstably Stratified Atmospheric Surface Layer. Part I: Velocity and Temperature Structure', *J. Atmos. Sci.* **41**, 3537–3550.

Wilczak, J. M. and Businger, J. A.: 1984, 'Large-scale Eddies in the Unstably Stratified Atmospheric Surface Layer. Part II: Turbulent Pressure Fluctuations and the Budgets of Heat Flux, Stress and Turbulent Kinetic Energy', *J. Atmos. Sci.* **41**, 3551–3567.

Wilczak, J. M. and Tillman, J. E.: 1980, 'The Three-dimensional Structure of Convection in the Atmospheric Surface Layer', *J. Atmos. Sci.* **37**, 2424–2443.

Wilson, N. R. and Shaw, R. H.: 1977, 'A Higher Order Closure Model for Canopy Flow', *J. Appl. Meteorol.* **16**, 1197–1205.

INDEX OF SUBJECTS
Volume 47

TABLE OF CONTENTS

Volume 47

ARTICLES

An Introduction to Boundary Layer Meteorology

by
ROLAND B. STULL

ATMOSPHERIC SCIENCES LIBRARY

1988, 680 pp.
Hardbound Dfl. 220.00/£64.00/$99.00   ISBN 90-277-2768-6
Paperback Dfl. 98.00/£29.00/$35.00   ISBN 90-277-2769-4

Part of the excitement in boundary-layer meteorology is the challenge associated with turbulent flow – one of the unsolved problems in classical physics. An additional attraction of the field is the rich diversity of topics and research methods that are collected under the umbrella-term of boundary-layer meteorology. The flavor of the challenges and the excitement associated with the study of the atmospheric boundary layer are captured in this textbook.

Fundamental concepts and mathematics are presented prior to their use, physical interpretations of the terms in equations are given, sample data are shown, examples are solved, and exercises are included.

The work should also be considered as a major reference and as a review of the literature, since it includes tables of parameterizations, procedures, field experiments, useful constants, and graphs of various phenomena under a variety of conditions.

It is assumed that the work will be used at the beginning graduate level for students with an undergraduate background in meteorology, but the author envisions, and has catered for, a heterogeneity in the background and experience of his readers.

Contents

Preface. **1.** Mean Boundary Layer Characteristics. **2.** Some Mathematical and Conceptual Tools: Part 1. Statistics. **3.** Application of the Governing Equations to Turbulent Flow. **4.** Prognostic Equations for Turbulent Fluxes and Variances. **5.** Turbulence Kinetic Energy, Stability, and Scaling. **6.** Turbulence Closure Techniques. **7.** Boundary Conditions and External Forcings. **8.** Some Mathematical and Conceptual Tools: Part 2. Time Series. **9.** Similarity Theory. **10.** Measurement and Simulation Techniques. **11.** Convective Mixed Layer. **12.** Stable Boundary Layer. **13.** Boundary Layer Clouds. **14.** Geographic Effects. **Appendices. A.** Scaling and dimensionless groups. **B.** Notation. **C.** Useful constants, parameters and conversion factors. **D.** Derivation of virtual potential temperature. Subject Index.

 Kluwer Academic Publishers

P.O. Box 989, 3300 AZ Dordrecht, The Netherlands
101 Philip Drive, Norwell, MA 02061, U.S.A.

Applications of Remote Sensing to Agrometeorology

Proceedings of a Course held at the Joint Research Centre of the Commission of the European Communities in the Framework of the Ispra-Courses, Ispra, Varese, Italy, April 6–10, 1987

edited by
F. TOSELLI

ISPRA COURSES ON REMOTE SENSING

1989, 344 pp.   ISBN 0–7923–0020–3
Hardbound Dfl. 200.00/£64.00/US$ 99.00

Agrometeorology has, over the last few years, had the benefit of the supplementation conventional ground investigative data by data derived from space, not only from the conventional dedicated earth resource satellites such as LANDSAT and SPOT, but also from other platforms such as METEOSAT and NOAA-TIROS. The use of such combined data sources leads, cost effectively, to a better characterization of natural systems.
The current work provides information on the state of the art in this area, as well as on future developments in the field. It is divided into three parts.

Contents
Preface. **1.** *Ph. Hartl* – Fundamentals of Remote Sensing. **2.** *Ph. Hartl* – Sensors. **3.** *G. Fraysse* – Platforms. **4.** *A. Hubaux* – The Luxuriant Image – An Introduction to Remote Sensing Data Processing. **5.** *J. Dejace* – Image Processing Techniques: Filtering, Exogenous Data, Geometrical Processing. **6.** *D. R. Nüesch* and *E. S. Kasischke* – Imaging Radar Fundamentals. **7.** *D. R. Nüesch* and *E. H. Meier* – Radar Post-Processing. **8.** *H. Grassl* – Extraction of Surface Temperature from Satellite Data. **9.** *B. Seguin* – Use of Surface Temperature in Agrometeorology. **10.** *E. Raschke* – Radiation Budget at the Ground from Satellite Data. **11.** *T. Schmugge* – Microwave Remote Sensing of Soil Moisture. **12.** *J. P. Malingreau* – The Vegetation Index and the Study of Vegetation Dynamics. **13.** *E. C. Barrett* – Satellite Remote Sensing of Rainfall.

New
Publication

KLUWER
ACADEMIC
PUBLISHERS

P.O. Box 322, 3300 AH Dordrecht, The Netherlands
P.O. Box 358, Accord Station, Hingham, MA 02018-0358, U.S.A.